Springer Proceedings in Mathematics & Statistics

Volume 467

This book series features volumes composed of selected contributions from workshops and conferences in all areas of current research in mathematics and statistics, including data science, operations research and optimization. In addition to an overall evaluation of the interest, scientific quality, and timeliness of each proposal at the hands of the publisher, individual contributions are all refereed to the high quality standards of leading journals in the field. Thus, this series provides the research community with well-edited, authoritative reports on developments in the most exciting areas of mathematical and statistical research today.

Matilde Bini · Antonio Balzanella ·
Lucio Masserini · Rosanna Verde
Editors

Advanced Methods in Statistics, Data Science and Related Applications

SIS 2022, Caserta, Italy, June 22–24

Società Italiana di Statistica

Springer

Editors
Matilde Bini
Dipartimento di Scienze Umane
Università Europea di Roma
Roma, Italy

Antonio Balzanella
Dipartimento di Matematica e Fisica
Università della Campania "Luigi Vanvitelli"
Caserta, Italy

Lucio Masserini
Dipartimento di Economia e Management
Università di Pisa
Pisa, Italy

Rosanna Verde
Dipartimento di Matematica e Fisica
Università della Campania "Luigi Vanvitelli"
Caserta, Italy

ISSN 2194-1009 ISSN 2194-1017 (electronic)
Springer Proceedings in Mathematics & Statistics
ISBN 978-3-031-65698-9 ISBN 978-3-031-65699-6 (eBook)
https://doi.org/10.1007/978-3-031-65699-6

Mathematics Subject Classification: 62P25, 62-06, 62C10, 62H11, 62-07, 62F35, 62M05, 62P20

This Springer imprint is published by the registered company Springer Nature Switzerland AG
The registered company address is: Gewerbestrasse 11, 6330 Cham, Switzerland

Preface

This volume contains a collection of contributions presented at the SIS 2022 scientific meeting held in Caserta, Italy. The meeting, the first entirely in-person event after the pandemic period, was widely attended by members of the Italian statistical scientific community, as well as many foreign colleagues, most of whom were co-authors of papers resulting from fruitful collaborations with Italian colleagues.

The contributions ranged from developments in the field of statistics, including Composite indicators, Regression estimators, Survey sampling, and PLS models, to recent contributions in the field of Data Science, where statistics plays a fundamental role, especially in interpretative and explanatory aspects, as well as in application areas with numerous innovative methodological insights in environmental, social, labor market, epidemiological, etc., fields.

The volume is organized into five parts, covering topics such as contributions to statistical models and data analysis techniques, applications to population and social studies, studies in economics, and sample design techniques. Most of these issues are relevant for the analysis of large amounts of data or complex data structures, typically within the Data Science framework. As is known, the high dimensionality of data and the complexity of new data require an increasingly appropriate adaptation of traditional data analysis techniques.

The first part, titled *Data Quality, Administrative Large Data Analysis and Composite Indicators*, containing six contributions, focuses on pre-treatment techniques for administrative database, data quality assessment on Big Data, composite indicators, and an aggregation method for the Human Development Index.

The second part, titled *New Dynamics in Labour Market and Population Migration*, contains four contributions, covering applications in the broad socio-demographic field, with an emphasis on economic analyses and insights into new dynamics in the labour market, also with reference to the pandemic period. It also includes contributions on subjective well-being studies, on the specific skills of foreign workers and analyses on the evolution of migration trends.

The third part, titled *Scaling, Correlation and Decomposition Approaches in Social Applications*, includes five contributions, which highlight novelties in the decomposition of inequality after an asymmetric shock, interrater agreement measures for quantitative data, and the use of PLS-SEM for confirmatory analysis in social applications.

The fourth part, *Non-linear Models and Spatial Analysis*, consists of six contributions, dealing with non-linear and spatial models for the analysis of scattered data, as well as microdata. The applications concern studies on the agricultural sector and industrial location; another contribution focuses on the analysis of disease occurrence in volcanic areas in Sicily.

The fifth part, titled *Experimental Design, Robust Methods, Bayesian Optimization, and Statistical Matching*, consists of four contributions, related to applicative studies, such as innovative measures for the strength assessment of randomized clinical trials; a robust factorial design in the experimental context; a Bayesian approach for the optimization of cloud resources by using machine learning techniques; and a statistical matching approach in the context of time series.

The *guest editors of this volume* wish to thank the colleagues who, through their review of the contributions, helpful comments, and careful revision work, made the publication of this collection of scientific contributions possible.

Caserta, Italy — Antonio Balzanella
Roma, Italy — Matilde Bini
Pisa, Italy — Lucio Masserini
Caserta, Italy — Rosanna Verde

Contents

Data Quality, Administrative Large Data Analysis and Composite Indicators

Administrative Database and Official Statistics: An IT and Statistical Procedure

Caterina Marini and Vittorio Nicolardi

Abstract The availability of a complete harmonised database that includes all data, both administrative and statistical, on the phenomenon under investigation is important to provide a precious analytical instrument that is very often unavailable because of issues related to important obstacles that involve both statistical and computational problems, and administrative and legal impediments. In this work, we depict the IT and Statistical procedure used to create the first Full Information Harmonised Real Estate Database in the analysis of the real estate economy.

Keywords Administrative data · Big data · Real estate economy · Harmonised database · Official statistics

1 Introduction

The debate about the administrative databases and their use to provide a statistical information that is not differently affordable is internationally considered noteworthy and relevant to face the nowadays scientific challenge. Worldwide, all the National Statistical Institutes are effectively considering the opportunity to use administrative data to complement survey data and delineating statistical regulations to facilitate merging and guarantee the quality of information. Nevertheless, the solution to the integration process is still on the horizon because of the vastness of the problem and the complexity of the merger application that cannot be implemented to all cases without adequate characterisations. Furthermore, the administrative procedures affect every aspect of the modern society, and many data are unintentionally

C. Marini · V. Nicolardi (✉)
Department of Economics and Finance, University of Bari Aldo Moro, Abbazia S. Scolastica, 70124 Bari, Italy
e-mail: vittorio.nicolardi@uniba.it

C. Marini
e-mail: caterina.marini@uniba.it

M. Bini et al. (eds.), *Advanced Methods in Statistics, Data Science and Related Applications*, Springer Proceedings in Mathematics & Statistics 467,
https://doi.org/10.1007/978-3-031-65699-6_1

collected. Therefore, many issues are involved and they are all fundamental to provide a complete depiction of the socio-economic scenario in a world that is becoming more complex to be analysed. It is generally believed that the administrative data are largely available and readily usable when it is assessed the need for their use in the analysis, but this is one of the most common errors that is unconsciously committed. The integrated use of administrative data needs to address several issues that are not encountered when working with official statistical data and, in the scientific context, many aspects are debate (Nordbotten [9], Kitchin [6], Thomsen et al. [11]). First, administrative data referring to the same phenomenon or correlated phenomena are very often produced by independent administrative offices, sometimes belonging to the same organisation, public authority or company, and they are differently collected for specific administrative purposes. Therefore, administrative data can be affected by material and human errors, such as duplication, missing values and erroneous information, and consequentially they are not controlled and statistically correct. Furthermore, in most cases, databases are structurally diverse because of well-established IT and administrative procedures that are not suitable for aligning all data. Second, administrative data are a little part of the more complex phenomenon known as Big Data. In literature, the Big Data aspect of the problem is widely discussed (Harford, [2], Kitchin [4, 5], Marini and Nicolardi [7]). Finally, it is crucial the harmonisation of the administrative data, which are raw information, with the official statistical data, as produced by the National Statistical Institutes (NSIs, hereafter), which are consolidated statistically. In the recent past, some experimental attempt to complement survey data by administrative data has been successful in the empirical analysis of some economic phenomenon (Pacheco and JuliáÃ‚Â¡n [10], Johnson et al. [3], Bryant and Graham [1], Marini and Nicolardi [8]). In some of our recent works (Marini and Nicolardi [7, 8]), we have provided an original fully comprehensive information statistical database for the analysis of the phenomenon of the real estate economy where administrative data and Italian NSI's data are perfectly aligned, and the Big Data crucial aspects solved. The Full Information Harmonised Real Estate Database (FIHRE-DB, hereafter) we created is the first attempt to align all the administrative and statistical data referred to the real estate economy, which still suffers the consequences of the dearth of a complete and harmonised data warehouse in Italy, as in many European countries. The potentialities of FIHRE-DB are vast, and Public Administrations, households and private agencies can take considerable advantages of its data in terms of financial and economic returns. In fact, Public Administrations can use the FIHRE-DB all-encompassing information to redefine their fiscal scheme more fairly in terms of tax rates based on a more exact economic evaluation of the socio-economic characteristics of the city areas. At the same time, through the FIHRE-DB data, households and private agencies can evaluate the economic profitability of an investment in the real estate market and its incomparable return in terms of business based on the location. Finally, the economic experiment we run on the sole city of Bari in South Italy can be an important starting point for a fair reform of the Italian National Registry of private and public properties as stated in the Italian National Recovery and Resilience Plan document, part of the Next Generation EU programme.

In this work, we describe the IT and statistical procedure we defined to create FIHRE-DB. The full procedure needed SAS software and QGIS software to produce geo-maps and define the overlapping steps.

The paper is organised as follows: the description of the dataset contents to create FIHRE-DB is reported in Sect. 2; Sect. 3 describes the method designed and employed through the IT computational code we defined; some concluding remarks are reported in Sect. 4.

2 The Data

In our work, the experiment is restricted to the city of Bari, in southern Italy, because Bari was part of a national research project. Databases used to create FIHRE-DB are essentially five key databases that include the fundamental data, and three instrumental databases that are helpful to harmonise all data.

The five key databases are independent administrative databases, generally managed by independent Public Administration (PA, hereafter) offices of the city of Bari, and provide an autonomous real estate information. Four of five key databases belong to the Real Estate Registry (RER, hereafter) and contain all data referring to the real estate units: technical and economic data as typology, size, economic value, which is normally used to calculate the council tax; RER identifiers; toponyms; council tax rates. The last key database belongs to the Real Estate Italian Observatory (REIO, hereafter) of the Italian Revenue Agency and includes all data referred to the monetary value of each real estate unit in the Italian real estate market.

The three instrumental databases are mutually fully independent because they belong to independent PAs: the first belongs to RER, includes the geo-references of each property and permits the final connection between the administrative data and the official statistical data from the Italian NSI; the second belongs to REIO and the last belongs to the Italian NSI, and both are the geo-localisation datasets used for the final overlapping action. In particular, the Italian NSI instrumental database contains the census sections that are generally used for all official territorial analyses.

3 The Method

The method we propose is a *step by step* procedure to obtain the final outcome, i.e. FIHRE-DB.

A first study of all databases highlighted that the four RER databases needed an inevitable prior action of cleaning because of many duplicated, missing and erroneous data. Afterwards, it was possible to proceed with the harmonisation of all data based on well identified key fields.

In the subsections that follow, all steps of the IT code, which is substantially the method of data alignment we designed, are depicted.

3.1 RER Databases

The administrative RER databases are fundamental to create the final harmonised database but, at the same time, require more effort in terms of cleaning. Furthermore, the analysis of the data framework highlighted also that a recoding action is necessary to correct some information that, in the past, was erroneously recorded because of some administrative territorial process that involved the inclusion of some little autonomous municipalities in the city of Bari. Therefore, the cleaning and recoding actions are important steps to homogenise information provided by RER and align the four RER databases.

The first RER database includes a list of records that contains all technical and economic information of each unit, i.e. property. Therefore, without affecting the statistical significance of the analysis, we deleted the incomplete and duplicated records in this first RER database. That is a multiple step procedure.

The first step is to sort data by means of two key fields that are *Id_Property*, which identifies each real estate in RER, and *Ordinal_Number*, which identifies the mutations corresponding to each unit:

Proc 1

```
proc sort data = name_Database1;
by Id_Property DESCENDING Ordinal_Number;
run;
```

The second step of the procedure is to clean the database. To delete the oldest data referring to each real estate mutation and, therefore, avoid duplication in the key field records, the key field is again *Id_Property*:

Proc 2

```
proc sort data = name_Database1 out = name_Database2 nodupkey;
by Id_Property;
run;
```

Successively, we proceeded with the removal of all real estate units without economic value, i.e. *Income*, because they lost the Certificate of Occupancy:

Proc 3

```
data mainDatabase1;
set name_Database2;
where (Income not in (.));
run;
```

This final step for the first RER database creates the *mainDatabase1*.

The second RER database comprises other real estate data, mainly the Urban Sections and the RER Sheet, Subordinate and Parcel reference information. This second RER database includes a much higher number of records compared with the first RER database because of both duplication, caused by some administrative change, and the presence of some real estate unit whose record has not been deleted though the building was demolished. Therefore, we deleted the incomplete and duplicated records in this second RER database and proceeded to recode some field. That is a multiple step procedure.

The first step is to sort data by means of two key fields that are *Id_Property* and *Urban_Section*:

Proc 4

```
proc sort data = name_Database3;
by Id_Property DESCENDING Urban_Section;
run;
```

The second step of the procedure is to clean the database. To delete the duplicated data for each real estate unit, we used as key field *Id_Property*:

Proc 5

```
proc sort data = name_Database3 out = name_Database4 nodupkey;
by Id_Property;
run;
```

Furthermore, we also deleted all records where the real estate units have an erroneous *Registry Sheet* code.

Proc 6

```
data name_Database5;
set name_Database4;
where (Registry_Sheet in
      (report the right "Registry_Sheet numbers/code" set));
run;
```

The third step of the procedure is to recode the sole field *Urban_Section* in the final database, that is *name_Database6*:

Proc 7

```
data name_Database6;
set name_Database5;
if (Urban_Section} in (report the old "Section_code" set))
    then Section = report the new "Section code";
...;
run;
```

Finally, to homogenise information between the two databases obtained through **Proc 3** and **Proc 7**, we deleted all the records that were in the second database and not in the first database. Therefore, we adopted a *right join* between the *mainDatabase1* and the *name_Database6*.

This final step for the second RER database creates the *mainDatabase2*.

The third RER database comprises the toponyms of each real estate unit. Toponyms are important to identify the exact localisation of each unit on the urban area. This third RER database shows many duplications because of the modifications of some toponym and/or building number. Therefore, a cleaning action is necessary, and a double step procedure was defined.

The first step is to sort data by means of two key fields that are *Id_Property* and *Ordinal_Number*, which identifies all the mutations in toponyms and/or building numbers corresponding to the real estate unit:

Proc 8

```
proc sort data = name_Database7;
by Id_Property DESCENDING Ordinal_Number};
run;
```

The second step of the procedure is to clean the database. To delete the oldest data referring to each real estate mutation and, therefore, avoid duplication in the key field records, the key field *Id_Property* is used:

Proc 9

```
proc sort data = name_Database7 out = name_Database8 nodupkey;
by Id_Property;
run;
```

This second step of the procedure creates the *name_Database8*.

Finally, to homogenise information between the two databases obtained through **Proc 3** and **Proc 9**, we deleted all the real estate records that were in the second database and not in the first database. Therefore, we adopted again a *right join* between the *mainDatabase1* and the *name_Database8.*

This final step for the third RER database creates the *mainDatabase3*.

The conclusion of the cleaning and recoding multiple procedure referred to the first three key RER databases is the creation of a sole omni comprehensive database that aligned data by means of the key field *Id_Property*:

Proc 10

```
data mainDatabaseRER;
merge mainDatabase1 mainDatabase2 mainDatabase3;
by Id_Property;
run;
```

This final step of the procedure referred to the first three key RER databases creates the *mainDatabaseRER*.

The last RER database contains all data referred to the real estate income and economic value, and the corresponding council tax rate. They are all administrative data. The cleaning and recoding problems we found in this last RER database are the same we encountered in the second RER database previously described. In this last RER database, the codification of each real estate unit is based on the taxpayer in a framework of Urban Sections, and RER Sheet, Subordinate and Parcel reference information. Duplication rises because more than one taxpayer can pay the council tax for the same building in a co-ownership status. Therefore, we followed a double step procedure and created the databases we need.

The first step of the procedure is to recode the fields *Urban_Section* and *Registry_Sheet*:

Proc 11

```
data name_DatabaseT1;
set name_Database9;
if (Urban_Section in (report the old
                      "Urban_Section code" set))
    then Urban_Section = report the new "Urban_Section code";
if (Registry_Sheet in (report the old
                       "Registry_Sheet code" set))
    then Registry_Sheet = report the new "Registry_Sheet code";
...;
run;
```

The second and final step of the procedure is to clean the database. To delete duplication in the key field records, we used as key fields *Urban_Section*, *Registry_Sheet*, *Registry_Parcels* and *Registry_Subordinate*:

Proc 12

```
proc sort data = name_DatabaseT1 out = mainDatabaseCT nodupkey;
by Urban_Section Registry_Sheet Registry_Parcels
   Registry_Subordinate};
run;
```

This final step for the fourth RER database creates the *mainDatabaseCT*.

In the end, we created the omni comprehensive RER database, which aligns all the administrative RER data, by merging the *mainDatabaseRER* and the *mainDatabaseCT* through the key fields *Urban_Section*, *Registry_Sheet* and *Subordinate*. We named the final RER database *FinalRER_DB*:

Proc 13

```
data FinalRER_DB;
merge mainDatabaseRER mainDatabaseCT;
by Urban_Section Registry_Sheet Subordinate;
run;
```

3.2 REIO Database

REIO database is the last key database we used in our work and one of the most reliable sources of data to analyse the real estate monetary value dynamics in the Italian real estate market. The REIO monetary values of each real estate unit are calculated based on the trade price per square meter of properties. They are open data on a biannual basis referred to the minimum and maximum price for all the different types of real estates at the level of the council area. Therefore, to use a univocal REIO value in the analysis, we calculated the midrange value for each record.

Furthermore, in the REIO dataset, each municipality is split into homogeneous areas that experience the same economic and socioenvironmental characteristics, i.e. the REIO zones. The latter, successively, are grouped into 5 territorial districts that delineate precise geographical portions of the urban space: Centre, Near-centre, Outskirts, Suburbs and Extra-Urban.

Finally, REIO data are statistical valued data and, therefore, cleaning and recoding actions were not necessary for our work.

4 Full Information Harmonised Real Estate Database

Once the RER databases are obtained in a homogenous format, the last part of the work is the creation of FIHRE-DB. The most crucial step in this last part of the procedure is to device a suitable method for connecting and merging all RER data with REIO data. In fact, *FinalRER_DB* cannot be connected with the REIO database by any field. Therefore, the instrumental databases, as described previously, play an important role to obtain FIHRE-DB. They are all geo-referred and, therefore, a geo-localised process is suitable to solve the problem. In fact, we defined a multiple procedure for merging all the five key databases through the geo-localised information.

First, the creation of a "linking table" named BRIDGEDB permitted to connect REIO zones with the Italian NSI census sections. In other words, the procedure connects the Italian NSI census sections, and indirectly all real estate units, with each REIO zone and, consequentially, the official statistical information with all the real estate market values.

Proc 14

```
data FinalREIO_DB;
merge mainREIO_DB BRIDGEDB;
by REIO_Zone;
run;
```

This first step of the procedure creates the *FinalREIO_DB*.

The second step of the procedure is to merge the Italian NSI census sections with *FinalRER_DB* through the key fields *Urban Section*, *Registry Sheet* and *Subordinate*:

Proc 15

```
data Final_Administrative_DB;
merge FinalRER_DB NSI_Census_DB;
by Urban_Section Registry_Sheet Subordinate;
run;
```

This second step of the procedure creates the *Final_Administrative_DB* where all administrative RER data are perfectly aligned with the official Italian NSI data. In other words, all official statistical data, geo-referred based on the census sections, can be easily connected to the *Final_Administrative_DB* data.

Finally, the last step of the procedure merged *Final_Administrative_DB* with *FinalREIO_DB* through the Italian NSI census sections:

Proc 16

```
data FIHRE_DB;
merge Final_Administrative_DB FinalREIO_DB;
by Census_Section;
run;
```

This is the final outcome of our work. In fact, this final step of the procedure creates the Full Information Harmonised Real Estate Database (FIHRE-DB) that contains, for each real estate unit, the following variables: RER property identifiers, toponyms, RER economic value, council tax, REIO market value, REIO zone and ISTAT census section. As stated in Sect. 1, all variables are aligned and harmonised.

5 Final Remarks

The IT and Statistical procedure we depicted in this work is a first attempt to provide an omni comprehensive database where all available data, both administrative and statistical, referring to the analysis of the real estate economy are included. As known in literature, administrative data and their use to complement official statistical data represent a new frontier of research. The objective is ambitious for the analysis and challenging because of its consequent complexity, at the same time. All the aspects of the nowadays society are characterised by innumerable activities that substantially represent data, and this is indisputable when referred to the administrative actions. Many are the issues that need to be solved when administrative data are used to integrate official statistical data, and solutions are not easy most of the time. The Full Information Harmonised Real Estate Database we created in this work in original and has vast potentialities to develop complete analyses of the real estate economy from the point of view of both statistical aspect and financial/economic perspective. In fact, all real estate data are aligned and harmonised in a sole database, and the final outcome is a complete and eclectic collection of information that is well suited to conduct all types of analysis. In the end, the experiment we run on the sole city of Bari solves the dearth of a complete information related to the real estate economy because of the practical non-existence of a full information harmonised database.

The Full Information Harmonised Real Estate Database we created can be an important input to replicate the procedure we defined to any other dimensional geographical area and any other economic analysis. The IT and Statistical procedure we describe to create the Full Information Harmonised Real Estate Database is quite complex but all issues, included those referred to the dimension of databases, are solved. Finally, we also consider our outcome an important starting point towards that fair reform of the Italian National Registry of private and public properties that the Italian government planned to reform through the Italian National Recovery and Resilience Plan document.

Acknowledgements The authors are grateful to the PA offices of the city of Bari for providing RER databases.

References

1. Bryant, J.R., Graham, P.: A Bayesian approach to population estimation with administrative data. J. Offic. Stat., Sciendo **31**(3), 475–487 (2015)
2. Harford, T.: Big data: a big mistake? Significance **11**, 14–19 (2014)
3. Johnson, B.W., Ludlum, M., Rib, T.: Using administrative records to improve official statistics produced by the statistics of income division. IRS. Stat. J. IAOS **34**(1), 25–32 (2018)
4. Kitchin, R.: Data, new epistemologies and paradigm shift. Big Data Soc. **1**, 2053951714522848O (2014)
5. Kitchin, R.: The Data Revolution: Big Data, Open Data. Data Infrastructures and Their Consequences. Sage Publications, London (2014)
6. Kitchin, R.: The opportunities, challenges and risks of bigdata for official statistics. Stat. J. IAOS **31**(3), 471–481 (2015)
7. Marini C., Nicolardi V.: Big Data and Economic Analysis: The Challenge of a Harmonized Database, Data Science and Social Research II. DSSR 2019. Studies in Classification, Data Analysis, and Knowledge Organization, Springer Nature (2020)
8. Marini, C., Nicolardi, V.: Administrative database and official statistics: The case of the real estate analysis. Statistica Applicata - Italian J. Appl. Stat. **33**(1), 83–95 (2021)
9. Nordbotten, S.: The use of administrative data in official statistics - past, present, and future - with special reference to the Nordic countries. J. Offic. Stat. 205–223 (2010)
10. Pacheco, A.F., Julián, M.P.: Use of administrative records in official statistics: a practical view. BEIO, Boletín de Estadística e Investigación Operativa **30**(1), 49–65 (2014)
11. Thomsen, I., Holmoy, A.M.K.: Combining data from surveys and administrative record systems. The Norwegian experience. Int. Stat. Rev. **66**(2), 201–221 (1998)

Improving Administrative Data Quality on Tourism Using Big Data

Antonella Bianchino, Armando d'Aniello, and Daniela Fusco

Abstract National Statistical Institutes (NSIs) have been using the Administrative Data for official statistics production for many years. However, they are characterized by typical errors of the source (e.g., coverage, delay in data registration) that could be solved by integrating other sources. The study shows how the use of Big Data can partially solve some of the errors. Particularly, the information about tourist accommodation obtained through web scraping, was integrated with regional Administrative Data to improve the final quality assessment.

Keywords Administrative-data · Big-data · Multisource · Tourism · Data-quality

1 Introduction

The NSIs ordinarily use the Administrative Data (AD) sources for statistical purposes. By AD we mean the information produced by the Public Administration for administrative purposes which, processed by research institutes, become usable for statistical purposes. The rapid development of Public Administrations (PA) Information Technology (IT) made this integration process possible, providing back structured and easily usable information on companies, institutions, individuals. However, non-sampling errors characterize the AD. For example, the time lag between the date of presentation of the deed and its reference period: the loss of information due to delay in updating the source and the presence of units cataloged

A. Bianchino · A. d'Aniello · D. Fusco (✉)
Istat, Rome, Italy
e-mail: dafusco@istat.it

A. Bianchino
e-mail: bianchin@istat.it

A. d'Aniello
e-mail: armando.daniello@istat.it

M. Bini et al. (eds.), *Advanced Methods in Statistics, Data Science and Related Applications*, Springer Proceedings in Mathematics & Statistics 467,
https://doi.org/10.1007/978-3-031-65699-6_9

with obsolete classifications. These aspects can reduce the informative power of administrative archives.

This study demonstrates how Big Data can improve the coverage of AD. Particularly we analyzed the Campania Region tourist accommodation affected by over coverage error. The Region uses two different AD sources to improve information about this kind of facilities: "Rilevatore Turistico Regionale" and "Turismo Web". The first one is a platform created for a direct communication of tourist movement between tourist accommodations and province tourist offices. Turismo Web is a portal dedicated to accommodation facilities located in the region for the on-line communication of prices to the municipalities of the region for the continuous census of accommodation facilities. Both sources contain identifying information about the structures (identification code, address, telephone number, e-mail, etc.).

The activity status of the structures was evaluated combining the data of the two AD Region sources with the Big Data source TripAdvisor and the Italian Business Register (Asia). The Statistical Register of Active Enterprises contains economic units that exercise arts and professions in industrial, commercial and service activities for businesses and households and provides identification (name and address) and structure (economic activity, employees and independent, legal form, start and end date of activity, turnover) information of these units.

TripAdvisor is among the largest travel platforms in the world, used by approximately 463 million travelers every month. It is used to consult over 859 million reviews and opinions on 8.6 million accommodations, restaurants, experiences, airlines and cruises. It is chosen as Big Data source for the number of accommodation facilities reported, similar with the number published by Italian statistical institute. The TripAdvisor data were extracted by web scraping, the reference period is March 2020. The data were obtained through two levels scraping, collecting information on name, address, municipality, contacts and type of facility. Regarding to Regional AD the reference period is December 2019.

The four different sources were integrated using record linkage models. The study shows the effect of the introduction of the Big Data source in the quality assessment of the integrated dataset.

2 The Quality Framework

2.1 Quality Issues Referred to Different Sources

Using a transformation function, the administrative objects are derived in Statistical units [14]; in this way the AD are aligned with statistical data firstly at the metadata level (through comparison and connection of definitions) and then at the data level. A mismatch between the statistical population of interest and the population identified by the administrative source may depend on several factors:

(a) Definitional difference between the target population and the population covered by the administrative source.
(b) Incorrect identification of statistical units.
(c) Delays and/or missing administrative registrations.

The three factors cause coverage errors (over or under).

Where is possible, more administrative sources should be used. This would allow to reduce both the under-coverage error (by integrating sources that cover different portions of the population) and the over-coverage error (having the possibility of more information to establish which units correctly belong to the desired field of observation) [1].

For quality assessment, Accuracy has a specific importance: for a long time, it was the main defining characteristic of the quality of a data product [3]. It represents the degree of closeness of computations or estimates to the exact or true values what the statistics were intended to measure. The bias and the variance of the estimates influence the accuracy of the final estimates. Theoretically, in multisource statistics, the use of additional sources should have a positive effect on the accuracy of the output, particularly on the coverage.

In this direction, Big Data play an important role. Big Data sources can generally be described as: "high volume, velocity and variety of data that demand cost-effective, innovative forms of processing for enhanced insight and decision making" [11]. In addition to generating new commercial opportunities in the private sector, Big Data are potentially a very interesting data source for official statistics, either for use on their own, or in combination with more traditional data sources such as sample surveys and administrative registers [4].

Among the differences between Big Data and traditional sources is the degree of control that NSIs have on the data acquisition and recording. In traditional surveys, NSIs can plan, design and carry out the acquisition procedures; for the use of administrative data sources, they may have agreements with the data providers [12]. Indeed, in a survey life cycle, coverage errors attain to the dimension of the representation line, i.e. to the target population or the set of units to be studied ("who") [6]. Instead, Big Data could not be representative of the whole population but just a fraction of it, which will probably have specific characteristics that differ from the broader population. In other words, the problem is similar to the one faced by studies dealing with non-probabilistic samples.

2.2 *The Sources Quality Evaluation*

Administrative data quality assessment was done using the method proposed by Statistics Netherlands [5]. The AD sources used in the study were evaluated using specific qualitative and quantitative indicators calculated on the three hyper-dimensions Source, Data and Metadata, in order to determine their statistical usability as a source of administrative data (Table 1). In the Source hyper-dimension, the

Table 1 Quality assessment of administrative data, year 2019

Indicators	Source	
	Turismo web	Rilevatore T. R
Contact	+	+
Purpose	+/o	+
Response burden (Effect of data source use on response burden)	+	+
Feedback	+	+
Security	+	+
Indicators	Metadata	
	Turismo web	Rilevatore T. R
Clarity	+	+
Comparability	+ /o	+
Unique key	+	+
Meta data treatment	–	–
Indicators	Data	
	Turismo web	Rilevatore T. R
Technical check	+	+
Over coverage	+	+
Under coverage	?	?
Data treatment	–	–
Not responding units	+	+
Missing partial response	+/o	o/–
Measurement	+/o	o/–
Sensibility	+	o

Source Our elaboration
Legend + Good, oReasonable, – Insufficient, ? Not clear, / Intermediate between values indicated

quality aspects related to the data source as a whole, the data source keeper, and the delivery of the data source to the NSI are studied. The Metadata hyper-dimension specifically focuses on the metadata related aspects of the data source. Clarity of the definitions and completeness of the meta information are some of the quality aspects included. The Data hyper-dimension focuses on the quality aspects of the data (facts) in the data source. The quality aspects of the Data hyper-dimension are mainly related to accuracy except for those included in the Technical Checks dimension [7].

Since each dimension contains several quality indicators, which are measured by one or more methods, the results shown were obtained by comparing the evaluation results for every measurement method for each quality indicator in each dimension. The symbols for the scores used in Table 1: good (+), reasonable (o), poor (−) and unclear (?); intermediary scores are created by combing symbols with a slash (/) as a separator.

Table 2 Number of tourist accommodation for source and Province, year 2020

Provinces	Source					
	Istat (2018)	TripAvisor	Google	Booking	Kayak	Hotels.com
Caserta	439	353	671	160	5,871	2,515
Benevento	653	244	415	63	160	343
Napoli	3,453	3,714	7,408	4,148	7,103	4,805
Avellino	403	231	399	406	108	1,660
Salerno	2,255	2,517	4,613	680	1,592	2,873
Total	7,185	7,059	13,506	5,457	14,815	12,196

Source Our elaboration on Istat and internet data

Regarding to Source and Metadata hyper-dimension, the sources considered do not present particular problems, except for the Meta Data treatment indicator, which is poor for both sources. Data hyper-dimension evaluation was totally different. Particularly, Turismo Web source has a poor evaluation for Missing partial response caused by the number of no-responses on the key variable Identification Code (90%). Furthermore, Measurement indicator is poor for the number of observations with errors, especially for the key variable Address (75%).

Under Coverage indicator is unclear because of the lack information about the percentage of units missing from the target population.

Currently, comprehensive analysis and research of quality standards and quality assessment methods for big data are lacking. When Big Data sources are used, evaluation plays an important role [2]. Most of the specificities of Big Data are related both to its quick pace of change and in terms of the population covered. Due to the complex nature of Big Data, was proposed, at European level, extensions to existing statistical data quality frameworks [13]. For this study, only hubs with relevant likability degree were considered. Likability is the ease with which the data can be linked or merged with other relevant datasets.

To evaluate population coverage of tourist accommodation hubs, they are compared with Italian statistical institute (Istat) official statistics (Table 2), used as benchmark. How table two shows, TripAdvisor has a number of units similar to Istat data, while other hubs have oversized or undersized the reference population compared to the benchmark.

3 Methodology and Results

The AD sources (Turismo Web and Rilevatore Turistico Regionale), the Big Data source (TripAdvisor) and the statistical source (Asia) are integrated applying record linkage models. The purpose of record linkage is to identify the same real-world entity that can be differently represented in data sources [8], even if unique identifiers are not available or data are affected by errors.

The main phases of a record linkage process are [10]:

1. Preprocessing
2. Matching
3. Analysis

Generally speaking, the preparation of input files is the first phase and requires 75% of the whole effort to implement a record linkage procedure, in fact data can be recorded in different formats, some items may be missing or with inconsistencies and errors. In this case, it was very onerous: the AD were not treated and Big Data were "dirty". It was necessary, for example: removing useless text, codifying municipality, removing special characters, modify variables as telephone number or e-mail address, identifying and removing the recurring words.

After the previous phase, it is important to choose matching variables that are as suitable as possible for the considered linking process. In this study, the variables chosen in the different models were the following: Identification code, municipality, structure name, owner name, address, telephone number, e-mail.

Before starting the actual linkage phase, it is generally necessary to reduce the search space in order to evaluate only a reduced set of pairs of records from the two datasets, thus reducing the computational burden of the operation and consequently the time required.

To reduce the search space, the functions used alternately in the models were:

- Blocking Union: reduce the search space only to the pairs that report the same values for the selected blocking variables (with Province and Municipality variables).
- SimHash: is a locality-sensitive hashing (LSH) technique used to find near-neighbors in high dimensional space (with address and name variables).

Comparison functions are used to compute the distance between records based on the values of the chosen matching variables. Equality and 3 g are used in approximate string matching by "sliding" a window of length q over the characters of a string s to create a number of q length grams (in this case we considered q equal to 3) for matching.

The decision model chosen was this: at first, the deterministic approach followed by the probabilistic one. It is important underline the main differences of these models. In the deterministic approach, a pair is a match if it agrees completely on all the matching variables chosen or satisfies a defined rule-base system, i.e., if it reaches a score that is beyond the selected threshold when applying the comparison function. The probabilistic approach, based on the Fellegi and Sunter model [9], requires an estimation of the model parameters that can be performed via the EM algorithm, Bayesian methods, etc. In particular, starting from two lists of records A and B, for each single pair of records (a, b), a in A and b in B, it can take one of the following decisions: link, possible link and non-link. The comparison between the matching variables of the two units (a, b) is made by means of a suitable comparison function. For each element of Ω containing all the pairs (a, b), the result of the comparison of the matching variables is summarized in the vector γ, called comparison pattern.

The probability models for linkage assume that the probability distribution of the comparison pattern comes from a mixture of two probability distributions. The first one comes from the pairs *(a, b)* that actually are the same unit, distribution *m*; the other one comes from the pairs *(a, b)* that actually represent different units, distribution *u*. Starting from the estimations of the two distribution, m(γ) and u(γ), it is possible to define the composite matching weight, given by the likelihood ratio: $r = \frac{m(\gamma)}{u(\gamma)} = \frac{P(\gamma|M)}{P(\gamma|U)}$, where M is the set of the pairs that are links and U is the set of the pairs corresponding to non-links, with $M \cup U = \Omega$, $M \cap U = \varnothing$. Fellegi and Sunter proposed an equation system to achieve the formulas for the estimates of $m(\gamma)$ and $u(\gamma)$ when the matching variables are at least three. Once the composite weight r is estimated, it is possible to classify a pair as a link if the corresponding weight r is above a certain threshold T_m, and as a non-link if the weight lays below the threshold T_u. Finally, for the pairs corresponding to weights falling into the range $I = (T_u, T_m)$, no-decision is made, and the pair is assigned to a clerical review analysis. According to the Fellegi and Sunter theory, a decision on the threshold levels has to be made in order to properly manage the trade-off between the need of a small number of expected no-decisions and small misclassified error rates for the pairs.

At the end of the process, 13,274 tourist accommodations were matched. Using the integrated dataset containing a new variable relating to the presence of each unit in each source called SOURCE, a model was built to establish the activity status of units.

The sources were classified considering the reliability: the Business Register is the primary source for the quality building process; between the two AD sources, Turismo Web resulted the most reliable, according with the results show in paragraph 2.2; lastly, TripAdvisor is the least reliable concerning Big Data's typical errors. It was then attributed a score on each unit considering their presence in each source (variable SOURCE): Asia 0.45, Turismo Web 0.25, Rilevatore Turistico Regionale 0.20, TripAdvisor 0.10.

Summarizing the scores on each unit, the thresholds for the activity status of units were defined as follows:

- More than 0.56: Active;
- Between 0.45 and 0.55: Eligible;
- Less than 0.44: Not eligible.

Results are shown in Table 3. Considering Istat official data as a benchmark, the total difference with the Eligible units into the integrate Dataset is of 36%. In particular, the difference was determined by the quantity of the structures in the province of Salerno, which, although decreased in number compared to the initial situation, is always twice the value found in the survey on the capacity of the accommodation facilities. This could depend on the presence of non-entrepreneurial managed structures that are not part of the Istat survey field of observation but could be present in the integrated dataset with score equal to 0.45.

Table 4 shows some quality indicators calculated on the integrated dataset. The coverage rate is calculated as the following ratio: *(Not eligible units+(1-alpha) Units solved)/(Units solved+Not solved units+alpha Not solved units)*. Units solved are the

Table 3 Integrated database tourist accommodation for typology

Province	DB units	Not eligible	Eligible	Actives	Eligible+actives	Istat
Caserta	566	88	199	279	478	439
Benevento	764	151	374	239	613	635
Napoli	5,111	1,898	1,104	2,109	3,213	3,453
Avellino	438	10	198	230	428	403
Salerno	6,395	1,289	2,952	2,154	5,106	2,255
Total	13,274	3,436	4,827	5,011	9,838	7,185

Source Our elaboration

Table 4 Integrated dataset quality indicators by province

Province	Coverage rate (%)	Eligibility rate (%)	Activity rate (%)
Caserta	15.5	35.2	49.3
Benevento	19.8	49.0	31.3
Napoli	37.1	21.6	41.3
Avellino	2.3	45.2	52.5
Salerno	20.2	46.2	33.7
Total	25.9	36.4	37.8

Source Our elaboration

number of units which was possible determine the eligibility; Not solved units are the unknowing eligibility units; alpha is the ratio of not solved units that may be eligible (Eurostat suggests to set alpha = 1).

The coverage rate for Campania region is +25.9% (over coverage), respect to +75.2% of over coverage obtained using only AD sources. If the integrated register is used as a frame, it could be accepted. It's important to underline that probably the over coverage depends on different referring data of the sources. This limit could be overcome by running the model in an ordinary process.

Eligibility rate (number of eligible structures out of the total) and activity rate (number of active eligible structures out of the total) show that the integrated register has made it possible to "discard" 26% of the administrative data sources structures, because of their low probability of existence.

4 Conclusion

Most EU Member States have been moving towards an increased use of administrative data sources for statistical purposes, as a substitution and/or as a complement to survey data. At the same time, the emergence of Big Data allows for a further increase of available sources for statistics [15]. As a result, statistical output is based on complex combinations of sources.

Moving in this direction, this paper analyses the appropriateness of the use of Big Data to improve the data accuracy. The results show how Big Data can be introduced in statistical productive processes as a partial solution to some administrative data limits. Thanks to the record linkage work carried out using enterprises details data, we have experienced how Big Data can contribute to reducing time lag errors or improving aspects relating to possible under or over coverage of administrative sources, based on the presence of individual accommodation facilities in different big data sources.

In this case, Big Data source was used as a secondary source, contributing to the definition of activity state of the accommodation structures. A step forward could be using Big Data as the main source, supplementing the administrative source. This would allow identifying the structures not yet registered due to delays in the registration of administrative acts by the PA.

To complete the quality control, a microanalysis of the structures deemed ineligible will be required, through sample checks with the help of the provincial tourism bodies with the specific role of data collectors.

References

1. Ascari, G., Alexander, P. J., Brancato, G., Burg, T., Zhang, L. C., Waldner, C.: Quality Guidelines for Multisource Statistics (QGMSS). Version 0.8.1. ESSnet Quality of Multisource Statistics. Komuso, Eurostat (2018)
2. Bogdanovits, F., Degorre, A., Gallois, F., Fischer, B., Georgiev, K., Paulussen, R., Quaresma, S. Scannapieco, M., Summa, D., Stoltze, P.: BREAL: Big Data Reference Architecture and Layers. Business Layer. ESSnet Big Data II. Work Package F. Process and architecture. Eurostat. Bruxelles (2019)
3. Brancato, G., Boggia, A., Barbalace, F., Cerroni, F., Cozzi, S., Di Bella, G., Di Zio, M., Filipponi, D., Luzi, O., Righi, P., Scanu, M.: Linee guida per la qualità dei processi statistici che utilizzano dati amministrativi. Istat (2016)
4. Cheung, P.: Big data, official statistics and social science research: emerging data challenges. Presentation at the December 19th World Bank meeting, Washington (2012)
5. Daas P.: On the quality of registers. Joint UNECE/Eurostat expert group meeting on register-based censuses. The Hague, The Netherlands, 10–11 May 2010 (2016)
6. Daas, P., Ossen, R., Vis-Visschers, R., Arends-Tóth, J.: Checklist for the quality evaluation of administrative data sources. Discussion paper (09042). Statistics Netherlands (2009)
7. Elmagarmid, A.K., Ipeirotis, P.G., Verykios, V.S.: Duplicate record detection: a survey. IEEE Trans. Knowl. Data Eng. **19**(1), 1–16 (2007)
8. Fellegi, I.P., Sunter, A.B.: A theory of record linkage. J. Am. Stat. Assoc. **64**(328), 1183–1210 (1969)
9. Groves, R.M., Fowler, F.J. Jr., Couper, M., Lepkowsky, J.M., Singer, E., Tourangeau, R.: Survey Methodology. Wiley, New York (2004)
10. Jabine, T.B., Scheuren, F.J.: Record linkages for statistical purposes: methodological issues. J. Off. Stat. **2**, 255–277 (1969); Fellegi, I.P., Sunter, A.B.: A theory of record linkage. J. Am. Stat. Assoc. **64**(328), 1183–1210 (1986)

11. Manyika, J., Chui, M., Brown, B., Bughin, J., Dobbs, R., Roxburgh, C., Hung Byers, A.: big data: the next frontier for innovation, competition, and productivity. Report of the McKinsey Global Institute, McKinsey & Company (2011)
12. Quaresma, S., Maślankowski, J., Salgado, D., Ascari, G., Brancato, G., Di Consiglio, L., Righi, P., Tuoto, T., Daas, P., Six, M., Kowarik, A.: Revised version of the quality guidelines for acquisition and usage of big data. ESSnet Big Data II. Eurostat (2020)
13. UNECE Big Data Quality Task Team.: A suggested framework for the quality of big data (2014).
14. Wallgren, A., Wallgren, B.: Register-Based Statistics: Administrative Data for Statistical Purposes, 2nd edn. Wiley, Chichester, UK (2014)
15. Vâju, S., Agafiţei, M., Gras, F., Kloek, W., Reis, F.: Measuring the quality of multisource statistics. Paper presented at NTTS 2015, Reliable Evidence for a Society in Transition. Brussels (2015)

A Regional Analysis of the Efficiency by Energy Producers in Italy

Gianna Greca, Giuseppe Cinquegrana, and Giovanni Fosco

Abstract The energy market in Italy is characterized by a competitive framework where regulatory Authority checks the conditions in which operate enterprises that produce, store, distribute and trade energy. The presence of government subsidies towards the energy's production Italian enterprises from two decades up now days in different ways (*Energy Account*, *Green Certificates*, *GRIN,* etc.) points out the issue if the efficiency of this industry was affected by these incentives. In this study we focus on the efficiency of the energy's production SMEs taking into account the territorial gap among the Italian Regions. An analysis on the impact of the subsidies to the energy producers has been conducted regarding the efficiency of the industry on the basis of data collected (Source ISTAT, GSE).

1 Introduction

The Italian energy market is regulated and supervised by the Energy, Networks and Environment Regulatory Authority (ARERA). The regulation covers the production, storage, distribution and trading of energy and affects companies of different sizes and industrial activities. The electricity production market has been characterized by a flow of public subsidies over the last two decades [6]. In particular, several funds have been granted to the renewable energy sector in order to promote clean energy production. Italy has employed various economic mechanisms, including *feed-in tariffs* and *feed-in premiums,* to incentivize electricity production from renewable sources. Such mechanisms have evolved over the years in response to changing market dynamics

G. Greca · G. Cinquegrana (✉)
Istat, Rome, Italy
e-mail: gicinque@istat.it

G. Greca
e-mail: greca@istat.it

G. Fosco
Università degli Studi della Campania, Capua, Italy
e-mail: giovanni.fosco@unicampania.it

M. Bini et al. (eds.), *Advanced Methods in Statistics, Data Science and Related Applications*, Springer Proceedings in Mathematics & Statistics 467,
https://doi.org/10.1007/978-3-031-65699-6_16

and declining costs of renewable energy generation [7]. However, these subsidies appear to have stimulated the expansion of the energy production market in Italy by broadening the renewable energy production base [3]. Some authors [8] argue that government incentives to companies can also create obstacles to competition, potentially leading to the fragmentation of the energy sector into numerous smaller entities with limited economies of scale. Therefore, the present study aims at shedding a light on the impact of governmental incentives to renewable energy production on the cost efficiency of companies within the sector.

First, we describe and analyze the mechanisms that have played a key role in renewable electricity generation. Specifically, the "Conto Energia—CE" (Energy Account) mechanism, which rewards photovoltaic systems for energy production, and the "Gestione Riconoscimento Incentivo—GRIN" (Incentive Recognition Management) mechanism, which has replaced green certificates for many renewable energy sources. The Conto Energia mechanism started in 2005, it underwent multiple phases each introducing new rules and tariff structures modification. This mechanism terminated once public allocated funds ended.

Then, we shift our attention to the GRIN mechanism, which was launched in 2016 and has been applied across a range of renewable energy sources, mainly encompassing wind, hydro, biomass and solar. The GRIN mechanism has displayed remarkable growth of disbursed incentives, with relevant amounts paid to eligible entities.

In the second part of our analysis, we employ a cost frontier function [2] to estimate the cost inefficiency of individual energy producers. Furthermore, we evaluate the impact of the GRIN incentive on cost efficiency using a quasi-experimental econometric approach, specifically employing the difference-in-differences (diff-in-diff) method. The results indicate that the GRIN incentive increases the cost inefficiency of energy producers, with a more significant effect observed in the northern regions. This can be attributed to the higher concentration of energy producers in the North and the dependence of these producers on the incentive for their profits, which may lead to a less efficient allocation of funds.

In summary, this study examines the complex landscape of government subsidies to energy producers in Italy, particularly in the renewable energy sector. It provides valuable insights into the impact of these subsidies on cost efficiency and highlights the disparities in the distribution of subsidies across regions. Using rigorous econometric analysis, the study sheds light on the effectiveness and consequences of such incentive mechanisms in promoting renewable energy production.

1.1 Government Subsidies to Energy Producers

The promotion and incentive of electricity produced from renewable sources has been achieved in Italy through multiple mechanisms that have followed one another over the years, taking into account both the market trend and the reduction in incentive levels in response to the reduction in costs of generation.

Several economic mechanisms have been implemented, in particular:

- "*feed in tariff*", incentive mechanism through which the producer is recognized, for the energy injected into the grid, with a single tariff which includes an incentive component and an energy valorisation component; the energy, therefore, does not remain available to the producer;
- "*feed in premium*", incentive mechanism through which the producer is granted an incentive for the energy produced or injected into the grid; the energy, therefore, remains available to the producer, who therefore also receives the revenue from the sale.

This paragraph analyses the mechanisms that most contribute, in terms of power, to the production of electricity from renewable sources and which benefit from greater subsidies in economic terms.

In particular, the following incentive mechanisms connected to "renewable and similar sources" are analysed:

- "Conto Energia—CE" (*Energy Account—EC*);
- "Gestione Riconoscimento Incentivo—GRIN" (*Incentive Recognition Management*, former *Green Certificates—CV*).

For plants that generate electricity through the conversion of solar energy, a specific incentive system called Conto Energia (CE) has been envisaged. The CE mechanism rewards the energy produced by photovoltaic systems with incentive tariffs for a period of twenty years (*feed in premium*). This mechanism, already provided for by Legislative Decree 387/2003, became operational following the entry into force of the interministerial decrees of 28 July 2005 and 6 February 2006 (I CE).

The II CE came into force, through which some critical issues were removed which were a brake on the construction of photovoltaic systems, by modifying and simplifying the rules for access to incentive tariffs (Ministerial Decree On 19 February 2007). Subsequently, the III EC, to be applied to systems, which entered into operation starting from 1 January 2011 and in continuity to the incentive mechanism, introduced specific tariffs for integrated photovoltaic systems with innovative characteristics and systems concentrated photovoltaics (Ministerial Decree 6 August 2010). With Law no. 129 of 13 August 2010 (so-called "Save Alcoa Law"), the tariffs for the year 2010 of the II CE were then confirmed for all plants capable of certifying the completion of the works by 31 December 2010 and enter into operation by 30 June 2011. After the issuing of Legislative Decree 28/2011, the Ministerial Decree was published. 5 May 2011 (IV CE) with the aim of aligning the level of tariffs with the evolution of the costs of photovoltaic technology and introducing a limit on the cumulative annual cost of incentives, set at 6 billion euros. As we approached the identified cost limit, the further Ministerial Decree was published. 5 July 2012 (5th CE).

The V CE partially confirmed the provisions already foreseen by the IV CE and introduced new rules. In particular, instead of a fixed incentive premium paid on the basis of the electricity produced, an incentive was defined consisting of two rates (on two different shares of the energy produced):

- as regards the self-consumed share of energy produced, a premium tariff has been envisaged;
- as regards, however, the share of net production injected into the network:
  - for plants with nominal power up to 1 MW, a feed in tariff (Tariffa omnicomprensiva—TO) has been established, determined on the basis—the power and type of system;
  - for plants with a nominal power exceeding 1 MW, recognition of the difference between a reference tariff and the hourly zonal price has been provided for (since the valorisation of electricity is the responsibility of the producer).

The incentive provisions for photovoltaic systems ceased to apply (meaning that operators other than those who had already obtained the right to the incentive could not access them) on 6 July 2013, after thirty days from the date of reaching the indicative cumulative annual cost of 6.7 billion euros. Beyond the aforementioned deadline, requests for recognition of incentives relating to plants affected by specific extensions remained entitled to be evaluated.

In the following table the amounts disbursed on CE mechanism are reported by decree and by year starting from 2007 up to 2020 (Table 1).

With reference to total disbursed amount, the main beneficiaries are "market enterprises" which received in 2021 incentives amounting to € 3.9 billion; these are more than 53,000 enterprises with the majority of capital companies, non-financial, not belonging to multinational groups, small and medium sized (belonging to the class of employees 0-249). Among the market enterprises, those receiving the largest amounts are those with registered offices in Lombardy, Trentino-Alto Adige, Lazio and Emilia-Romagna (Fig. 1) [4, 5]. Since the majority of enterprises are small and medium-sized in terms of employees, it can be assumed that the registered office corresponds to the place of production. In terms of number, the biggest beneficiaries are, instead, the consumer households, with more than 350,000 units to which incentives amounting to about 800 million euros have been granted.

Another relevant incentive mechanism for the production of electricity is GRIN mechanism that allows eligible subjects to take advantage of the incentives provided for by the D.M. 06/07/2012 for all plants powered by renewable sources which had

Table 1 Disbursed amounts through the Conto Energia mechanism by decree from 2007 up to 2019–2020 (millions of euros)

Decree	2007	2019	2020
I CE	17	72	75
II CE	2	2,16	2,978
III CE	...	584	596
IV CE	...	2,187	2,227
V CE	...	210	219
Total	19	5,969	6,096

Source GSE

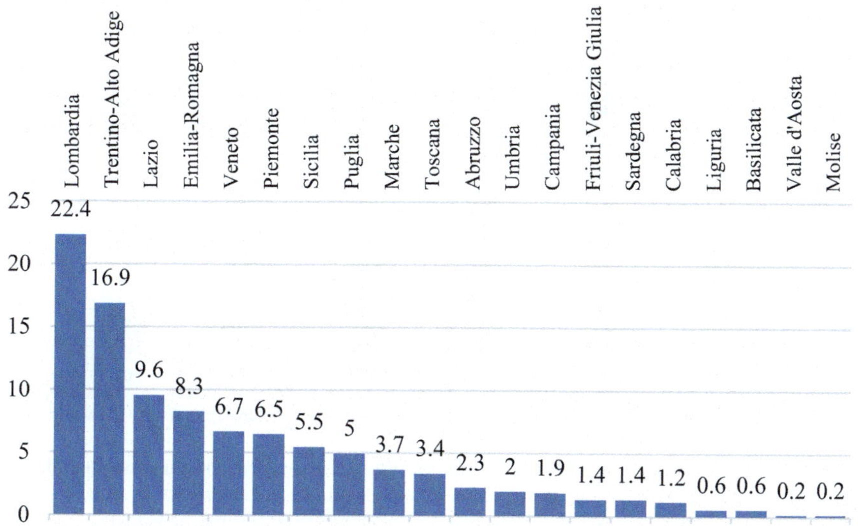

Fig. 1 Disbursed amounts through the Conto Energia mechanism to market enterprises by region (percentage values)—Year 2021. *Source* Istat, elaborations Istat thematic research Project "I meccanismi di incentivazione energetico-ambientale (Energy-environmental incentive mechanisms)"

been qualified to be subsided by Green Certificates up to 2015. Since 2016, the Green Certificates mechanism has been replaced by this new form of incentive.

Unlike the CE mechanism which concerns the production of electricity only from solar source, the GRIN mechanism concerns the production of electricity form multiple energy resources: wind, hydroelectric, biomass, solar, etc. This work focuses on GRIN mechanism in order to evaluate the impact of these subsidies on the energy sector.

The analysis of disbursed amounts for GRIN incentive mechanism shows that in the period 2016–2021 an extraordinary growth of 77.1% was recorded with a peak of 3.2 billion euros reached in 2017 (Fig. 2).

If we consider the balanced panel of enterprises present in the period 2016–2021, it may be observed that the disbursed amounts amount to approximately 1.1 billion euros and that most of the incentives are paid to companies with fewer than 50 employees (85.6% of the total) (Table 2). Furthermore, it should be noted that the higher amounts are intended for beneficiaries with registered offices in the North, in particular in Trentino-Alto Adige, Lombardia and Emilia-Romagna (respectively 25.2, 12.1 and 11.9%) (Tables 3 and 4). However, among the southern regions, Calabria and Puglia should be noted, which received respectively 9.5 and 8.6% of the total incentives. On the other hand, enterprises with registered offices in Abruzzo and Basilicata did not receive any incentive.

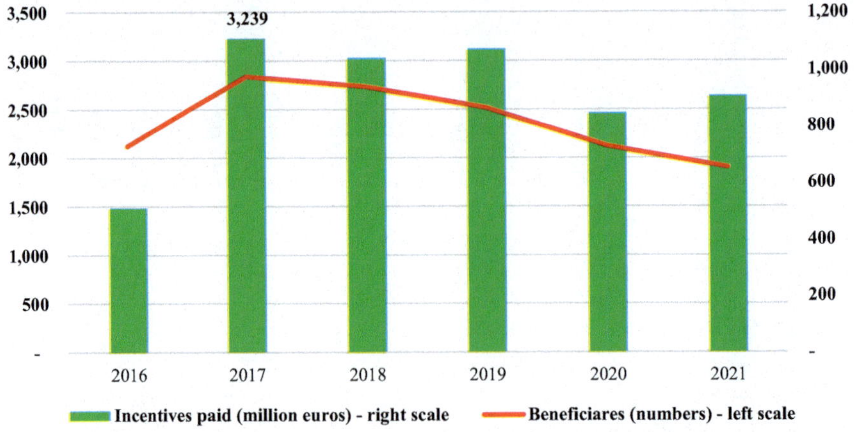

Fig. 2 GRIN mechanism: disbursed amounts (millions of euros) and beneficiaries (number)—Years 2016–2021. *Source* Istat, elaborations Istat thematic research Project "I meccanismi di incentivazione energetico-ambientale (Energy-environmental incentive mechanisms)"

Table 2 Disbursed amounts through the GRIN mechanism to the panel of enterprises by employee class (millions of euros)—Years 2016–2021

Size by employee	Total
0–9	490
10–49	415
50–249	86
Oltre 250	69
Total	1.060

Source Istat, elaborations Istat thematic research Project "I meccanismi di incentivazione energetico-ambientale (Energy-environmental incentive mechanisms)"

Table 3 Disbursed amounts through the GRIN mechanism to the panel of companies by geographical area (millions of euros)—Years 2016–2021

Geographical areas	Total
Central Italy	79
Islands	7
Southern Italy	241
North-Western Italy	192
North-Eastern Italy	541
Total	1.060

Source Istat, elaborations Istat thematic research Project "I meccanismi di incentivazione energetico-ambientale (Energy-environmental incentive mechanisms)"

Table 4 Ranking of regions receiving incentives through the GRIN mechanism (millions of euros and position)—Year 2021

Regions	Total	Ranking position
Trentino-Alto Adige	266.8	1
Lombardia	128.3	2
Veneto	126.1	3
Emilia-Romagna	112.3	4
Puglia	100.5	5
Calabria	90.9	6
Lazio	48.4	7
Campania	42.5	8
Friuli-Venezia Giulia	35.6	9
Piemonte	33.7	10
Toscana	28.6	11
Valle d'Aosta	15.5	12
Liguria	14.6	13
Abruzzo	7.1	14
Sicilia	6.5	15
Marche	2.0	16
Sardegna	0.3	17
Total	1,060	–

Source Istat, elaborations Istat thematic research Project "I meccanismi di incentivazione energetico-ambientale (Energy-environmental incentive mechanisms)"

2 Data

We collect data on the entire population of firms categorized under the Italian Classification of Economic Activities coded as 3511, which pertains to energy production. Our data sources come from various databases in the Istat repository. Specifically, from the Frame SBS, an integrated system of administrative and statistical data for each enterprise, we collect balance sheet and economic data. Meanwhile, from Enerdata, an outcome of the Istat thematic research project "Energy Environmental Incentive Mechanisms," we gather data about the administrative units that receive the GRIN incentive. Finally, we create a balanced panel dataset consisting of 3,910 energy producers observed over a seven-year period spanning from 2014 to 2020, during which we could obtain comparable data series.

Table 5 reveals that our data predominantly comprises small and medium-sized enterprises. In Table 6, where we present the distribution of energy producers across Italian macro areas, it becomes evident that energy firms are primarily located in the northern macro areas. The GRIN incentive starts to be implemented in the year 2016. Specifically, the distribution of firms that receives it during the period 2016–2020, which we call treated firms, are displayed in the Table 7 was 347 in 2016, 373 in

Table 5 Number of energy producers' units by size included in the panel 2014–2020

Size by employee	Number of units
Up to 9	3819
10–49	85
50–249	6
Total	3910

Source our elaborations on Istat data

Table 6 Number of energy producers' units by Italian macro areas in the panel 2014–2020

Macro areas	Number of units
North-East	1522
North-West	1144
Centre	611
South	470
Islands	163
Total	3910

Source our elaborations on Istat data

Table 7 Number of energy producers' units treated between 2016–2020

Year	Number of treated
2016	347
2017	373
2018	358
2019	334
2020	294

Source our elaborations on Istat data

2017, 358 in 2018, 334 in 2019, and 294 in 2020. In the next section, we present the econometric model used to estimate cost efficiency for each macro area in a way that allows us to obtain comparable estimates.

3 Empirical Model

Our econometric analysis consists of two main steps. In the first step, we estimate the cost frontier function, which allows us to compute time-varying cost efficiencies for each energy producer, providing also different models for different Italian macro areas. In the second step, we aim to assess the impact of the GRIN incentive on cost efficiency through a quasi-experiment econometric methodology.

Our primary objective is to estimate a cost frontier function, which defines the minimum production cost based on a given output level, input prices, and technological constraints. By measuring the difference between the fitted frontier cost and the observed cost for each firm, we can evaluate cost efficiency, which encompasses both technical and allocative efficiency.

To estimate the cost efficiency for each energy producer, we employ the stochastic frontier model developed by Battese and Coelli [2], a parametric approach designed for panel data. In comparison to non-parametric methods such as Data Envelopment Analysis (DEA), parametric models offer the advantage of distinguishing inefficiency from statistical noise resulting from data errors and omitted variables. Moreover, they enable us to verify the statistical significance of the covariates in the model.

Battese and Coelli's [2] stochastic frontier model assumes that the inefficiency terms are independently distributed as truncations of normal distributions with constant variance, but with means that are linear functions of observable variables. Consequently, this model offers the advantage of providing a structural representation for the mean inefficiency, conditioning it through other covariates. Typically, the inefficiency function includes a trend variable with its polynomial to account for time-varying inefficiency.

Our frontier cost function assumes the Cobb–Douglas functional form as the technology underlying the production function. We use the classical form with one output (Q) and two inputs, namely labor (L) and capital (K). Because of its good mathematical properties, Cobb–Douglas facilitates the handling and interpretation of the parameters, allowing us also to assess the returns to scale of the production function in a simple way. Thus, our cost function depends on two prices: the price of labor input (p_w) and the price of the capital input (p_r).

Furthermore, we impose the linear homogeneity condition on our cost function, normalizing the total cost, labor price, and capital price by the energy price, enabling us to obtain coefficients and estimated cost efficiencies within the range of 0–1. The following equations describes our frontier cost function,

$$\begin{aligned}\ln\left(\frac{TC}{p_e}\right)_{i.t,c} =& \alpha_i + \gamma_t + \delta_c + \ln Q_{i.t,c} + \beta_r \ln\left(\frac{p_r}{p_e}\right)_{i.t,c} + \beta_w \ln\left(\frac{p_w}{p_e}\right)_{i.t,c} \\ &+ \beta_1 t + \beta_r t^2 + U_{i,t} + V_{i,t}\end{aligned}$$

TC represents the total costs derived from balance sheet information for the firm i in the period t located in the municipality c. While p_e indicates the energy price based on the deflator of value added in the energy sector. P_r denotes the capital price, calculated as the ratio between the sum of depreciation and finance charges over the value of fixed assets from the balance sheet. P_w points out the average wage per employee, computed as the total salary costs divided by the number of employees per firm. Q represents the chained value-added deflated by the price index of 2015, whereas t and its polynomial are trend variables intended to capture technological dynamics or progress. As such, there are fixed effects for the firm α_i, year γ_t, and municipality δ_c. β_q indicates the marginal cost, while β_k and β_w represent the price elasticities of capital and labor. In this specification, the error term comprises two uncorrelated components: the non-negative inefficiency component U and the statistical disturbance term V, which is assumed to follow a normal distribution. Furthermore, the specification by Battese and Coelli [2] allows for modelling the path of inefficiency by including explanatory variables:

$$U_{i,t} = \varphi_0 + \varphi_1 t + \varphi_2 t^2 + Z_{i,t}$$

In the first step of our analysis, we include only a trend variable with its polynomial to obtain a dynamics of the inefficiency estimates that change over time. This equation also accounts for Z as its error term for inefficiency. In the second step of our analysis, which is the central focus of our study, we aim to assess the impact of the GRIN incentive on the cost efficiency of energy producers who receive it. We break down the analysis by macro-region, dividing firms that receive the treatment from those that do not. In this second step, we modify the cost inefficiency equation as follows:

$$U_{i,t} = \varphi_1 + \varphi_1 post_{2016} + \varphi_2 GRIN_i + \varphi_3 GRIN_i \cdot post_{2016} + Z_{i,t}$$

Using the same frontier cost function, we implement the difference-in-differences (diff-in-diff) method to estimate the impact of GRIN subsidies [1]. In this modified formulation of cost inefficiency, "post2016" is an indicator that represents years after 2015, which is when the GRIN incentive was first implemented. "GRIN" is a dummy indicator identifying the recipients of GRIN subsidies. Therefore, the interaction between these two indicator variables allows us to estimate the effect of the GRIN incentive on the cost inefficiency of energy producers.

4 Results

Table 7 presents the estimates for the frontier cost and the model for inefficiency cost, providing a model for each Italian macro areas. The lambda parameter, which represents the ratio between the standard deviation of the efficiency term U and the standard deviation of the error term V, provides insight into the relevance of the inefficiency term. The lambda value indicates that energy firms operating in the northern regions of Italy exhibit lower inefficiency compared to those in the southern regions.

Additionally, when we assess the impact of the variable trend and its polynomial, we observe a convex pattern of inefficiency over time for energy producers located in the South and Islands, as opposed to the northern regions. This dynamic is clearly illustrated in Fig. 3, where we chart the cost inefficiency of energy producers averaged across Italian macro-regions from 2014 to 2020. The graph reveals that energy firms in the northern regions display concave inefficiency trends over time, while those in the South of Italy exhibit convex inefficiency trends. This suggests that energy

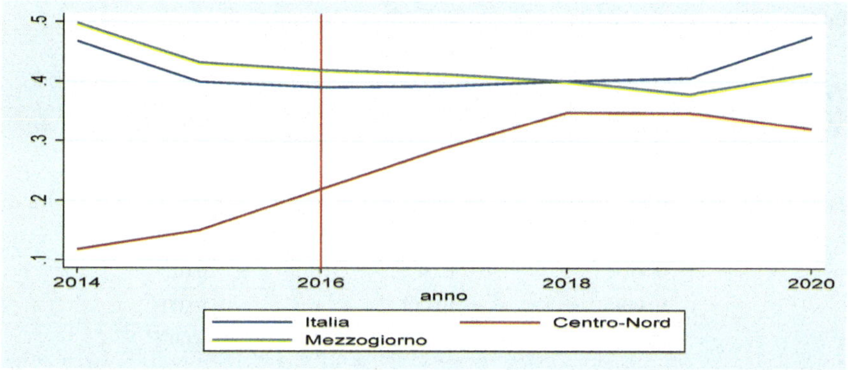

Fig. 3 Cost efficiency by Italian macro-regions (Years 2014–2020). Source: our elaborations on Istat data

producers in southern regions tend to experience a worsening of cost efficiency over time.

Furthermore, the marginal cost parameters show that the additional cost per unit of output produced is relatively similar in all Italian macro areas, as is the elasticity of total costs to changes in labour prices. However, the elasticity for the cost of capital is slightly higher in the South and Islands (0.10) than in the Centre-North (0.04). This suggests that the frontier cost in the southern regions is more sensitive to positive changes in capital prices (Table 8).

Table 9 presents the results of the applied difference-in-differences (diff-in-diff) analysis concerning the GRIN incentive. The outcomes related to the frontier parameters closely mirror the previous findings, where the inefficiency term in the frontier cost function considered the trend variable and its polynomial. However, the newly calculated lambda values exhibit some changes compared to previous ones. Notably, inefficiency seems more pronounced among energy producers in the northern regions. Nevertheless, the most significant result pertains to the coefficient interaction between GRIN and post. The interaction's coefficient, which is not only statistically significant but also economic relevant, indicates that the GRIN incentive on average amplifies energy producers' cost inefficiency across all macro areas, with a more prominent impact observed among energy producers in the northern regions. This result can be attributed to two main factors. Firstly, energy producers receiving this type of incentive, which is similar to a subsidy, often make their profits too dependent on the incentive itself, leading to a less efficient allocation of the funds received into their production processes. Second, there is a higher concentration of energy producers in the northern regions than in the southern regions, resulting in more incentive recipients in the north.

Figure 4 illustrates the average cost inefficiency conditional on the GRIN incentive for the period 2014–2020. In this scenario, we observe a convex relationship between cost inefficiency and time for each macro area. Even if it is not surprising, energy producers situated in the southern regions exhibit lower cost inefficiency compared to

Table 8 Frontier cost (panel data, years 2014–2020)

	Italy	Center-North	South-Islands
Frontier			
Y	0.96***	0.91***	0.95***
	[0.00]	[0.00]	[0.01]
p_r	0.04***	0.04*	0.10***
	[0.01]	[0.02]	[0.03]
p_w	0.03***	0.04***	0.02***
	[0.00]	[0.00]	[0.00]
t	0.13***	−0.05**	0.09***
	[0.02]	[0.01]	[0.03]
t^2	−0.02***	0.00	−0.01**
	[0.00]	[0.00]	[0.00]
Mu			
t	−2434.72**	3.88e + 08***	−1825.89
	[799.23]	[0.00]	[4199.07]
t^2	310.72**	−3.52e + 07***	165.98
	[96.45]	[0.00]	[380.08]
Lambda	243.70	197.58	238.59

$^{*}p < 0.05$, $^{**}p < 0.01$, $^{***}p < 0.001$
Source our elaborations on Istat data

those in the northern regions, especially from 2016 onwards, which is the year when the incentive was implemented. Furthermore, we can note that the overall country-wide pattern and the northern region converge because most firms benefiting from the GRIN incentive are located in the north Italy.

5 Conclusions

The existence of subsidies to energy producers in Italy has been widely debated over the last decade. In particular, the debate has attempted to address the choice of policy makers to allocate resources to firms in the sector within the EU framework setting strict environmental targets to reach.

In this paper, our econometric analysis using diff-in-diff methods highlights that the cost inefficiency has been amplified by the GRIN incentive to small firms and points out that there are differences in the impact of government subsidies across regions. In particular, the most important effect concerns the northern firms, as most of the energy producers are located in the Centre-North of Italy.

Table 9 Frontier cost with GRIN incentives (panel data, years 2014–2020)

	Italy	Center-North	South-Islands
Frontier			
Y	0.96***	0.96***	0.99***
	[0.00]	[0.00]	[0.01]
p_r	0.04***	0.03**	0.11***
	[0.01]	[0.01]	[0.03]
p_w	0.03***	0.04***	0.02***
	[0.00]	[0.00]	[0.00]
T	0.09***	0.09***	0.08
	[0.01]	[0.01]	[0.04]
t^2	−0.01***	−0.01***	−0.01*
	[0.00]	[0.00]	[0.00]
Mu			
Post	−410.81	−148.02	−813.69
	[279.52]	[462.83]	[479.60]
GRIN	−4799.30***	−9073.60***	−2451.59
	[1091.90]	[2108.50]	[2133.82]
Post*GRIN	4318.72***	8355.83***	512.69
	[1134.26]	[2168.50]	[2189.47]
Lambda	175.29	212.89	164.81

$^{*}p < 0.05$, $^{**}p < 0.01$, $^{***}p < 0.001$
Source our elaborations on Istat data

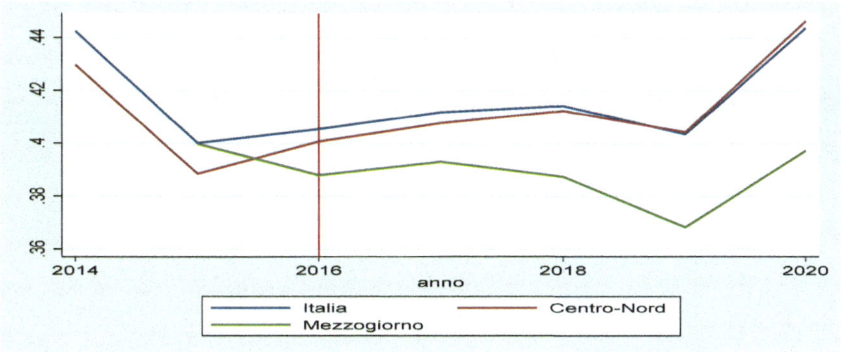

Fig. 4 Cost efficiency by Italian macro-regions with treatment for firms with GRIN incentives (Years 2014–2020) *Source* our elaborations on Istat data

The small companies seem to allocate the additional financial resources obtained by the GRIN incentive inefficiently and the energy production process does not benefit from this subsidy.

Nevertheless, in spite of cost efficiency considerations the described cost incentive mechanism allowed Italy to develop a renewable energy production chain thus reaching EU-set environmental targets.

References

1. Angrist, J.D., Pischke, J.-S.: Mostly Harmless Econometrics: An Empiricist Companion. Princeton University, Princeton (2009)
2. Battese, G.E., Coelli, T.J.: A model for technical inefficiency effects in a stochastic frontier production function for panel data. Empir. Econ. **20**, 325–332 (1995)
3. Greca, G.: La filiera dell'energia elettrica: il valore aggiunto e le sue principali caratteristiche strutturali, Relazione sulla situazione energetica nazionale del 2018, Ministero dello sviluppo economico (2019)
4. Greca, G.: #Incentiviallerinnovabili #EfficienzaenergeticaeCO2 #Mercatodell'energiaecompetitività #ENERDATA", "I meccanismi di incentivazione energetico-ambientale", Poster scientifico, 14ª Conferenza nazionale di Statistica (2021)
5. Greca, G.: I meccanismi di incentivazione energetico-ambientale (*Energy-environmental incentive mechanisms*). ISTAT Thematic Research Project, Report Finale (2022)
6. Greca, G.: Transizione energetica. "Energia pulita e accessibile": uno strumento di monitoraggio degli obiettivi, Paper, XLIII Conferenza Annuale AISRE, Milano 5–7 settembre 2022 (2022)
7. Greca, G.: Il valore aggiunto del settore energetico, "Relazione sulla situazione energetica nazionale del 2022", Ministero dell'ambiente e della sicurezza energetica, luglio (2023)
8. Necoechea-Porras, P.D., López, A., Salazar-Elena, J.C.: Deregulation in the energy sector and its economic effects on the power sector: a literature review. Sustainability **13**, 3429 (2021)

Working with Non-compensatory Composite Indicators: A Case Study Based on SDG for Mediterranean Countries

Francesca Mariani, Mariateresa Ciommi, Maria Cristina Recchioni, Giuseppe Ricciardo Lamonica, and Francesco Maria Chelli

Abstract In recent years there has been an increasing interest in measuring phenomena through the use of composite indicators. There are at least two main motivations: on the one hand, composite indicators are able to summarize in a single number the many faces of the analyzed phenomena. On the other hand, any composite indicator induces a ranking that allows easy comparisons both over time and space. Here, we propose a two-stage aggregation procedure to construct four overall composite indicators. In the first stage, the aggregation is made at domain level and then domains are aggregated to get the overall indicator. The four methods are based on the idea of mitigating the compensability hypothesis. Starting from the simplest non-compensatory index, computed through the geometric mean, we modify it taking into account the horizontal variability and the vertical variability, obtaining the so-called Penalized Geometric Mean Indicator and Weighted Geometric Mean Indicator, respectively. Then, by combining both horizontal and vertical variability we define the Penalized Weighted Geometric Mean indicator. We apply the above-mentioned methods to measure sustainability according to the 17 targets of the Sustainable Development Goals (SDGs). Thus, for each method, we calculate an overall index and, among the 193 countries we put attention on the results provided by 17 Mediterranean countries.

Keywords Composite indicators · Horizontal and vertical penalty · SDG · Mediterranean

---

F. Mariani (✉) · M. Ciommi · M. C. Recchioni · G. Ricciardo Lamonica · F. M. Chelli
Department of Economic and Social Sciences, Università Politecnica delle Marche, Ancona, Italy
e-mail: f.mariani@univpm.it

M. Ciommi
e-mail: m.ciommi@univpm.it

M. C. Recchioni
e-mail: m.c.recchioni@univpm.it

G. Ricciardo Lamonica
e-mail: g.ricciardo@univpm.it

F. M. Chelli
e-mail: f.chelli@univpm.it

M. Bini et al. (eds.), *Advanced Methods in Statistics, Data Science and Related Applications*, Springer Proceedings in Mathematics & Statistics 467,
https://doi.org/10.1007/978-3-031-65699-6_2

1 Introduction

The 17 United Nations Sustainable Development Goals (SDGs) are targets for global development adopted in September 2015, set to be achieved by 2030, according to the Agenda for Sustainable Development.

All 193 United Nations member states have committed to achieve sustainable development and annual progress are monitored.

To track progress across several goals with multiple indicators and track overall progress, it is important to aggregate the sub-indicators. So it becomes essential to have a tool to make synthesis. Literature accounts for several initiatives. Sachs et al. [9] propose the so-called *Sustainable Development Goal Index* (SDG-I), computed as a simple aritmethic mean among SDGs. After this first attempt of aggregation, Lafortune et al. [5] compare different methods such as the arithmetic mean (CES function), the minimum (Leontief production function) and the geometric mean (Cobb-Douglas production function) for aggregating SDGs. Guijarro [4] proposes a parametric weighting scheme for the calculation of the SDG Index based on the multicriteria Goal Programming (GP) approach. Finally, Biggeri et al. [2] introduce an adjusted SDG Index based on the Multidimensional Synthesis of Indicators (MSI) method with the twofold objective of overcoming the perfect substitutability problem of the arithmetic mean and avoiding the tendency of geometric mean approach to collapse to zero.

The aim of this paper is twofold. On the one hand, we are interested in investigating the role of weights in the aggregation and their effect on the ranking. On the other hand, we believe that the Sustainable Development Goals represent an extraordinary challenge that merit to be in-depth analysed.

Thus, we propose a two-stage procedure that allows to aggregate within the 17 Goals and then among Goals to get overall indices. In particular, since we believe that compensability should be avoided in the measurement of economic, environmental and social indicators, we start with the simplest non-compensatory index, namely the geometric mean, (GM). Mariani and Ciommi [6] define a new aggregation method, called the penalized geometric mean, (GM_p). The method introduces a penalty in the weighted geometric mean in terms of the (horizontal) variability of the normalized indicators transformed via the zero-order Box-Cox function.

Similarly, Ciommi et al. [3] propose a new method to capture the vertical variability, based on the Gini inequality index, (GM_w) believing that a more unequal distribution of an elementary indicator among countries implies a greater weight for that indicator. Thus, for each sub-indicator (first step) and then for each goal (step 2), we compute the corresponding Gini inequality index and we use it to weight each unit.[1] Stiglitz et al. [11] justify this choice because in line with the recommendations by the Commission on the Measurement of Economic Performance and Social Progress that proposes to take into account the distribution of indicators. In the litera-

---

[1] We observe that even if the Gini index is usually defined for transferable variables, Becchetti et al. [1] affirm that the possibility of redistribution is not essential for the definition and the measurement of inequality.

ture on the construction of composite indicators, there are several works introducing weights reflecting specific characteristics of the distribution. For instance, Mazziotta and Pareto [8] propose the use of weights obtained by applying Principal component Analysis in formative model. In this case, the weights depend on the correlations among indicators.

Finally, we combine horizontal and vertical variability by computing the so-called weighted penalized geometric mean, (GM_{wp}). The idea behind this last index is to capture simultaneously the horizontal variability, as for the penalized geometric mean, and the vertical variability, as for the Gini weighted geometric mean.

The paper is organized as follows: in Sect. 2, we discuss the methods used and describe the data, Sect. 3 collects the results, whereas, Sect. 4 concludes and suggests possible extensions.

2 Methods and Data

2.1 Notations

Let $Z = \{z_{ijl}\}$ denote the matrix of normalized values of the indicator $j, j = 1, \ldots, k_l$ across domain $l, l = 1, \ldots, s$ relative to unit $i, i = 1, \ldots, n$.

Although the normalization procedure is one of the crucial aspects in the construction of a composite indicator, here we concentrate on the next step of the construction of a composite indicator (herafter, CI), namely the aggregation step. Therefore, we simply assume that the data has been normalized in the range [0, 1] and transformed with positive polarity[2].

We compute the overall index following a two stage procedure.

Firstly, for each domain we aggregate the normalized values according to four different methods.

The first method tries to solve the problems of compensation among variables by assuming a multiplicative relationship among them. From a mathematical point of view this means to compute the composite indicator by applying the geometric mean. Thus, we refer to it as GM method and for a given domain l it is defined as follows:

$$GM_{i \cdot l} = \left(\prod_{j=1}^{k} z_{ijl} \right)^{1/k}. \tag{1}$$

Then, we modify the previous formula to capture the vertical variability. This can be made by adding weights. Following Ciommi et al. [3], the weights are set such as a more unequal distribution of an elementary indicator among countries implies

[2] As usually, the term polarity of a variable represents the sign of the relation between the variable and the phenomenon to be measured. Since here we are measuring sustainable development, we have positive polarity for all indicators respect to the corresponding Goals as well as for Goals respect to the overall index.

a greater weight for that indicator. Thus, to capture the inequality, we use the Gini index w_{ijl} of the normalized indicators. We refer to this method, as the *weighted Geometric Mean*, *GMw*. Thus, the corresponding CI is computed as follows:

$$GMw_{i \cdot l} = \prod_{j=1}^{k} z_{ijl}^{w_{ijl}} \tag{2}$$

However this is just a way of adding weights. Mariani and Ciommi [6] propose a new method that accounts for the (horizontal) variability of the normalized indicators. We refer to this method as the *Penalized Geometric Mean*, *GMp*. Formally:

$$GMp_{i \cdot l} = GM_{i \cdot l} e^{-\tilde{S}_{il}^2} \tag{3}$$

where

$$\tilde{S}_{il}^2 = \frac{1}{k} \sum_{j=1}^{k} \left(\ln \frac{z_{ijl}}{GM_{i \cdot l}} \right)^2 \tag{4}$$

represents the error made by approximating the normalized indicators z_{ijl} with $GM_{i \cdot l}$ coincides with the (biased) sample variance of z_{ijl}. $GMp_{i \cdot l}$ is the product between the geometric mean $GM_{i \cdot l}$ and a penalty factor $e^{\tilde{S}_{il}^2}$ that allows us to discriminate between units with the same geometric mean but different geometric mean reliability. The penalty factor gives smaller value to the units for which the geometric mean is less reliable.

Finally, we define the penalized weighted geometric mean (GM_{wp}) relative to the i-th unit as follows:

$$GMwp_{i \cdot l} = GMw_i \, e^{-w\tilde{S}_{il}^2}, \quad i = 1, 2, \ldots, n \tag{5}$$

$$w\tilde{S}_{il}^2 = \sum_{j=1}^{k} w_{ijl} (\ln z_{ijl} - \ln(GMw_{il}))^2. \tag{6}$$

As for GM_p, the size of (6) determines the reliability of the estimate $\ln GMw_{il}$ and, as a consequence, the weighted geometric mean GMw_{il}.

As second stage, we aggregate across dimensions to obtain the four overall indices. For each of the methods we apply Eq. (1), (2), (3) and (5) by substituting the values of the normalized sub-indicators z_{ijl} with the values of the indicators computed at the first stage.

Thus, in the case of geometric mean, the overall index Geometric Mean Indicator is:

$$overall\ GM_i = \left(\prod_{l=1}^{s} GM_{i \cdot l} \right)^{1/s} \tag{7}$$

2.2 Data on SDG

To illustrate the appealing of this penalized geometric mean, we focus on Sustainable Development Goals. Data comes from the Europe Sustainable Development Report (2022) [10], hereafter SDR2022. Authors use official SDG indicators endorsed by the UN Statistical Commission and data come from a mix of official and non-official data sources.[3] Data refer to 2022.

According to Sachs et al. [10], the procedure for computing the SGD Index accounts for three steps: (i) data manipulation, (ii) normalization and (iii) aggregation. According the first step, data require to be treated, this implies two phases, a first one devoted to the choice of performance thresholds for each sub-indicator and a second one aiming at removing extreme values. These steps are quite standard in the literature of the construction of composite indicators. For instance, in the procedure to compute the Human Development Index, for the economic dimension, computed in terms of the per capita Gross Nation Income at purchasing power parity (GNIpc) the minimum and maximum values to normalize data are \$100 and \$75, 000, respectively. Countries having GNIpc values greater than \$75, 000, are arbitrarily set equal to this maximum value, as in the case of Liechtenstein (\$131, 032), Qatar (\$92, 418), and Singapore (\$88, 155) (see Mariani and Ciommi [6]). For the second step, the normalization one, Sachs et al. [10] use a modified version of the so-called min-max method,[4] transforming data in the range [0, 1]:

$$z_{ijl} = \frac{x_{ijl} - \underline{x}_{ijl}}{\overline{x}_{jl} - \underline{x}_{jl}} \tag{8}$$

where $\overline{x}_{jl}$ and $\underline{x}_{jl}$ denote the upper and lower bounds, respectively.[5]

The rescaling step ensures that z_{ijl} were expressed as ascending variables (higher values denoted better performance), that is, we have all data with positive polarity.

The dataset obtained after step 1 and step 2 is our starting point. The original dataset accounts for 17 Goals, 95 sub-indicators collected for 193 countries.[6] After

---

[3] See Sachs et al. [10], Sect. A.3 Methodology (overview) for more details.

[4] We observe that in the original equation of the max-min method, $\overline{x}_{jl}$ and $\underline{x}_{jl}$ are the maximum an minimum across units, respectively. However, this procedure has several drawbacks, firstly, it does not handle outliers very well. Moreover, for each sub-indicators there is always at least one zeros and this could lead problems in the aggregation approaches since several aggregative methods work only with positive non-zeros values.

[5] The original normalization proposed by Sachs et al. [10] rescales data in [0, 100]. Authors justify their choice to an easiest interpretation since a country that scores 50 on a variable is half-way towards achieving the optimum value, or, similarly, it has covered one-half of the distance from worst to best.

[6] Data can be freely downloaded at the following link: https://www.sdgindex.org/reports/europe-sustainable-development-report-2022/.

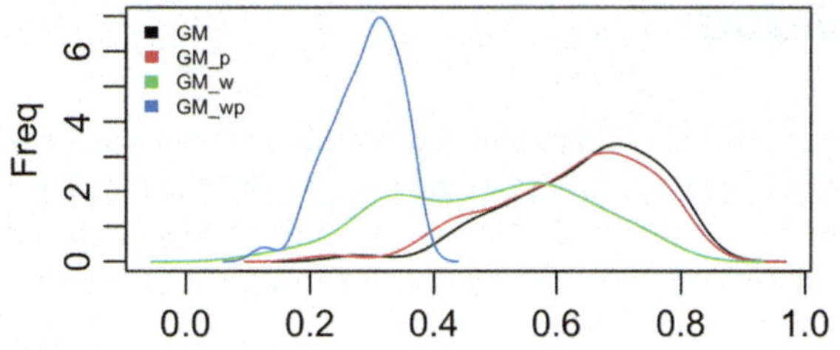

Fig. 1 Distributions for all countries

Table 1 Summary statistics of the four overall methods

	Overall GM	Overall GM_p	Overall GM_w	Overall GM_{wp}
Min.	0.2566	0.2220	0.0996	0.1156
1st Qu.	0.5645	0.5360	0.3502	0.2515
Median	0.6670	0.6482	0.4914	0.2969
Mean	0.6446	0.6221	0.4858	0.2884
3rd Qu.	0.7332	0.7225	0.6045	0.3292
Max.	0.8498	0.8405	0.7761	0.3839
IRQ.	0.1687	0.1865	0.2543	0.0777

applying two criteria to eliminate both units and sub-indicators with more than 50% of missing values, the final dataset collects data for 163 countries and 92 sub-indicators.[7]

Finally, as explained before, for the aggregation step, we first estimate a country's score on each goal using one of the above mentioned aggregation method of its score on the goal indicators. Then, these scores are again aggregated across all 17 SDGs to obtain four overall country's scores, one for each method.

3 Results

Figure 1 reports the distribution of the four indices. The overall indices computed via Geometric Mean and Penalized Geometric Mean exhibit a similar distribution, as confirmed by summary statistics reported in Table 1.

Although we compute the four overall indices for all countries, we focus on 17 Mediterranean Countries, partitioned into 9 European Mediterranean countries,

[7] In brackets the number of sub-indicators for each Goals: Goal 1: No Poverty (2); Goal 2: Zero Hunger (8); Goal 3: Good Health and Well-being (14); Goal 4: Quality Education (4); Goal 5: Gender Equality (4); Goal 6: Clean Water and Sanitation (5); Goal 7: Affordable and Clean Energy (4); Goal 8: Decent Work and Economic Growth (6); Goal 9: Industry, Innovation and Infrastructure (6); Goal 10: Reduced Inequality (2); Goal 11: Sustainable Cities and Communities (4); Goal 12: Responsible Consumption and Production (7); Goal 13: Climate Action (3); Goal 14: Life Below Water (5); Goal 15: Life on Land (5); Goal 16: Peace and Justice Strong Institutions (10); Goal 17: Partnerships to achieve the Goal (3).

namely Croatia, Cyprus, France, Greece, Italy, Malta, Portugal, Slovenia and Spain and 8 non-European Mediterranean countries, namely Algeria, Egypt, Israel, Jordan, Lebanon, Morocco, Tunisia, and Turkey.

Table 2 reports summary statistics for Mediterranean Countries. Also in the sub-group of Mediterranean countries, GM_w, that is the method that takes into account the vertical variability, displays the larger variability measured as the Interquartile difference (IRQ), that is (0.1612). This value significantly reduces when an horizontal penalty is simultaneously applied to the vertical one ($GM_w = 0.0460$).

On average, Mediterranean Countries perform better than non-Mediterranean Countries for all indices, as shown in Fig. 2. Moreover, the values of the Overall Weighted methods, both in their penalized and non-penalized version, are lower than the corresponding values of the geometric mean, meaning that adding weights reflect-

Table 2 Summary statistics of the four overall methods for Mediterranean countries

	Overall GM	Overall GM_p	Overall GM_w	Overall GM_{wp}
Min.	0.6465	0.6252	0.3851	0.2790
1st Qu	0.6731	0.6571	0.5497	0.3098
Median	0.7237	0.7104	0.5846	0.3136
Mean	0.7264	0.7140	0.6141	0.3276
3rd Qu.	0.7860	0.7769	0.7110	0.3558
Max.	0.7967	0.7910	0.7675	0.3686
IRQ	0.1129	0.1339	0.1613	0.0460

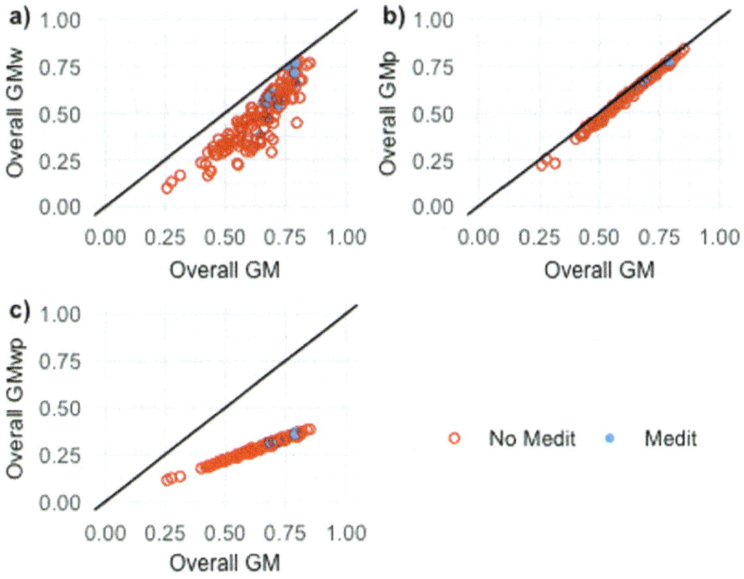

Fig. 2 Comparisons

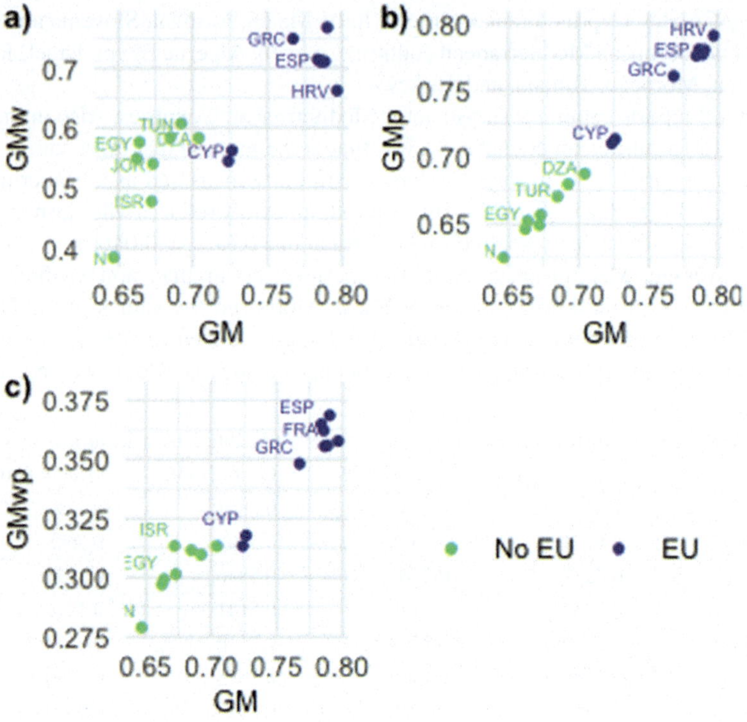

Fig. 3 Mediterranean countries

ing vertical variability (GM_w) and both horizontal and vertical variability (GM_{wp}) reduce the value of the indices.

However, not all Mediterranean Countries perform at the same way, In general, there are differences between EU Countries and non-EU Countries (Fig. 3) since the former perform better than the latter.

Table 3 reports the results of the four methods and the corresponding rankings.

Lebanon is the country with the worse performance according to all methods. Taking into account the vertical variability, countries modify their ranking, meaning that countries perform very differently for each Goal.

Finally, to compare the role of the weights, we compute the relative contribution of the weights of GM_p, GM_w and GM_{wp}, in terms of their relative differences in values with GM, that is, we compute:

$$e_{GM_w} = \frac{GM_w - GM}{GM} \quad e_{GM_p} = \frac{GM_p - GM}{GM} \quad e_{GM_{wp}} = \frac{GM_{wp} - GM}{GM} \tag{9}$$

Table 4 reports the values of efficiency (as a percentage) as well as rank differences for Mediterranean Countries. By computing the average values, we find that there is a reduction of 1.75%, on average, moving from Geometric mean to the penalized

Table 3 Values of the four overall methods and corresponding rankings for Mediterranean countries

ISO	Country	GM	rank GM	GM_p	rank GM_p	GM_w	rank GM_w	GM_{wp}	rank GM_{wp}
CYP	Cyprus	0.7261	43	0.7136	40	0.5621	56	0.3178	52
DZA	Algeria	0.7042	52	0.6879	52	0.5839	43	0.3134	58
EGY	Egypt	0.6639	76	0.6527	73	0.5767	48	0.2994	71
ESP	Spain	0.7841	19	0.7760	17	0.7144	10	0.3649	7
FRA	France	0.7866	17	0.7785	15	0.7116	12	0.3556	17
GRC	Greece	0.7677	24	0.7609	22	0.7487	7	0.3482	23
HRV	Croatia	0.7967	8	0.7910	7	0.6629	22	0.3578	14
ISR	Israel	0.6721	74	0.6496	75	0.4783	79	0.3136	57
ITA	Italy	0.7860	18	0.7808	13	0.7110	13	0.3622	9
JOR	Jordan	0.6731	71	0.6571	69	0.5406	66	0.3015	69
LBN	Lebanon	0.6465	88	0.6252	87	0.3851	102	0.2790	92
MAR	Morocco	0.6621	79	0.6466	77	0.5497	60	0.2971	74
MLT	Malta	0.7237	44	0.7104	42	0.5444	64	0.3133	59
PRT	Portugal	0.7903	14	0.7803	14	0.7675	2	0.3686	5
SVN	Slovenia	0.7887	15	0.7769	16	0.7109	14	0.3558	16
TUN	Tunisia	0.6925	62	0.6801	58	0.6081	38	0.3098	63
TUR	Turkey	0.6850	63	0.6709	63	0.5846	42	0.3117	60

Table 4 Efficiency and rank differences

ISO	Country	e_{GM_p}	e_{GM_w}	$e_{GM_{wp}}$	$rdif_{GM_p}$	$rdif_{GM_w}$	$rdif_{GM_{wp}}$
CYP	Cyprus	−1.72	−22.59	−56.23	−3	13	9
DZA	Algeria	−2.32	−17.08	−55.49	0	−9	6
EGY	Egypt	−1.68	−13.13	−54.9	−3	−28	−5
ESP	Spain	−1.03	−8.89	−53.46	−2	−9	−12
FRA	France	−1.04	−9.54	−54.8	−2	−5	0
GRC	Greece	−0.88	−2.47	−54.65	−2	−17	−1
HRV	Croatia	−0.72	−16.8	−55.09	−1	14	6
ISR	Israel	−3.35	−28.83	−53.35	1	5	−17
ITA	Italy	−0.65	−9.54	−53.91	−5	−5	−9
JOR	Jordan	−2.38	−19.69	−55.21	−2	−5	−2
LBN	Lebanon	−3.3	−40.44	−56.85	−1	14	4
MAR	Morocco	−2.34	−16.99	−55.13	−2	−19	−5
MLT	Malta	−1.83	−24.77	−56.7	−2	20	−15
PRT	Portugal	−1.27	−2.89	−53.36	0	−12	−9
SVN	Slovenia	−1.5	−9.87	−54.88	1	−1	1
TUN	Tunisia	−1.79	−12.19	−55.27	−4	−24	1
TUR	Turkey	−2.07	−14.66	−54.5	0	−21	3

version. The highest difference is observed with the weighted penalized indicator (−54.9%). Computing the (absolute) average of rank differences, we find that to the Penalized Geometric mean corresponds a change in about two positions. In this case, the higher (absolute) average position is between the weighted method and the classical one (that is, 13 positions).

4 Conclusions and Further Research

In this paper we propose a two-stage procedure to construct an overall SDG score using four different methods, namely the Geometric Mean GM, the Weighted Geometric Mean GM_w, the penalized Geometric Mean GM_p and the Weighted Penalized Geometric Mean GM_{wp}.

The proposed modified versions of the classical geometric mean have several advantages. First, they allow for a discrimination among units with same generalized mean; second, they account for the degree of horizontal and/or vertical variability experienced by each unit. Finally, as proved by Mariani et al. [7], they are based on the minimum information loss principle, usually used for constructing composite indicators.

The empirical results for the 17 Mediterranean countries show that northern Mediterranean countries better perform compared to southern and eastern countries.

The analysis of weights reveals that their introduction in the geometric mean has a great impact with respect to introducing them in the penalized version of the geometric mean. This could be interpreted as a strength for the penalized geometric mean. In fact, the simple geometric mean requires to be weighted to account for inequality, while the penalized method already accounts for inequality across indicators. That is, the ranking deriving from penalised method remains unchanged when weights are included to account for inequality across countries.

Further research will be conducted on several directions: firstly, we are interested in analysing the properties of the three modified version of the classical geometric mean. Second, here we have used Gini as measure of inequality but this is just one of the long list of inequality measures. So, we retain important to investigate alternative inequality indices to build the weights and account for the differences in rankings produced. Related to the analysis of ranking, a possible extension of the work concerns the use of different methods to track for ranking differences. Finally, from an empirical point of view we think that it is of potential of interest to compute the overall indices over time to analyse if there are differences in the role of weights.

References

1. Becchetti, L., Massari, R., Naticchioni, P.: The drivers of happiness inequality: suggestions for promoting social cohesion. Oxf. Econ. Pap. **66**(2), 419–442 (2014)
2. Biggeri, M., Clark, D.A., Ferrannini, A., Mauro, V.: Tracking the SDGs in an 'integrated' manner: A proposal for a new index to capture synergies and trade-offs between and within

goals. World Dev **122**, 628–647 (2019)
3. Ciommi, M., Gigliarano, C., Emili, A., Taralli, S., Chelli, F.M.: A new class of composite indicators for measuring well-being at the local level: an application to the Equitable and Sustainable Well-being (BES) of the Italian Provinces. Ecol. Ind. **76**, 281–296 (2017)
4. Guijarro, F.: A multicriteria model for the assessment of countries' environmental performance. Int. J. Environ. Res. Public Health **16**(16), 2868 (2019)
5. Lafortune, G., Fuller, G., Moreno, J., Schmidt-Traub, G., Kroll, C.: SDG index and dashboards detailed methodological paper, sustainable development solutions network (SDSN) (2018). www.sdgindex.org/reports/sdg-index-and-ashboards-2018/
6. Mariani, F., Ciommi, M.: Aggregating composite indicators through the geometric mean: a penalization approach. Computation **10**(4), 64 (2022). https://doi.org/10.3390/computation10040064
7. Mariani, F., Ciommi, M., Recchioni, M.C.: A new class of composite indicators: the penalized power mean. Eur. J. Oper. Res. **317**(3), 1015–1035 (2024)
8. Mazziotta, M., Pareto, A.: Use and misuse of PCA for measuring well-being. Soc. Indic. Res. **142**, 451–476 (2019). https://doi.org/10.1007/s11205-018-1933-0
9. Sachs, J., Schmidt-Traub, G., Kroll, C., Durand-Delacre, D., Teksoz, K.: SDG Index and Dashboards – Global report Bertelsmann Stiftung and Sustainable Development Solutions Network, New York (2016)
10. Sachs, J., Kroll, C., Lafortune, G., Fuller, G., Woelm, F.: Sustainable Development Report 2022. Cambridge University Press (2022). https://s3.amazonaws.com/sustainabledevelopment.report/2022/2022-sustainable-development-report.pdf Cited 09 dic 2022
11. Stiglitz, J.E., Sen, A., Fitoussi, J.P.: Report by the Commission on the Measurement of Economic Performance and Social Progress Commission on the Measurement of Economic Performance and Social Progress, Paris (2009). https://portal.cor.europa.eu/europe2020/news/Documents/Sen-Stiglitz-Fitoussi%20Report.pdf

The Construction of a Heat Vulnerability Index by Means of the Composite Indicator Approach: A Case Study for Friuli Venezia Giulia Region, Italy

Laura Pagani, Maria Chiara Zanarotti, and Anja Habus

Abstract Last decades have seen increasing consensus towards the issue of climate change and rising awareness of the implied responsibility of human activity. With growing global warming, extreme climate events like heat waves have increased in duration, frequency and intensity leading to higher heat-related morbidity and mortality rates. In this context, heat vulnerability assessments play an important role supporting decision-makers in implementing targeted mitigation and prevention actions. With this motivation, this work develops a heat vulnerability index by means of the Composite Indicator techniques to depict heat vulnerability in the Friuli Venezia Giulia region at the census tract level. The results show that heat vulnerability follows a spatial pattern, where most vulnerable census tracts are located in urbanised and densely populated areas, lower risk is observed in rural areas and lowest danger in mountainous areas. The Performance Interval approach confirms that these results do not depend on the aggregation method used to construct the index.

Keywords Climate change · Heat waves · Composite indicator · Heat vulnerability index

1 Introduction

Anthropogenic greenhouse gases emissions have increased since the pre-industrial era largely driven by economic and population growth, lifestyle, energy and land use patterns, technology and climate policy, and are now higher than ever. Their

L. Pagani (✉)
Department of Economics and Statistics, University of Udine, Udine, Italy
e-mail: laura.pagani@uniud.it

M. C. Zanarotti
Department of Statistical Sciences, Catholic University of Milan, Milan, Italy
e-mail: maria.zanarotti@uniucatt.it

A. Habus
Directorate General Statistics, European Central Bank, Frankfurt am Main, Germany
e-mail: anja.habus@ecb.europa.eu

M. Bini et al. (eds.), *Advanced Methods in Statistics, Data Science and Related Applications*, Springer Proceedings in Mathematics & Statistics 467,
https://doi.org/10.1007/978-3-031-65699-6_5

effects have been detected throughout the climate system and are the dominant cause of global warming and climate changes [7]. Among others, heat extremes have increased in frequency, duration and intensity since 1950. The most immediate and direct impact of rising temperature on human health is the increase in heat-related mortality and morbidity [8]. The European heat wave (HW) of 2003 caused more than 70,000 additional deaths, where France and Italy totalised 19,490 and 20,089 excess deaths, respectively [25]. In 2006, the Californian HW resulted in 16,000 additional emergency room visits and 1,200 additional hospitalisations [11]. The 2010 HW in Russia was responsible of killing an estimated 55,000 people [3]. Heat stress adverse effects go beyond deaths, also including strains on man-made infrastructure and increasing rates and intensities of wildfires, hence impacting both the natural and built environment. Specific categories of individuals are at greatest risk of being negatively affected by HWs, such as people with pre-existing medical conditions, elderly, and children [11, 21, 26]. HWs have a much bigger health impact in cities than in surrounding suburban and rural areas due to the urban heat island (UHI) effect [9]: impervious and heat-absorbing surfaces such as buildings, roads and paved surface store heat during the day leading to higher daytime temperatures and slowly release it during overnight, keeping urban lands warmer than surrounding rural areas. With increasing urbanisation, population aging, land cover and land use, this effect is expected to exacerbate.

Several studies have shown that adverse heat effects can be minimised with targeted prevention programmes and economic investments [10, 19, 20]. Many cities worldwide have developed heat warning systems to inform decision-makers and the general public to impending dangerously hot weather and to provide guidance for avoiding negative health outcomes associated with hot weather extremes. Early alerts have been recognised as an effective tool in reducing negative health impacts. Despite in the international context Italy has the highest heat-related effects on daily mortality, Italian cities have recorded a decreasing trend in heat-related mortality risk after the introduction of the national heat prevention plan which issues warnings and activates prevention measures when extreme temperatures are foreseen [31].

This study aims to construct a heat vulnerability index (HVI) for Friuli Venezia Giulia (FVG) region at the census tract level by means of the Composite Indicator (CI) techniques [23]. Section 2 defines the theoretical framework for HV measurement by outlining (i) the concept of heat vulnerability (HV), (ii) the most common manifest indicators used in the literature to determine the association with HV and (iii) the main measurement methods used in the literature to assess HV. In Sect. 3 the HVI for FVG is constructed based on climate, health, demographic, and environmental data; the Performance Interval (PI) approach [17] is used to validate the index against the impact of the chosen aggregation function. Thereafter, in Sect. 4 the HVI is mapped to identify areas and populations that are vulnerable to heat stress in FVG. Finally, some concluding remarks are presented in Sect. 5.

2 Heat Vulnerability

HV is a latent concept as its direct measurement is not possible due to the lack of a readily available and unique instrument. To carry out a meaningful measurement of this complex phenomenon, a sound theoretical framework needs to be designed following specific steps [14, 16]: (i) conceptual definition of HV and identification of its determinants, (ii) development of a complete set of non-interchangeable manifest measures, also called basic indicators (BIs), for a comprehensive representation of HV and (iii) definition and modelling of relationships between the latent concept of HV and the manifest measures.

2.1 Definition of Heat Vulnerability

In the natural hazards and climate change field vulnerability is broadly referred to as the potential for harm or loss of a system due to exposure to environmental hazards [5]. The Intergovernmental Panel on Climate Change (IPCC) defines vulnerability as

> [...] the degree to which a system is susceptible to, or unable to cope with, adverse effects of climate change, including climate variability and extremes. Vulnerability is a function of the character, magnitude, and rate of climate change and variation to which a system is exposed, its sensitivity, and its adaptive capacity [18].

In addition, it outlines the three determinants of vulnerability as follows:

- Exposure (E): the nature and degree to which a system is exposed to significant climatic variations.
- Sensitivity (S): the degree to which a system is affected, either adversely or beneficially, by climate-related stimuli.
- Adaptive capacity (AC): the ability of a system to adjust to climate change (including climate variability and extremes), to moderate potential damages, to take advantage of opportunities or to cope with the consequences.

In other words, systems that are highly exposed, highly sensitive and less able to adapt to the hazard of interest are more vulnerable. From the perspective of heat extremes, vulnerability corresponds to the adverse effects of excessive heat stress on human health and life and is also conceived as a function of the aforementioned dimensions. In fact, the generalized IPCC framework is widely applied in HV assessments [9, 15, 20, 29, 32] and it is also adopted for this analytical framework.

2.2 Risk Factors

The IPCC framework highlights the multifaceted nature of HV, but no manifest indicators are yet indicated to measure it. The manifest indicators, called risk factors, are identified through consulting relevant experts, studying epidemiological literature and heat episode studies that investigate and quantify the heat-health relationship. Heat exposure and climatic factors, health conditions, behavioural and demographic characteristics, socioeconomic status and built environment are the most common risk factors available in the literature.

Exposure to heat stress depends on frequency, magnitude (daily maximum, mean and minimum temperature), duration (number of consecutive hot days defined as the days with maximum temperature over a certain threshold), timing in season (HWs early in the summer are more dangerous than late in the season) and variability (temperature range) of HWs [8]. Furthermore, heat exposure depends on the climate and synoptic weather conditions of a HW (e.g. radiation and elevation, thermal, wind and humidity conditions) which are exacerbated by additional physical factors and processes of the built environment. In the literature the most widespread metrics for outdoor heat load are (i) air temperature measured in the canopy layer by meteorological stations and (ii) land surface temperature (LST), i.e. satellite images of thermal emission. Thermal indices are also measures of exposure: they combine air temperature with humidity and/or wind speed to capture heat exchange between the human body and the surrounding thermal environment (e.g. Thom index [28]). Built environment factors play an important role in mitigation or aggravation of HV of populations, such as land cover and land use, urban greening and characteristics of urban material (e.g. reflective, storage and transportation properties). Urban and metropolitan areas are particularly affected by HWs due the UHI effect: temperatures are on average 1–3°C higher than in the surrounding rural areas during daytime and potentially up to 12°C warmer at night [4]. Hence, researchers represent heat exposure in urban settings indirectly through measures of impervious surfaces, land cover, lack of vegetation and open spaces, combined daytime maximum and nighttime minimum temperatures, housing and population density.

Sensitivity refers to risk factors that indicate the extent to which individuals can absorb heat exposures and the range of associated impacts; these include health, demographic and socioeconomic factors. Infants, elderly and people suffering from pre-existing medical and chronic conditions (e.g. diabetes, obesity, cardiovascular or respiratory diseases) and taking medications are at greatest risk of serious complications due to the intrinsic less efficient thermoregulatory system and reduced sweating capacity [11, 12]. People with disabilities and psychological disorders might have limited mobility and reduced capacity to assess risk, seek for a cooler environment or assistance. Socioeconomic conditions (e.g. poverty level, unemployment and type of occupation) reflect the capability of individuals to have access to enough information, support and resources, such as air-conditioning and thermo-isolated houses [6, 22].

Adaptive capacity is the potential of a population to modify its features and behaviour to better cope with existing and anticipated stresses. Behavioural measures (e.g. wearing light clothes), at-home strategies (e.g. opening windows overnight) and access to government intervention programmes (e.g. urban greening and warning systems) are typical indicators of adaptation but not easily available since they require the collection of survey data at household level. Therefore, air-conditioning data are commonly used to proxy adaptive capacity in HV assessments: having a working air-conditioner or being able to visit other air-conditioned environments have been identified as protective against heat-related death and illness during HWs [24]. However, adaptive capacity is often excluded from empirical studies [6, 20]. This occurs for instance because prevalence of air-conditioning is interpreted to reflect the thermal properties of buildings and indoor environments (i.e. exposure) [30].

HV determinants, exposure, sensitivity and adaptive capacity, are interactive elements therefore there is no uniform classification of the risk factors in the literature. While climate data are unambiguously indicated as metrics of exposure, data related to land cover, vegetation coverage, indoor temperature and housing do not find a clear-cut categorisation.

2.3 *Measures of Heat Vulnerability*

The published literature on vulnerability to heat recognizes a wide range of potential measures and methods. Heat stress is often represented via proxy indicators, for instance emergency calls, hospitalisations, morbidity and mortality counts or rates [25]. However, most commonly the identified risk factors are combined to create a single index of vulnerability, which is then used as a tool for translating research into policy action via spatial representations of vulnerability [6, 20, 22, 24, 30]. In this context, different methods are typically applied to compute the final HVI, the most common are CI techniques but also multivariate statistical tools, such as the Principal Components Analysis (PCA) [2].

Building a CI is a conceptual and mathematical operation that consists in aggregating a set of individual indicators on the grounds of a well-defined measurement model that takes into account the multi-dimensional structure of the phenomenon of interest [16]. CIs benefit from their ability to summarise complex multidimensional constructs and their ease of interpretation, which in turns facilitates communication with ordinary citizens and supports decision-makers in their political agenda. On the other hand, the construction of CIs encompasses multiple stages in which subjective decisions are made: BIs selection, outliers and missing data treatment, BIs normalisation, weights definition and BIs aggregation. The multi-step nature of this process underlies the main criticism pointed out to this technique: CIs construction involves several alternatives and subjective decisions that may affect the quality and reliability of the results [16]. OECD and JRC European Commission [23] provide an extensive guide on the CI construction process and a comprehensive description of different methods within each stage, with their advantages and disadvantages. In the field of heat extremes, the most common approach to construct the HVI is to linearly aggregate equally weighted normalised BIs, i.e. the HV risk factors [1, 20]. Indicators selection for inclusion in the vulnerability index can also be supported by correlation levels with heat-related mortality, or other health-outcomes [6, 13]. Uncertain interactions among risk factors and their influence on HV lead researchers to assume that all the indicators are of equal importance and thus weight them equally. Sometimes weighting schema are introduced by means of regression coefficients where the health-outcome is the dependent variable and risk factors are the independent ones [13]. A relevant number of HV studies exclude measures of exposure (e.g. air temperature or LST) from the HVI [24]. Whereas in their multiplicative model, Zhu et al. [32] consider the occurrence of the HW (i.e. exposure) as a necessary condition for the existence of vulnerability.

PCA is probably the most popular multivariate statistical technique for reducing multidimensional data and thus commonly used to construct HVIs [6, 20, 22, 24]. PCA-based HVIs can be distinguished into those obtained by summing up equally weighted [4, 22] versus variance weighted components [30]. Most studies apply PCA across all indicators, regardless the dimension they belong to, whereas fewer [9] apply PCA within each HV dimension (exposure, sensitivity, and adaptive capacity).

The choice of the measurement method should be grounded on a well-defined measurement model that considers the relationships between the latent concept of HV and the manifest measures. This aspect is rarely made explicit in the literature on HV assessments, where usually the difference between the formative and reflective model is not considered [16] and PCA and CI techniques are used interchangeably. The measurement framework underlying HV is in fact formative and not reflective [27]. In a formative context the latent construct does not exist as an independent entity, but it is a composite measure determined by (i.e. *formed from*) a set of non-interchangeable manifest indicators, whereas in a reflective setup manifest indicators are seen as functions of the latent concept, whereby changes in the former make manifest (i.e. *reflect*) changes of the latter [14, 16]. HV follows a formative approach as it does not exist as an independent entity but rather it is defined by its dimensions (exposure, sensitivity, and adaptive capacity) and the related risk factors, which determine the vulnerability to HWs of the geographical area or population of interest. In this context, Mazziotta and Pareto [16] highlight the common misuse of PCA in well-being studies, specifying that composite indices based on PCA are more suited for a reflective approach than a formative one.

3 Data and Methods

The IPCC framework and the CI techniques are embraced to evaluate and build a synthetic measure of vulnerability associated with heat stress in Friuli Venezia Giulia. The geographical scale selected for the assessment is the census tract: the smallest statistical unit for which *Istituto nazionale di statistica* (ISTAT) publishes geographical and census data, in other words the smallest entity of the municipality based on which the data collection of census surveys is organised.

3.1 Study Area

Friuli Venezia Giulia is Italy's north-easternmost region. Within an area of sole 7,847 km^2, its landscape spans from the Alps to the Adriatic Sea, determining the subdivision of the region into four main parts: mountainous-alpine terrain in the north, hilly to the south of the mountains and in the central part of the eastern border with Slovenia, and plain from the centre to the coastal area in the south. Mediterranean, continental and alpine climate conditions can be observed across the region due to the complex orography. FVG counts 8,287 census tracts which are nested in 218 municipalities[1] and 4 provinces.

3.2 System of Basic Indicators

Table 1 depicts the system of non-interchangeable BIs employed in the construction of the HVI. These have been selected based on the literature review outlined in Sect. 2.2 and in line with the IPCC framework.

Socio-demographic, health and environmental data are provided by *Regione Autonoma Friuli Venezia Giulia*.[2] Source of socio-demographic data is the 15th national population and housing census[3] conducted in 2011. People aged 65 years or more and 5 years or less correspond to individuals born before 1946 (inclusive) and after 2006 (inclusive), respectively. Raw health data are dated 2019 and provided at municipal level. To turn them at the census tract scale, health data are weighted by the municipal population share of each census tract. As count data are affected by population size and thus by the size of the census tract, socio-demographic and health data are normalised by the surface of the census tract in terms of km^2.

Environmental indicators are an elaboration of high-resolution imperviousness data collected in 2015 by a set of dedicated satellites (the Sentinel families) serving Copernicus.[4] The Largest Class Patch Index[5] (LCPI) corresponds to the largest continuously built area of each census tract. It takes 0 when there are no artificially built objects and it takes 1 when the census tract is completely covered by a continuous and unique built area, i.e. there is no discontinuity given by vegetation or open green spaces. The Impermeability Index (II) captures the portion of impermeable cover of

[1] As of 1 January 2019, Friuli Venezia Giulia counts 215 municipalities due to administrative amendments.

[2] Specifically by (i) *Servizio programmazione, pianificazione strategica, controllo di gestione, statistica e sicurezza sul lavoro*, (ii) *Direzione centrale salute, politiche sociali e disabilità* and (iii) *Direzione centrale infrastrutture e territorio*.

[3] ISTAT, *15° Censimento generale della popolazione e delle abitazioni*.

[4] European Union's Earth Observation Programme. Copernicus delivers data, information and services based on satellite Earth Observation data and ground (in-situ) sensors. More details are available at https://www.copernicus.eu/en.

[5] Elaborated by *Istituto Superiore per la Protezione e la Ricerca Ambientale* (ISPRA).

Table 1 System of basic indicators of heat vulnerability index in FVG

Class	BI	Description	Dimension	Range
Socio-demographic[a]	Under5	No. of individuals aged 5 years or less	S	[0 – 15992]
	Over65	No. of individuals aged 65 years or more	S	[0 – 25942]
	NoAC	No. of dwellings without air-conditioning	E, AC	[0 – 37100]
Health[a]	Car	No. of individuals with cardiovascular diseases	S	[0 – 51956]
	Res	No. of individuals with respiratory diseases	S	[0 – 6582]
	End	No. of individuals with diabetes	S	[0 – 5118]
	Psy	No. of individuals with psychological disorders	S	[0 – 11622]
	MultiPat	No. of individuals with at least two of the above diseases	S	[0 – 25921]
Environmental	II	Degree of soil imperviousness	E, AC	[1 – 4]
	LCPI	Largest patch of continuously built area	E, AC	[0 – 1]
	LAI	Leaf Area Index	E, AC	[1 – 6]
Climate	ThomMax	10-years average of daily Thom Index maxima	E	[21 – 30]
	ThomAvg	10-years average of daily Thom Index means	E	[13 – 23]

[a]Normalised by census tract's area (km^2)

soil. The lowest the value, the lowest the level of soil imperviousness and thus its vulnerability to heat and vice versa. The Leaf Area Index (LAI) is a dimensionless measure defined as the one-sided green leaf area per unit ground surface area in broadleaf canopies. It is estimated from remote sensing and includes all the green contributors such as the understorey when existing under forests canopies. The LAI has an inverse relationship with the HVI: the higher the LAI the lower the vulnerability to heat. The II and the LAI refer to sub-portions of the census tract, they are normalised for the share area of the census tract they represent and aggregated at census tract level.

Climate data, in the form of Thom Index, are provided by *ARPA FVG Osservatorio Meteorologico Regionale* (OSMER) at the census tract level and are based on elaborations of hourly humidity and temperature data recorded during summertime months from 2010 to 2019 by weather stations located around the region. Averages across ten yearly maximum of daily Thom maxima and ten yearly averages of daily Thom means are computed to get ThomMax and ThomAvg, respectively. These are proxies of the number of days where the daily ThomAvg is greater than 23 and those in which ThomMax is greater than 26. According to ARPA FVG OSMER no discomfort is perceived by the human body when ThomAvg is under 23 and ThomMax is under 26.

Sensitivity to heat is measured by data on population structure and healthiness condition. Climate BIs are indicators of exposure as the Thom Index represents the combined effect of both temperature and humidity on the warming and discomfort level perceived by the human body. Instead, availability of air-conditioning and environmental variables do not find a clear-cut classification across literature: they are either proxies of exposure or adaptive capacity, depending on the subjective choice of the analyst.

Data skewness is addressed by applying a log-transformation, whereas missing values are filled in based on the hypothesis that the behaviour of a variable in a census tract is similar to the one of contiguous census tracts located in the same municipality or in the neighbouring municipalities. Census tracts where climate indicators always signalled a situation of no discomfort are removed.

Figure 1 displays the correlation matrix of the system of BIs chosen for the HVI for FVG. Due to its inverse relationship with HV, the LAI has a negative correlation with all the other indicators. Climate indicators are highly correlated between them but uncorrelated with the other indicators, which further highlights the importance of including measures of hazard exposure in the index. The II and the LCPI are also highly correlated between them (they also share the same data source), however they display non negligible correlation levels with the LAI, socio-demographic and health

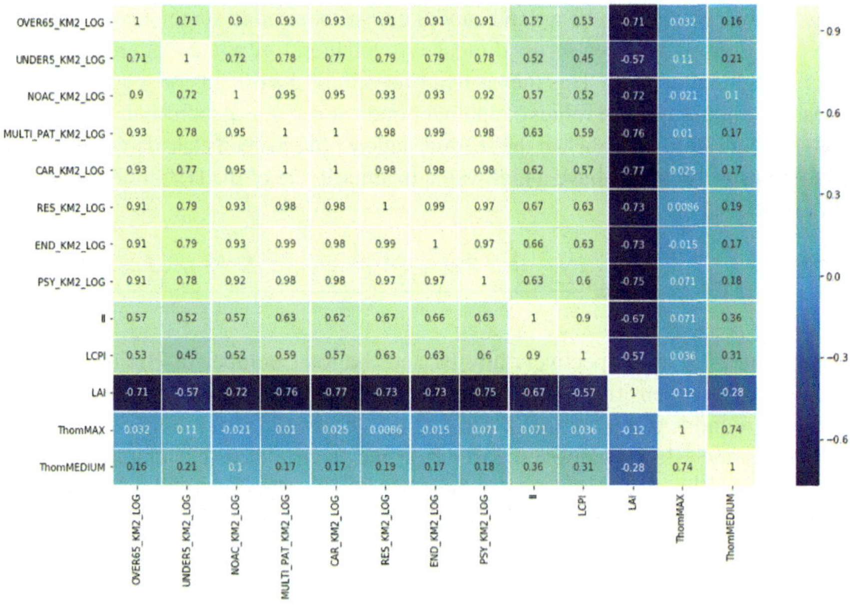

Fig. 1 Correlation heat map of basic indicators of HVI in FVG

indicators. The third group of data is made up of socio-demographic and health BIs that display high correlations. Despite highly correlated, the selected BIs represent different dimensions and features of HV and correlations might be driven by other factors.

3.3 Heat Vulnerability Index

Before constructing the index, polarity of the LAI is reversed and BIs are normalised with the min-max method to transform those expressed in different units into indices between 0 and 1 [23]. To maintain the dimensional structure of the IPCC framework, BIs are aggregated in two stages. At the first stage, a cubic mean with equal weights is applied within HV dimensions. The adoption of a non-linear aggregation function allows the introduction of a partially compensatory approach. This choice is driven by the intention to preserve the impact of those features that make a census tract, and the related inhabitants, vulnerable to HWs according to any dimension of HV. In other words, in case of imbalance between BIs, those with a high value do not compensate the ones with a low value. For instance, a census tract characterised by high density of elderly is not fully compensated by a healthy population and thus is still highly sensitive to HWs. As the HVI is an indicator with negative polarity, meaning that an increase of the index corresponds to a worsening of the phenomenon, an upward penalisation must be used [17]. As regards the intensity, a high penalisation is chosen by selecting the cubic mean. Nevertheless, a quadratic mean (low upward penalisation) would also be a good candidate. At the second stage, a compensatory approach is instead chosen across dimensions and therefore a linear aggregation with equal weights is applied. This means that low values of a dimension, e.g. sensitivity, linearly compensate high values of another dimension, e.g. exposure and vice versa. A non-compensatory approach at the second stage would instead classify as vulnerable all the census tracts that have a high value in at least one of the HV dimensions. The HVI is defined as:

$$HVI_{(S,E,AC)} = \frac{(\sum_{s}^{S} s^3 \frac{1}{\#S})^{\frac{1}{3}} + (\sum_{e}^{E} e^3 \frac{1}{\#E})^{\frac{1}{3}} + (\sum_{a}^{AC} a^3 \frac{1}{\#AC})^{\frac{1}{3}}}{3}, \quad (1)$$

where s and $\#S$ are respectively any BI and the total number of BIs representing the dimension of sensitivity. Similarly applies for exposure (e and $\#E$) and adaptive capacity (a and $\#AC$).

Since some BIs of adaptive capacity do not have a clear-cut classification, an alternative HVI, called $HVI_{(S,E)}$, is built based on a classification of BIs whereby only the dimensions of exposure and sensitivity are represented. That is:

$$HVI_{(S,E)} = \frac{(\sum_{s}^{S} s^3 \frac{1}{\#S})^{\frac{1}{3}} + (\sum_{e}^{E} e^3 \frac{1}{\#E})^{\frac{1}{3}}}{2}. \quad (2)$$

Both HVIs are rescaled to range between 0 and 1, so that census tracts with the highest index are considered to be the most vulnerable, whereas census tracts with the lowest index are considered the least vulnerable to heat. The two-dimensional index, $HVI_{(S,E)}$, is preferred to the three-dimensional one, $HVI_{(S,E,AC)}$, based on a more symmetric distribution and larger coefficient of variation (i.e. higher discrimination power across census tracts) of the former compared to the latter.

The CI techniques imply multiple subjective decisions that might influence the resulting index. To assess the impact of the chosen aggregation function, the PI approach proposed by Mazziotta and Pareto [17] is employed in this analysis. The researchers suggest computing, for each statistical unit, an interval of possible values rather than a single one. The range of the PI depends on the level of compensability of BIs and their imbalances, and it generates a lower (LB) and upper (UB) bound for the HVI. Since the HVI has negative polarity, the LB corresponds to the hypothesis of full compensability (arithmetic mean of BIs), whereas the UB corresponds to the hypothesis of non-compensability (maximum across the BIs). The midpoint (MP) of the interval represents the case of partially compensatory BIs. Taking into account that the HVI is computed in two steps, the PI approach is adopted at the second level of aggregation, i.e. investigating different levels of compensability between the dimensions of exposure and sensitivity, whereas within the two HV dimensions the cubic mean is maintained.

4 Results and Discussion

This section depicts the spatial patterns of HV in FVG by means of the selected index $HVI_{(S,E)}$. Figure 2 illustrates vulnerability patterns to heat extremes in FVG region.[6] Blank areas refer to census tracts having no inhabitants or that never recorded a discomfort from heat as per the Thom Index and hence are dropped from the analysis: these are mostly located in the mountainous area, as in line with the orography of the region. The map clearly shows that most urbanised and densely populated areas are the ones that record the highest values of HV: Trieste, Udine, Pordenone, Gorizia, Monfalcone and Sacile. Moreover, HVI increases in the surroundings of SR252, the regional road that connects Palmanova to Codroipo. Also some census tracts in the touristic and coastal cities of Lignano Sabbiadoro and Grado display high levels of HV, whereas other less urbanised census tracts in the same area do not record the same values, e.g. Lignano Riviera. This is a representative case of the impact of invasive tourism and high urbanisation that could serve as an example to decision-makers for avoiding similar policies when planning new touristic spots. In the mountainous areas only the towns of Tolmezzo and Gemona display high levels of HV due to elevated levels of urbanisation. In line with expectation, remaining census tracts located in the north are overall less vulnerable to HWs, whereas census tracts in the central part of the region record intermediate levels of HV.

[6] 6, 835 census tracts in total.

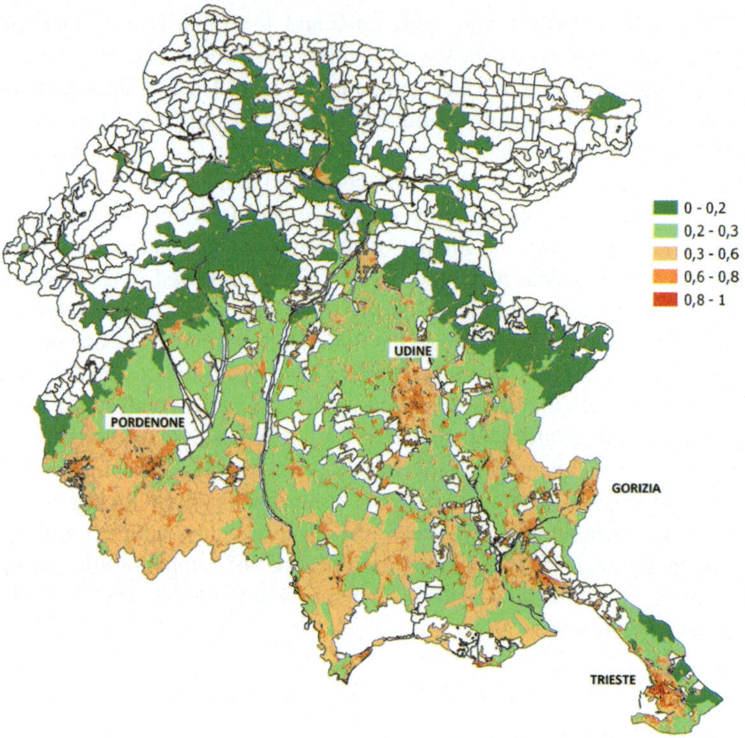

Fig. 2 Heat vulnerability map for FVG

Additional insights on the relationship between heat vulnerability and its drivers and on the features of the HVI are obtained by examining the maps of the two dimensions of HV, exposure (Fig. 3) and sensitivity (Fig. 4), which follow very different patterns. While for exposure the inverse relationship with latitude is the predominant spatial pattern, urbanisation is the one that models the spatial pattern of sensitivity. The pattern reflected in Fig. 3 is mainly driven by climate indicators, nevertheless for the most populated cities (e.g. Trieste, Udine) environmental indicators play an important role as well. As regards the sensitivity dimension, urban census tracts record much higher levels of young, old or ill individuals per km^2 than rural census tracts. Correlations of sensitivity and exposure with the HVI, respectively 0.96 and 0.85, tell that both the dimensions are meaningful to the final HVI. The fully compensatory dynamic introduced between HV dimensions through the linear aggregation is reflected in the HVI map of FVG in Fig. 2. In fact, high exposure values observed for the plain area are heavily offset by low sensitivity values.

To allow for a better interpretability of the results, the PI approach is applied at a higher administrative level, i.e. the municipality. Figure 5 displays the PIs of the top

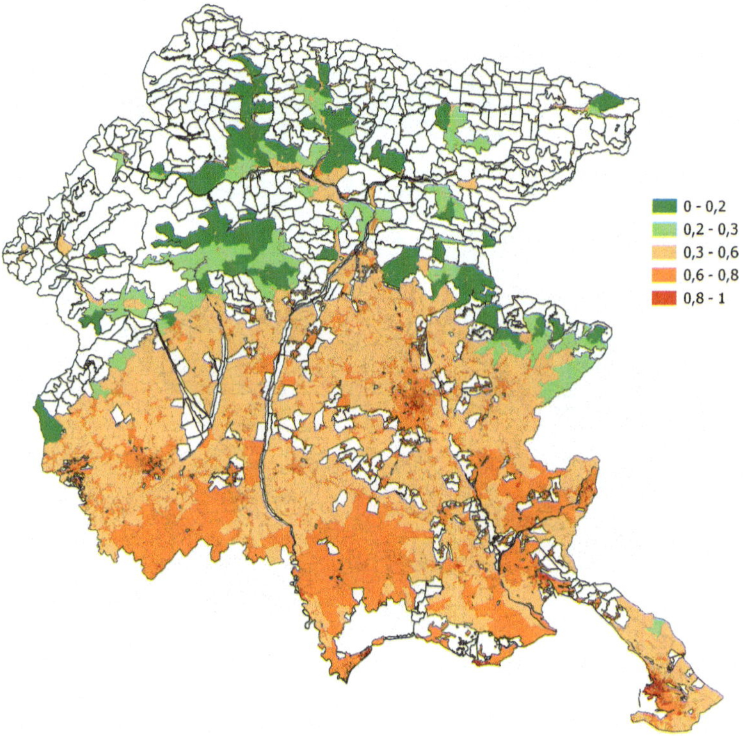

Fig. 3 Heat exposure in FVG

and the bottom 10 municipalities when ranked according to the MP.[7] It can be noticed how the ranking changes depending on the chosen aggregation function. Grado is ranked 5th based on a MP ranking but it is 1st based on UB ranking. Beyond the impact of the aggregation function, Fig. 5 further confirms that the biggest cities in FVG display the highest values of vulnerability: Trieste, Monfalcone, Pordenone, Udine and Gorizia are all ranked in the first 10 positions, which confirms the validity of the built HVI and the observations resulting from Fig. 2. Conversely, municipalities with the lowest levels of vulnerability are all located in the Alps.

5 Conclusions

This study is the first attempt to compute a HVI and geographically illustrate vulnerability pattern to heat in FVG. Communities with higher vulnerabilities are observed in urbanised and highly populated areas, specifically in the cities of Trieste, Udine,

[7] FVG counts 218 municipalities, determining the impossibility to effectively visualise all of them.

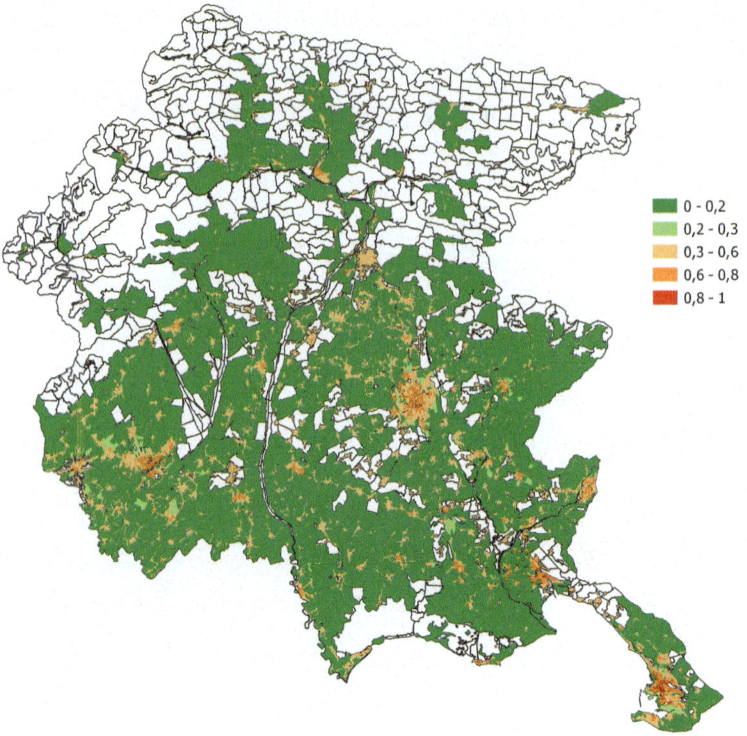

Fig. 4 Heat sensitivity in FVG

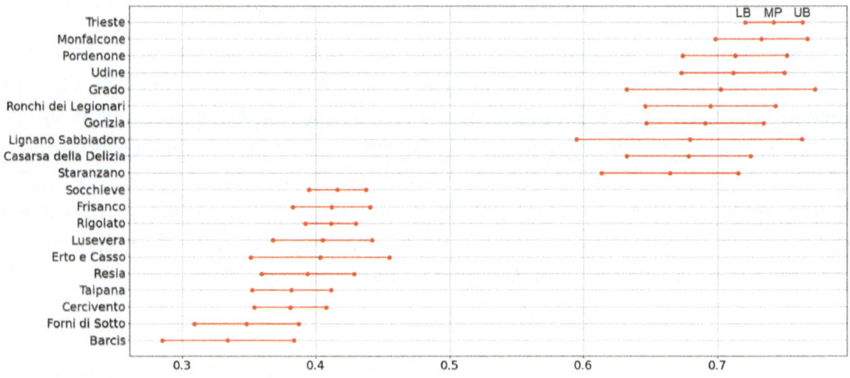

Fig. 5 Performance intervals – Top and bottom 10 municipalities

Pordenone, Gorizia and Monfalcone. Some census tracts at risk are also located in the coastal touristic towns such as Lignano Sabbiadoro and Grado. These outcomes are in line with expectations, as the synergies of climate and the UHI effect lead to higher daytime temperatures and reduced night-time cooling capacity of cities with respect to rural areas. This behaviour is driven by reduced availability of green areas and higher imperviousness levels of the urban landscape. Furthermore, Trieste, Udine and Pordenone, compared to the surrounding areas, record higher at risk population density, i.e. elderly, children and ill individuals. Low and very low vulnerability values are observed in census tracts located in the rural and mountainous part of the Region characterised by lower population density and/or reduced temperatures. These results are validated and confirmed through the PI approach: regardless the use of more or less compensatory techniques in aggregating the BIs, the set of risky areas previously detected remains the same.

The HVI constructed in this study provides policy makers a new and accessible instrument to identify areas vulnerable to heat in FVG and implement targeted prevention measures. Policy makers are suggested to redesign urban plans with the aim to change the way cities develop and grow. The current structure of urban areas should also be reshaped by introducing parks, tree-lined avenues or vegetated rooftops and incentives to steer economic development outside urban areas should be introduced in order to control the urbanisation process.

Future developments of this research could consider other metrics of vulnerability to be included in the index, for instance low income, population living alone and low education. Comparative analysis of temperature patterns, health outcome (mortality, emergency calls or hospitalisations), and their relationship with the constructed HVI would enhance the index reliability and contribute valuable insight in to how the HVI may detect spatial patterns of heat vulnerability.

References

1. Aubrecht, C., Özceylan, D.: Identification of heat risk patterns in the US National Capital Region by integrating heat stress and related vulnerability. Environ. Int. (2013). https://doi.org/10.1016/j.envint.2013.03.005
2. Bao, J., Li, X., Yu, C.: The Construction and Validation of the Heat Vulnerability Index, a Review. Int. J. Environ. Res. Public Health (2015). https://doi.org/10.3390/ijerph120707220
3. Barriopedro, D., Fischer, E., Luterbacher, J., Trigo, R., García-Herrera, R.: The hot summer of 2010: redrawing the temperature record map of Europe. Science (New York, N.Y.). (2011).https://doi.org/10.1126/science.1201224
4. Bradford, K., Abrahams, L., Hegglin, M., Klima, K.: A heat vulnerability index and adaptation solutions for Pittsburgh. Environ. Sci. Technol. (2015). https://doi.org/10.1021/acs.est.5b03127
5. Cutter, S.L., Boruff, B.J., Shirley, W.L.: Social vulnerability to environmental hazards. Soc. Sci. Q. (2003). https://doi.org/10.1111/1540-6237.8402002
6. Conlon, K.C., Mallen, E., Gronlund, C.J., Berrocal, V.J., Larsen, L., O'Neill, M.S.: Mapping human vulnerability to extreme heat: a critical assessment of heat vulnerability indices created using principal components analysis. Environ. Health Perspect. (2020). https://doi.org/10.1289/EHP4030
7. Core Writing Team, Pachauri, R.K., Meyer, L.A.: Climate Change 2014: Synthesis Report. Contribution of Working Groups I, II and III to Fifth Assessment Report of the Intergovernmental Panel on Climate Change. IPCC, Geneva (2014)

8. Hajat, S., Kovats, R.S., Atkinson, R.W., Haines, A.: Impact of hot temperatures on death in London: a time series approach. J. Epidemiol. Commun. H. (2002). https://doi.org/10.1136/jech.56.5.367
9. Inostroza, L., Palme, M., De La Barrera, F.: A heat vulnerability index: spatial patterns of exposure, sensitivity and adaptive capacity for Santiago de Chile. PLoS ONE (2016). https://doi.org/10.1371/journal.pone.0162464
10. Ito, K., Lane, K., Olson, C.: Equitable access to air conditioning: a city health department's perspective on preventing heat-related deaths. Epidemiology (2018). https://doi.org/10.1097/EDE.0000000000000912
11. Knowlton, K., Rotkin-Ellman, M., King, G., Margolis, H.G., Smith, D., Solomon, G., Trent, R., English, P.: The 2006 California heat wave: impacts on hospitalizations and emergency department visits. Environ. Health. Persp. (2009). https://doi.org/10.1289/ehp.11594
12. Kovats, S., Hajat, S.: Heat stress and public health: a critical review. Annu. Rev. Publ. Health. (2008). https://doi.org/10.1146/annurev.publhealth.29.020907.090843
13. Loughnan, M.E., Tapper, N.J., Phan, T., Lynch, K., McInnes, J.: A spatial vulnerability analysis of urban populations during extreme heat events in Australian capital cities. National Climate Change Adaptation Research Facility, Gold Coast (2013)
14. Maggino, F.: Developing indicators and managing the complexity. In: Maggino, F. (ed.) Complexity in Society: From Indicators Construction to their Synthesis, pp. 87–114. Springer International Publishing, Cham (2017)
15. Mallen, E., Stone, B., Lanza, K.: A methodological assessment of extreme heat mortality modeling and heat vulnerability mapping in Dallas. Texas. Urban Clim. (2019). https://doi.org/10.1016/j.uclim.2019.100528
16. Mazziotta, M., Pareto, A.: Use and misuse of PCA for measuring well-being. Soc. Indic. Res. (2019). https://doi.org/10.1007/s11205-018-1933-0
17. Mazziotta, M., Pareto, A.: Composite indices construction: the performance interval approach. Soc. Indic. Res. (2020). https://doi.org/10.1007/s11205-020-02336-5
18. McCarthy, J.J., Canziani, O.F., Leary, N.A., Dokken, D.J., White, K.S.: Climate Change 2001: Impacts, Adaptation, and Vulnerability. Contribution of Working Group II to Third Assessment Report of the Intergovernmental Panel on Climate Change. Cambridge University Press, Cambridge (2001)
19. McGregor, G.R., Bessmoulin, P., Ebi, K., Menne, B.: Heatwaves and Health: Guidance on Warning-System Development. WMO, Geneva (2015)
20. Méndez-Lázaro, P., Muller-Karger, F.E., Otis, D., McCarthy, M.J., Rodríguez, E.: A heat vulnerability index to improve urban public health management in San Juan. Puerto Rico. Int. J. Biometeorol. (2017). https://doi.org/10.1007/s00484-017-1319-z
21. Naughton, M.P., Henderson, A., Mirabelli, M.C., Kaiser, R., Wilhelm, J.L., Kieszak, Stephanie M., Rubin, C.H., McGeehin, M.A.: Heat-related mortality during a 1999 heat wave in Chicago. Am. J. Prev. Med. (2002). https://doi.org/10.1016/s0749-3797(02)00421-x
22. Nayak, S.G., Shrestha, S., Kinney, P.L., Ross, Z., Sheridan, S.C., Pantea, C.I., Hsu, W.H., Muscatiello, N., Hwang, S.A.: Development of a heat vulnerability index for New York State. Public Health (2018). https://doi.org/10.1016/j.puhe.2017.09.006
23. OECD and JRC European Commission: Handbook On Constructing Composite Indicators: Methodology and User Guide. OECD publishing, Paris (2008)
24. Reid, C.E., O'Neill, M.S., Gronlund, C.J., Brines, S.J., Brown, D.G., Diez-Roux, A.V., Schwartz, J.: Mapping community determinants of heat vulnerability. Environ. Health Perspect. (2009). https://doi.org/10.1289/ehp.0900683
25. Robine, J.M., Cheung, S.L., Le Roy, S., Van Oyen, H., Griffiths, C., Michel, J.P., Herrmann, F.R.: Death toll exceeded 70,000 in Europe during the summer of 2003. C. R. Biol. (2008). https://doi.org/10.1016/j.crvi.2007.12.001
26. Stafoggia, M., Forastiere, F., Agostini, D., Caranci, N., De'Donato, F., Demaria, M., Michelozzi, P., Miglio, R., Pandolfi, P., Picciotto, S., Rognoni, M., Russo, A., Scarnato, C., Perucci, C.: Vulnerability to Heat-Related Mortality: A Multicity, Population-Based, Case-Crossover Analysis. Epidemiology. Cambridge, Mass. (2006). https://doi.org/10.1097/01.ede.0000208477.36665.34

27. Teshome, M.: Rural households' agricultural land vulnerability to climate change in Dembia woreda. Northwest Ethiopia. Environ. Syst. Res. (2016). https://doi.org/10.1186/s40068-016-0064-3
28. Thom, E.C.: The discomfort index. Weatherwise (1959). https://doi.org/10.1080/00431672.1959.9926960
29. Wilhelmi, O.V., Hayden, M.H.: Connecting people and place: a new framework for reducing urban vulnerability to extreme heat. Environ. Res. Lett. (2010). https://doi.org/10.1088/1748-9326/5/1/014021
30. Wolf, T., McGregor, G.: The development of a heat wave vulnerability index for London. United Kingdom. Weather. Clim. Extremes. (2013). https://doi.org/10.1016/j.wace.2013.07.004
31. World Health Organization and United Nations: Climate Change and Health Country Profile: Italy. World Health Organization, Geneva (2018)
32. Zhu, Q., Liu, T., Lin, H., Xiao, J., Luo, Y., Zeng, W., Zeng, S., Wei, Y., Chu, C., Baum, S. et al.: The spatial distribution of health vulnerability to heat waves in Guangdong Province. China. Glob. Health Action. (2014). https://doi.org/10.3402/gha.v7.25051

An Alternative Aggregation Function for the UNDP Human Development Index

Manuela Scioni and Paola Annoni

Abstract Several composite indices have been proposed in literature to measure multidimensional phenomena such as well-being, quality of life, development, societal progress and so on. A fundamental point in the construction of composite indices in general, and in the well-being context in particular, is the degree of compensability allowed between individual indicators or dimensions. When building measures of well-being, the deficiency on one aspect shall not be offset by the surplus on another. In this paper we take the Human Development Index (HDI) as a relevant example of composite index and propose an alternative aggregation function between its dimensions. Having received quite few criticisms about compensability, in 2010 the HDI developers switched from the linear, unweighted arithmetic mean to geometric one. In practice, however, both approaches, the geometric and arithmetic mean, lead to similar 2021 HDI scores for most countries. The "Generalized Mean-Min" method we propose introduces an additive penalisation to country's unbalanced profiles, thus mitigating the compensability among the HDI dimensions. Furthermore, in this paper we provide guidelines for choosing the level of penalisation in an objective way. Thus, the new method allows for adjusting the penalisation of unbalanced country profiles across the HDI dimensions.

Keywords Composite indicator · Aggregation method · Unbalance penalization · Compensability · Human development · HDI

M. Scioni (✉)
Department of Statistical Sciences, University of Padua, via Cesare Battisti 241, 35134 Padua, Italy
e-mail: manuela.scioni@unipd.it

P. Annoni
European Commission, Directorate-General for Communication Networks, Content & Technology, Digital Economy, Recovery Plan and Skills Unit, Brussels, Belgium
e-mail: Paola.Annoni@ec.europa.eu

M. Bini et al. (eds.), *Advanced Methods in Statistics, Data Science and Related Applications*, Springer Proceedings in Mathematics & Statistics 467,
https://doi.org/10.1007/978-3-031-65699-6_17

1 Introduction

Based on the need to go beyond conventional economic measures and beyond the Gross Domestic Product (GDP), several composite indices have been proposed in the last decades to measure multidimensional phenomena such as well-being, quality of life and social progress. Summarizing this kind of concept into a single number poses various methodological challenges, including the choice of indicators, normalisation, weighting and aggregation. As advocated by [25], the issue of aggregation in the search for a scalar measure of quality of life requires the valuation and aggregation of the achievements in various important domains of life. This issue is "often perceived as the single most important challenge faced by quality-of-life research" [15 pg. 56]. Indeed, a fundamental point in the construction of composite indices in this context is the degree of compensability allowed among individual indicators. Compensability refers to the existence of trade-offs, i.e., the possibility of offsetting a disadvantage on some indicators by a sufficiently large advantage on another indicator [15]. When building measures of well-being, the deficiency on one aspect shall not be offset by the surplus on another. The different aspects of quality of life, social progress or human development are all important and "we should not let changes in any of them go unnoticed" [19, 26 pg. 27].

A fully compensatory approach involves the use of additive methods, such as the arithmetic mean, which assumes constant compensation among indicators and/or dimensions [9]. A non-compensatory or partially compensatory approach generally requires the use of non-linear functions such as the geometric mean, or Multi-Criteria approaches, which consider preferential relationships from the pairwise comparisons of the indicators [16]. A third category, recently developed in the literature, includes the so-called Mixed Strategies, which are hybrid methodologies using a combination of different approaches to address compensability issues [11].

Compensability can happen when aggregating indicators/dimensions with very different values observed on a specific statistical unit (country in the HDI case). In these cases, the country is said to have an unbalanced profile. Various approaches, belonging to the Mixed Strategies, which apply an additive penalty function to correct for the unbalance have been proposed in the literature [1, 4, 11, 13, 14]. The Mazziotta-Pareto index (MPI) employs as aggregation function the arithmetic mean adjusted by a penalization coefficient that accounts for the variability of indicators in relation to their mean [13, 14]. The mean-min function, developed by [4] is another approach subtracting a penalty component from the arithmetic mean. The penalty they apply depends on two parameters. Taking inspiration from their work, [1] proposed the "Generalized Mean-Min" method, which applies an additive penalisation to the generalized mean of order 0.5.

This paper employs the "Generalized Mean-Min" method [1] as an alternative method for the Human Development Index (HDI), a well-known composite index designed by the United Nations Development Programme that serves as a representative example, having been the symbol of the "dilemma" between compensatory and non-compensatory methods since its aggregation function switched from the arithmetic to the geometric mean in 2010. Data come from the Human Development Data Center for 2021 and refer to the three classical dimensions of HDI [27].

In the context of the 'Beyond GDP' discussion, it is important to promote alternative indicators to better reflect societal development, since these metrics of quality of life are key elements for setting objectives, monitoring implementation and benchmarking performances. In this perspective, the HDI was recently listed among the 10 composite indices recommended by the European Parliament as a non-exclusive selection of composite indices for policy-making [17].

Firstly, we will show that both the geometric and the arithmetic mean lead to similar 2021 HDI scores for most countries and, therefore, the non-compensatory effect of the geometric mean is overshadowed. Then, we investigate the reasons behind this behaviour and propose possible solutions based on the unbalance penalisation approach. Furthermore, we provide guidelines for choosing the level of penalisation in an objective way. Finally, we investigate the impact on country scores and rankings of the alternative method proposed.

This paper is an extended version of the paper presented at the 51st Scientific Meeting of the Italian Statistical Society on June, 2022 [23]. The paper unfolds as follows: Sect. 2 describes in detail the HDI and lists some alternative aggregation methods that have been proposed in the literature, Sect. 3 introduces the family of generalized means and the "Generalized Mean—Min" method focusing on the choice of the parameters controlling the level of penalisation. Section 4 describes and discusses results. Concluding remarks are in Sect. 5.

2 The HDI

Since 1990, the United Nations Development Programme (UNDP) has published a series of annual Human Development Reports (HDRs) in which the human development index (HDI) is computed for each country. The HDI was created to emphasize that people and their capabilities should be the ultimate criteria for assessing the development of a country, not economic growth alone.

The HDI is a summary measure of three dimensions: (D1) Life expectancy at birth, (D2) Schooling (arithmetic mean of expected years of schooling for adults aged 25 years and expected years of schooling for children of school entering age) and (D3) Gross National Income (GNI) per capita (natural logarithm). Having received quite few criticisms [7, 8, 10, 20, 22], the developers of the HDI switched the aggregation function from linear—unweighted arithmetic mean—to geometric in 2010. In their yearly report, they stated that 'Poor performance in any dimension is now directly reflected in the HDI, and there is no longer perfect substitutability across dimensions. This method captures how well rounded a country's performance is across the three dimensions. As a basis for comparisons of achievement, this method is also more respectful of the intrinsic differences in the dimensions than a simple average is. It recognizes that health, education and income are all important, but also that it is hard to compare these different dimensions of well-being and that we should not let changes in any of them go unnoticed' [26 pg. 15].

Nevertheless, one effect of replacing the arithmetic mean with the geometric one was a relatively higher penalisation of less developed countries [19], especially for the 2010 version of the HDI. This effect diminished for more recent releases of the index as the developers adjusted the theoretical minimum and maximum values for normalisation, as will be shortly discussed (Sect.2.1). The higher penalisation of less developed countries is related to two intrinsic characteristics of the geometric mean [5]. The first one, called decreasing penalization, means that for a given unbalance among dimensions, the lower the human development level of a country, the greater its penalization. The second one is called constraint at zero and concerns the fact that the geometric mean produces distorting effects for small positive values, since it tends to the minimum as any of the input values approaches zero. This can be considered a suitable property in the context of well-being composite indicators, in which countries with lower scores in a given dimension shall not be able to compensate fully in other dimensions [11].

2.1 HDI Computation Methodology

The individual indicators included in the HDI are normalised and converted to a common scale with range [0,1] by the Min-Max method. Since the geometric mean was introduced as aggregation function for the HDI in 2010, some changes were applied to avoid values too close to zero, that causes mathematical instability when included in a geometric mean.

Details about these changes are shown in Table 1. The theoretical minimum and maximum values for normalization have been modified with the HDI versions. The

Table 1 Theoretical minima and maxima for the HDI since 2010

Year	Health	Education	Income
2010	Life expectancy at birth Maximum values taken from dataset Minimum values set at 20	Arithmetic mean between the mean years of schooling and expected years of schooling (adult literacy rate no longer included) Maximum values taken from dataset Minimum values set at 0	Logarithm (base e) of GNI/capita Maximum set at $108211/capita Minimum set at $163/capita
2011	As in 2010	As in 2010	Maximum set at $107721/capita Minimum set at $100/capita
2012	Maximum set at 85 Minimum set at 20	As in 2011	Maximum set at 87478 $/capita Minimum set at 100 $/capita
2013–2021	As in 2012	Maximum values set at 15 and 18 Minimum values set at 0	Maximum set at $75000/capita Minimum set at $100/capita

current minimum and maximum values act respectively as the natural zeros and utopian targets and they haven't changed since 2013. In this way it is possible to track absolute performance of countries across time together with cross-country relative comparison.

In general, minimum values are far from the observed minima. For example, one change was to lower from 163 to 100 the minimum of GNI per capita in the min-max normalization, whilst the observed minimum, considering 2021 HDI, is about 731.

The choice of the theoretical minimum and maximum values for normalisation can greatly affect the impact of the aggregation function on the final score of a composite index. The values chosen for the HDI had the effect of squeezing the distribution of the normalised indicator values towards the right-hand side of their range (See Fig. 1).

This in turns overshadows the low compensatory effect of the geometric mean that leads to final scores almost identical to the ones calculated by using a simple, unweighted linear aggregation, as shown for the 2021 HDI in Fig. 2 and highlighted also by [18].

Alternative aggregation procedures for the HDI have been proposed in literature [12, 18, 19]. Ravaillon [19] offered an alternative aggregation function based

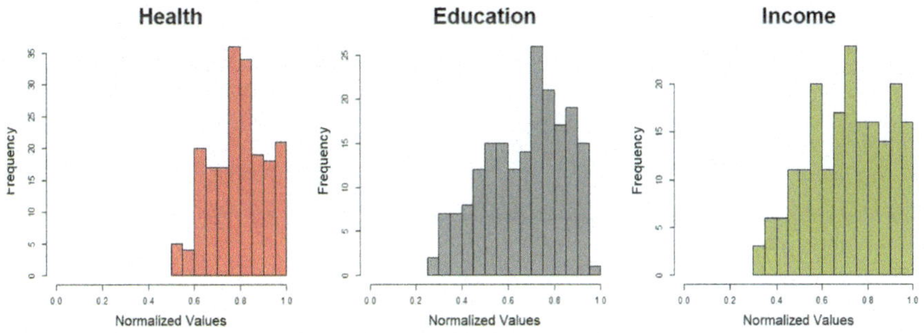

Fig. 1 HDI 2021 case study: normalized values of the three dimension

Fig. 2 Difference between 2021 HDI and Arithmetic Mean

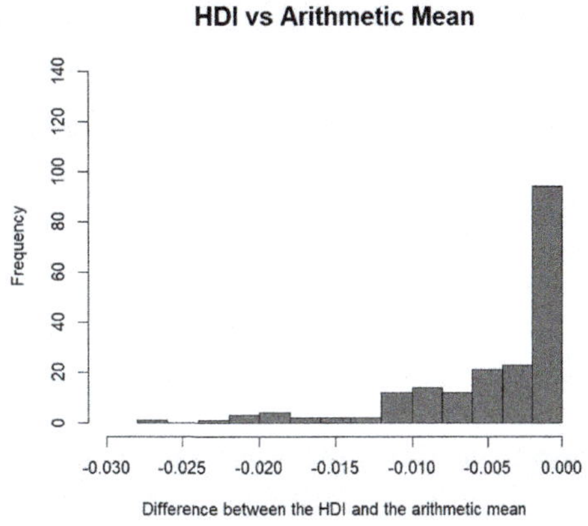

on the generalized aggregation formula by [6], which led to more sensible tradeoffs across the dimensions compared to the geometric mean. Mariani and Ciommi [12] proposed a non-compensative approach by means of a penalization factor capturing the unbalance among indicators. Similarly to the Mazziotta Pareto Index [13, 14], they propose a composite indicator defined as the product between the geometric mean and a penalisation term, that depends on the variance of the unbalance of the log-transformed and normalised indicators. Pinar [18] uses a flexible non-additive aggregation model-the Choquet integral-to account for potential synergies and redundancies of the HDI dimensions also penalizing the unbalanced values of the indicators. To this aim, they consider five hypothetical set of weights that allow for varying positive interactions among the HDI dimensions.

3 Controlling the Level of Compensability

The alternative aggregation function we propose for the HDI is the "Generalized Mean-Min" method. Its starting point is the generalized mean of order $\beta = 0.5$ that is added to a penalisation depending on the country's indicators unbalance. This penalty element is tuned according to two parameters. Section 3.1 describes the generalized mean and the family of the "Generalized Mean—Min" functions, whilst Sect. 3.2 is focused on the choice of the two parameters controlling the penalty.

3.1 *The Generalized Mean and the Generalized Mean—Min Functions*

Let x_{ji} denote the score of indicators (or dimensions) j $(j = 1, ..., q)$ for unit i $(i = 1, ..., n)$, the generalized mean of order β can be expressed as follows:

$$I_i^\beta = \begin{cases} \left(\frac{1}{q} \sum_{j=1}^{q} w_j x_{ji}^\beta \right)^{1/\beta} & \beta \neq 0 \\ \left(\prod_{j=1}^{q} w_j x_{ji} \right)^{1/q} & \beta = 0 \quad \text{geometric mean} \end{cases} \tag{1}$$

where $w_1, ..., w_q$ are non-negative weights normalised to have unitary sum and β is a constant, possibly negative. The indicators (dimensions) are all assumed to be in $\Re+$ and positively oriented with the phenomenon to be measured, for example in the case of the HDI, the higher the dimension's value, the higher the level of human development. The family of the generalised means of order β allows for controlling the level of compensability between the indicators/dimensions in a composite index [2, 4, 9, 21, 24]. I_i^β depends on the value of β. Special cases include the arithmetic mean for $\beta = 1$ and the minimum function (MIN) for the limit value $\beta \rightarrow -\infty$ [4, 9]. The former is a fully compensatory aggregation function whilst the latter is a fully non-compensatory aggregation function. By adjusting the value of β the analyst can obtain his own ideal, theoretical level of compensation across the values to be aggregated. If $\beta = 0$ the generalised mean coincides with the geometric mean, a partially compensatory function. While linear aggregation assumes constant trade-offs among the indicators, geometric aggregation offers inferior compensability for indices with lower values (diminishing returns) [28]. So, in terms of compensability, the family of generalised means of order $\beta \in (0, 1)$ occupy an intermediate position between the arithmetic and the geometric mean.

β plays the important role of balancing the achievements between two indicators. As already mentioned, compensability refers to the existence of trade-offs, i.e., the possibility of offsetting a disadvantage on some indicators by a sufficiently large advantage on another indicator [15]. This trade-off is the Marginal Substitution Rate (MSR_{st}) which for the generalized mean is equal to

$$MSR_{st}(x_t, x_s) = \left. \frac{\frac{\delta I}{\delta x_s}}{\frac{\delta I}{\delta x_s}} \right|_x = \left(\frac{w_s}{w_t}\right)\left(\frac{x_t}{x_s}\right)^{1-\beta} \tag{2}$$

The elasticity of the MSR_{st}, so called elasticity of substitution σ, helps us understanding the role played by β in defining the inequality adverse level of the generalized mean for $\beta > 0$. It is easy to show that the family of generalized means has constant elasticity of substitution between every pair of variables [3]:

$$\sigma = \frac{1}{1-\beta} \tag{3}$$

that includes the cases $\sigma = \infty$ for $\beta = 1$ (arithmetic mean), corresponding to complete substitutability between the two indicators, and $\sigma = 0$ for $\beta \rightarrow -\infty$ (minimum function), corresponding to null substitutability. The value of $\beta = 0.5$ chosen here is characterized by a constant positive elasticity of substitution $\sigma = 2$.

We apply here the "Generalized Mean—Min" function (GMM) as an alternative aggregation method of the HDI dimensions that introduces a penalty proportional to the unbalance of each country score across the index dimensions [1]. The GMM function, that is an adaptation of the "mean—min" method originally proposed by [4], is a two parameters function and uses the generalized mean of order $\beta = 0.5$ plus a penalty applied for the countries with very different values across the three HDI dimensions. Using the generalized mean of order $\beta = 0.5$ as a starting point of our aggregation function allows us to start from an intermediate position between the arithmetic and the geometric mean.

For each country i, the penalised HDI score—HDI_GMM—is computed as follows:

$$\text{HDI\_GMM}(x_1, x_2, x_3) = \bar{x}_{GM} - \alpha\left(\sqrt{(\bar{x}_{GM} - \min x)) + \gamma^2} - \gamma\right) \tag{4}$$

where (x_1, x_2, x_3) are the HDI dimension scores for country i , $\bar{x}_{GM}$ is the generalised mean of order $\beta = 0.5$ among the dimensions, $\min x = \min(x_1, x_2, x_3)$, the parameter α is the intensity of penalisation for unbalance $(0 \leq \alpha \leq 1)$ and $\gamma \geq 0$ is the level of substitutability between the dimensions. As desired, the penalisation term vanishes exactly when the dimensions are perfectly balanced and it is strictly positive elsewhere. The GMM function incorporates the two extreme cases of penalisation:

a. $\alpha = 0$ (γ irrelevant): in this case is the generalized mean of order $\beta = 0.5$ (no penalisation)
b. $\alpha = 1$ and $\gamma = 0$: in this case HDI_GMM = $\min x$, that is maximum penalisation.

All the others are intermediate cases between these two extremes. Parameters α and γ are key elements for the choice of the intensity of the penalisation. Casadio Tarabusi and Guarini [4] proposed $\alpha = \gamma = 1$ as "reasonable values" corresponding "to an intermediate easy case of adjustment with incomplete and progressive compensability" [4, pg. 31], but guidelines for the selection of these parameter values are not provided in their original proposal. Mariani and Ciommi [2] proposed a data-driven method to select the value of the parameters. In the next subsection we provide guidelines for choosing the level of penalisation in an objective way, presenting the procedure in detail and considering the 2021 HDI as case of study.

3.2 Choosing the Intensity of the Penalisation

Between the two parameters, α is the one with the most limited variability and the weakest impact on compensability effects. In this analysis we set $\alpha = 0.75$. The values $\alpha = 0.5$ and $\alpha = 1$ were also tested but discarded because they led to iso-curves either too linear ($\alpha = 0.5$) or too sharp-cornered ($\alpha = 1$). Results are not shown for sake of brevity. If $\gamma = 0$, $\Delta(\gamma) = \sqrt{(\bar{x}_{GM} - \min x)^2 + \gamma^2} - \gamma = (\bar{x}_{GM} - \min x)$ and it monotonically decreases as γ increases. When $\gamma >> (\bar{x}_{GM} - \min x)$, it prevails over the unbalance making it irrelevant in the penalisation term. Therefore, the choice of γ is case dependent and must be chosen taking into account the order of magnitude of the unbalance term $(\bar{x}_{GM} - \min x)$. To select the value of γ we examine two types of graphs:

1. The iso-curves of the penalisation component $\Delta(\gamma)$ and of the unbalance $(\bar{x}_{GM} - \min x)$. We set the range of γ values in between 0 and $\max(\bar{x}_{GM} - \min x) = 0.176$ in the 2021 HDI case (Fig. 3, left).
2. The distribution of the unbalance term across all the countries (Fig. 3, right).

The penalisation iso-curves inform us on the choice of γ. For a given value of γ shown on the y-axis, the iso-curves allow us to assess how the penalisation component Δ changes for different values of the unbalance (on the x-axis).

The unbalance distribution depends solely on the 2021 HDI indicators and not on parameters α and γ,but helps assessing the entity and distribution of the unbalance to select possible values for γ. The two graphs shall be jointly read:

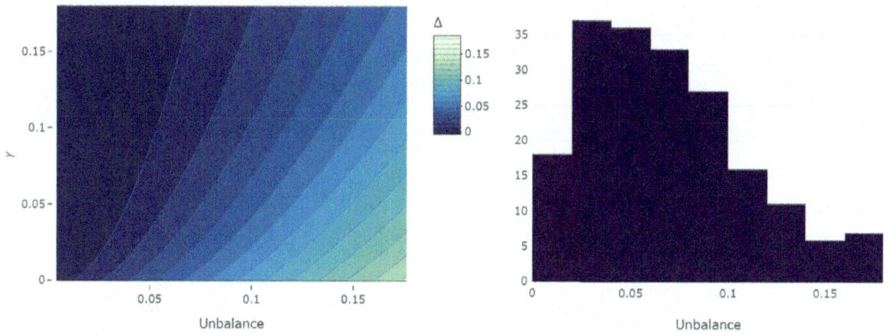

Fig. 3 HDI 2021 case study: Iso-curves of the penalization components and histograms of the unbalance distribution across the all the countries included in the index

a. In the case of a symmetrical unbalance distribution, we select a value of γ to reach a compromise between two extreme situations: the case with minimum or null penalty, Δ close to 0, and the case with maximum penalty. The former case corresponds to the darker area in Fig. 3 (left-hand side), whilst the latter to the lighter area.
b. For a positively (negatively) skewed distribution of the unbalance, we would choose a γ value closer to 0 (closer to the maximum), to differentiate as much (as less) as possible the penalisation value in left-hand side of the distribution.

These considerations led us to choose $\gamma = 0.05$.

An additional step is then needed to check whether the level substitutability between each pair of the 2021 HDI dimensions is limited. If the aim is to get a level of compensability that is in between maximum penalisation and no penalisation, the shape of the observed iso-curves shall be in between the L-pipe shape and the straight line. In the 2021 HDI case that shows a positively skewed unbalance distribution, we want to get a level of compensability more towards the maximum penalty. To this purpose we aggregate pairwise each 2021 HDI dimensions using the Generalised Mean-Min aggregation function (2) with $\alpha = 0.75$ and $\gamma = 0.05$ and analyse their corresponding iso-curves. The value of γ will be eventually confirmed if the iso-curves have a shape in between the straight line (fully compensatory) and the L-pipe (fully non compensatory).

Figure 4 shows the iso-curves for each pair of HDI dimensions. The shape of the iso-curves is in between the two extremes for all cases confirming the validity of the choice of the value of $\gamma = 0.05$.

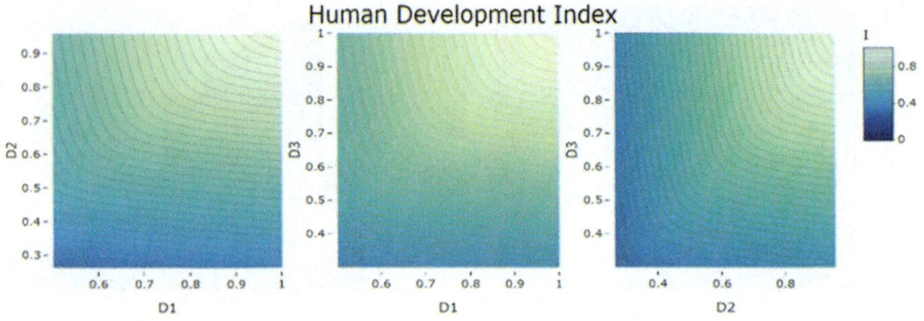

Fig. 4 HDI 2021 case study: Iso-curves for the Generalized Mean—Min aggregating function (GMM) for each pair of HDI dimensions

4 Results

Figure 5 illustrates the effect that the GMM method has on the 2021 HDI scores. The map shows the distribution across world countries of the percentage difference between the benchmark (HDI) and the penalised scores (HDI_GMM). The histogram shows its distribution across all the countries.

The percentage difference goes from a minimum of 0 to a maximum of 17.6%. African countries above the Equator, such as, Dijbouti, Mali and Senegal, are the

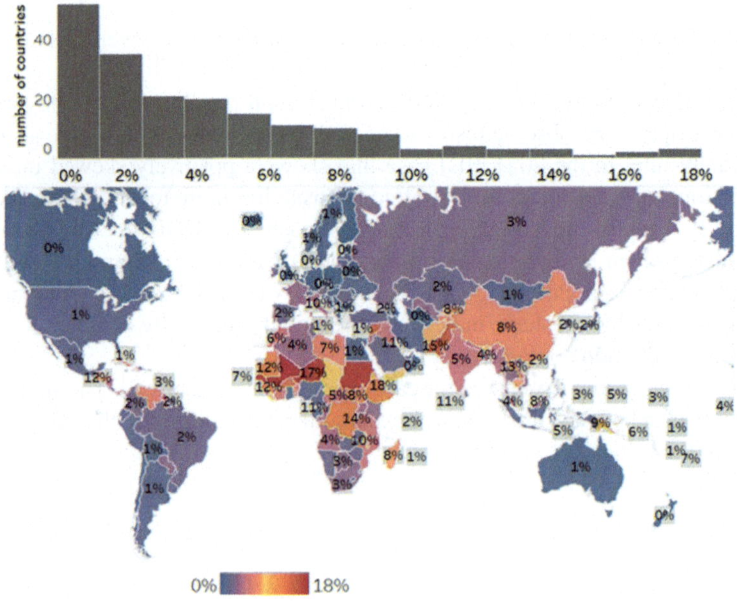

Fig. 5 HDI 2021 case study: Percentage difference between the benchmark (HDI) and the penalised (HDI_GMM) country scores

most penalised in percentage terms. On the contrary, countries higlighted in blue, such as the USA or UK, are the least penalised.

As expected, the proposed method has the effect of an overall penalisation, with the new scores smaller than the original ones for all the countries. The basic statistics of the score difference between the original and new method all confirm this effect (see Table 2). The original 2021 HDI average is 0.721, on a $0-1$ scale, in the original case and 0.694 in the penalised one. The average difference between HDI and penalised scores is 0.027 for the whole set of countries going from a minimum of 0 to a maximum of 0.094, with an interquartile range between 0.008 and 0.037. The average difference between HDI and arithmetic mean-based scores is much lower, 0.004, with an interquartile range between 0 and 0.007 and maximum value equal to 0.028 (Table 2).

The correlation between the scores from the two methods is 0.992 with a p-value < 0.001. Correlation between 2021 HDI scores and the scores obtained through the unweighted linear mean is 1.000 (p-value < 0.001) (Table 2).

Figure 6 shows the 2021 HDI scores versus the penalised one (HDI_GMM) for all the countries. In general, countries with poor HDI performances are more penalised, but important differences can also be seen for countries with higher scores. The intensity of the penalty does not depend on the index value but on the balance between the three dimensions. If it is usually more frequent for poor performers to

Table 2 HDI 2021 case study: descrptive statistics of the 2021 HDI and the HDI_GMM

Basic statistics		Value
HDI	Mean (s.d.)	0.721 (0.151)
HDI_GMM	Mean (s.d.)	0.694 (0.162)
Difference between HDI and HDI_GMM	Mean (s.d.)	0.027 (0.023)
	Minimum	0
	IQR	0.008–0.037
	Maximum	0.094
Difference between HDI and arithmetic mean	Mean (s.d.)	0.004 (0.005)
	Minimum	0
	IQR	0–0.007
	Maximum	0.028
Percentage difference between HDI and HDI_GMM	Mean (s.d.)	4.2 (4.1)
	Minimum	0
	IQR	1.1–6.1
	Maximum	17.6
Correlation between HDI and HDI_GMM	Pearson ρ (p-value)	0.992(<0.001)
Correlation between HDI and the arithmetic mean	Pearson ρ (p-value)	1.000(<0.001)
Correlation between ranks obtained with HDI and HDI_GMM	Spearman ρ (p-value)	0.992(<0.001)
Correlation between ranks obtained with HDI and the arithmetic	Spearman ρ (p-value)	1.000(<0.001)

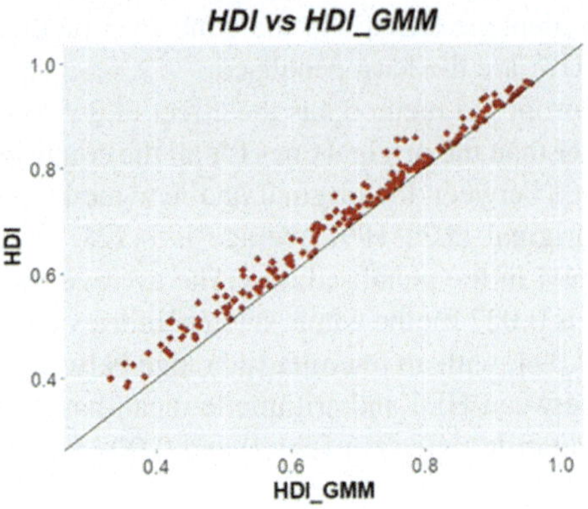

Fig. 6 HDI versus the penalised (HDI_GMM) country scores

have an unbalanced profile across the HDI dimensions, it is also a fact that unbalanced profiles can be observed among the top performers as well. Highlighting this pattern is highly important particularly for countries that, excelling on some aspects of human development, compensate poor scores on others.

On the contrary, the penalisation has a minor impact on countries with a balanced profile. Amongst the most stable ones, we find Netherlands and Ghana. They score very differently on the 2021 HDI, 0.941 for Netherlands and 0.632 for Ghana, but the original and modified scores are almost identical (Fig. 7, first two rows). The

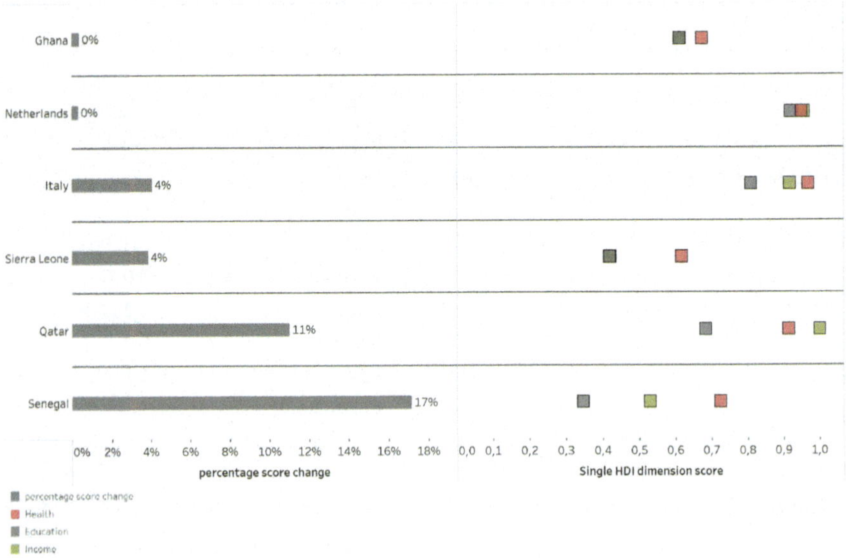

Fig. 7 Percentage difference between the benchmark (HDI) and the penalised (HDI_GMM) for some countries chosen as example and corresponding dimensions' scores

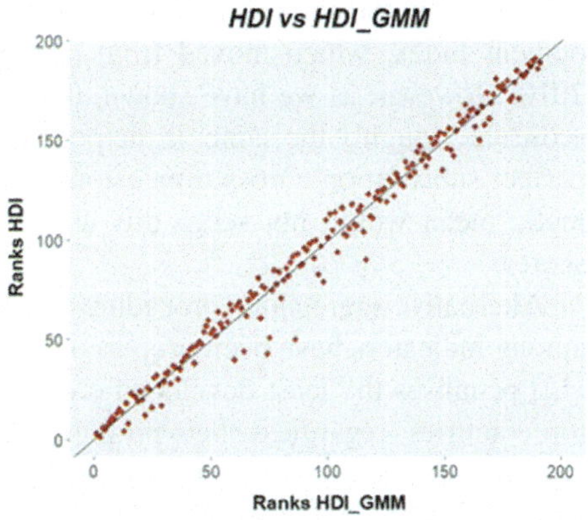

Fig. 8 HDI vs the penalised (HDI_GMM) country ranks

percentage score difference is intermediate—4%—for Italy (HDI = 0.895) and Sierra Leone (HDI = 0.477) and their profiles are penalised approximately in the same way. Italy scores well on the health and income dimension but poorly on education; whilst, Sierra Leone has low income and education scores and a relatively better health score. Qatar (HDI = 0.855, percentage difference = 11%) and Senegal (HDI = 0.511, percentage difference = 17%) are amongst the most penalised ones (Fig. 6, last two rows). Qatar performs poorly on the Education dimension, but it performs rather well on all the other dimensions. Senegal, on the other hand, is characterised by three different values across the HDI dimensions (a low value of education, an intermediate score of income and a higher performance on health).

Correlation between the ranks from the two methods is 0.991 with a p-value < 0.001. Correlation between 2021 HDI scores and the scores obtained through the unweighted linear mean is 1.000 (p-value < 0.001) (Table 2). However, despite the high correlation between rankings there are cases of clear rank differences (Fig. 8). The two countries at the bottom of the ranking according to the method are Kuwait and Qatar, which lose respectively 36 and 31 positions. In particular, Kuwait went from the 43rd to the 74th position whilst Qatar was relegated from the 50th to the 81st position. As the case of Qatar, has already been described above we focus here on Kuwait. The three normalized HDI dimensions for Kuwait take the following values: 0.903, 0.670 and 0.947. In other words, geometric mean allows countries to achieve higher positions in the ranking by having relatively good achievements in just one or two dimensions. On the other hand, the "generalized mean—min" rewards countries with more balanced scores across the HDI dimensions.

5 Conclusions

Building metrics of well-being and quality of life poses various methodological challenges. Our interest is here in the aggregation function and its effect on the level of compensability across the index's components.

A well-known composite indicator that serves as case study is the Human Development Index, which moved from unweighted arithmetic to geometric mean in 2010. However, as we have shown, the geometric and arithmetic mean 2021 HDI scores are roughly the same. If the policymakers in the UNDP were trying to avoid perfect substitution across dimensions, switching from the geometric to the arithmetic mean would not serve this aim as both methods mostly result in similar scores.

Alternative aggregation procedures for the HDI taking into account the unbalance among indicators have been proposed recently in literature [12, 18]. The method of [12] penalises the least developed countries while it does impact significantly the top countries, regardless their profile unbalance. On the other hand, [18] includes interactions among indicators that experts shall quantify. As the number of dimensions/indicators increases, however this procedure becomes more difficult, as well as requiring the opinion of experts.

We propose an alternative aggregation function, the "Generalized Mean-Min" method that introduces a penalty proportional to the unbalance of the country scores across the index dimensions. This additive element depends on two parameters that represent the intensity of the penalisation and the level of compensability, and we propose a method to make an informed, data-driven choice of these parameters. Our proposal is a simple and intuitive method that allows us to control the level of compensability between the indicators/dimensions by varying the two parameters tuning the amount of the penalization. We apply it to the 2021 HDI and compare original and modified scores. All countries see their performance deteriorating by the penalisation method. This effect is particularly evident for countries with an unbalanced profile across the index dimensions, regardless of the index value. This is the effect we desire for a metric of well-being where all the aspects are equally important and a good performance on one aspect shall not compensate for a poor one in another.

Our proposal contributes to the debate on the development of aggregate metrics, in particular in the field of well-being, quality of life and societal progress, where compensability across dimensions/indicators is not desirable. The goal was to provide an instrument to choose and limit, when necessary, the level of compensability between the index components, applying a non-arbitrary data informed term of penalisation. The approach proposed is purely methodological but has important implications from the political point of view as well, as if these metrics are more and more used for monitoring policies effectiveness.

Parameters α and γ are key elements for the choice of the intensity of the penalisation. Their choice is guided by the inspection of the iso-curves of the penalisation component and of the unbalance, and confirmed by looking at the iso-curves for each pair of the HDI dimensions. Possible improvements of the method include the development of a tool for implementing this type of graphs.

References

1. Annoni, P., Scioni, M.: The unbalance penalisation method for metrics of social progress. Soc. Indic. Res. **162**, 1093–1115 (2022). https://doi.org/10.1007/s11205-021-02876-4
2. Annoni, P., Weziak-Bialowolska, D.: A measure to target anti-poverty policies in the European Union regions. Appl. Res. Qual. Life. **11**, 181–207 (2016). https://doi.org/10.1007/s11482-014-9361-z
3. Arrow, K.J., Chenery, H.B., Minhas, B.S., Solow, R.M.: Capital-labor substitution and economic efficiency. Rev. Econ. Stat. **43**(3), 225–250 (1961). https://doi.org/10.2307/1927286
4. Casadio Tarabusi, E., Guarini, G.: An unbalance adjustment method for development indicators. Soc. Indic. Res. **112**, 9–45 (2013). https://doi.org/10.1007/s11205-012-0070-4
5. Casadio Tarabusi, E., Guarini, G.: Level dependence of the adjustment for unbalance and inequality for the human development index. Soc. Indic. Res. **126**, 527–553 (2016). https://doi.org/10.1007/s11205-015-0918-5
6. Chakravarty, S.R.: A generalized human development index. Rev. Dev. Econ. **7**(1), 99–114 (2003). https://doi.org/10.1111/1467-9361.00178
7. Chowdhury, S., Squire, L.: Setting weights for aggregate indices: an application to the commitment to development index and human development index. J. Dev. Stud. **42**(5), 761–771 (2006). https://doi.org/10.1080/00220380600741904
8. Davies, A.: Human development and the optimal size of government. J. Socio. Econ. **38**(2), 326–330 (2009). https://doi.org/10.1016/j.socec.2008.07.015
9. Decancq, K., Lugo, M.A.: Weights in multidimensional indices of wellbeing: an overview. Econom. Rev. **32**, 7–34 (2013). https://doi.org/10.1080/07474938.2012.690641
10. Desai, M.: Human development: concepts and measurement. Eur. Econ. Rev. **35**(2–3), 350–357 (1991). https://doi.org/10.1016/0014-2921(91)90136-7
11. Greco, S., Ishizaka, A., Tasiou, M., Torrisi, G.: On the methodological framework of composite indices: a review of the issues of weighting, aggregation, and robustness. Soc. Indic. Res. **141**, 61–94 (2019). https://doi.org/10.1007/s11205-017-1832-9
12. Mariani, F., Ciommi, M.: Aggregating composite indicators through the geometric mean: a penalization approach. Comput. **10**(4), 64–80 (2022). https://doi.org/10.3390/computation10040064
13. Mazziotta, M., Pareto, A.: On a generalized non-compensatory composite index for measuring socio-economic phenomena. Soc. Indic. Res. **127**, 983–1003 (2016). https://doi.org/10.1007/s11205-015-0998-2
14. Mazziotta, M., Pareto, A.: Measuring well-being over time: the adjusted Mazziotta-Pareto index versus other non compensatory indices. Soc. Indic. Res. **136**, 967–976 (2018). https://doi.org/10.1007/s11205-017-1577-5
15. Munda, G.: Choosing aggregation rules for composite indicators. Soc. Indic. Res. **109**(3), 337–354 (2012). https://doi.org/10.1007/s11205-011-9911-9
16. Munda, G., Nardo, M.: Noncompensatory/nonlinear composite indicators for ranking countries: a defensible setting. Appl. Econ. **41**(12), 1513–1523 (2009). https://doi.org/10.1080/00036840601019364
17. Pichon, E., Widuto, A., Dobreva, A., Jensen, L.: Ten composite indices for policy-making. In: European Parliamentary Research Service, European Parliament. PE 696.203 (2021). https://www.europarl.europa.eu/thinktank/en/document/EPRS_IDA(2021)696203
18. Pinar, M.: Choquet-integral aggregation method to aggregate social indicators to account for interactions: an application to the Human Development Index. Soc. Indic. Res. **149**, 1–53 (2022). www.s11205-021-02726-3
19. Ravaillon, M.: Troubling tradeoffs in the Human Development Index. J. Dev. Econ. **99**, 201–209 (2012). https://doi.org/10.1016/j.jdeveco.2012.01.003
20. Ray, A.K.: Measurement of social development: an international comparison. Soc. Indic. Res. **86**(1), 1–46 (2008). https://doi.org/10.1007/sll205-007-9097

21. Ruiz, N.: Measuring the joint distribution of household's income, consumption and wealth using nested Atkinson measures. OECD Working Paper, vol. 40, pp. 1–37 (2011). https://doi.org/10.1787/18152031
22. Sagar, A.D., Najam, A.: The human development index: a critical review. Ecol. Econ. **25**(3), 249–264 (1998). https://doi.org/10.1016/S0921-8009(97)00168-7
23. Scioni, M., Annoni, P.: An alternative aggregation function for the UNDP Human Development Index. In: Balzanella, A., Bini, M., Cavicchia, C., Verde, R. (eds.) Book of Short Papers SIS 2022, pp. 964–969. Pearson, Milan (2022)
24. Seth, S.: Inequality, interactions, and human development. J. Hum. Dev. Capab. **10**(3), 375–396 (2009). https://doi.org/10.1080/19452820903048878
25. Stiglitz, J.E., Sen, A., Fitoussi, J.P.: Report by the Commission on the Measurement of economic performance and social progress (2009). https://ec.europa.eu/eurostat/documents/8131721/8131772/Stiglitz-Sen-Fitoussi-Commission-report.pdf
26. UNDP: Human Development Report 2010: The Real Wealth of Nations—Pathways to Human Development. UNDP, New York (2010). http://hdr.undp.org/en/content/human-development-report-2010
27. UNDP: Human Development Report 2021—Beyond Income, Beyond Averages, Beyond Today: Inequalities in Human Development in the 21st Century. UNDP, New York (2021)
28. Van Puyenbroeck, T., Rogge, N.: Geometric mean quantity index numbers with benefit-of-the doubt weights. Eur. J. Oper. Res. **256**(3), 1004–1014 (2017). https://doi.org/10.1016/j.ejor.2016.07.038

New Dynamics in Labour Market and Population Migration

The Manual, Communicative and Quantitative Abilities of Native and Foreign Workers According to Their Level of Education in Italy

Camilla Pangallo, Oliviero Casacchia, and Corrado Polli

Abstract Many studies address the labour market consequences of immigration flows. A branch of them argues that native workers respond to immigration by specializing in jobs where they have a comparative advantage, thus protecting themselves from labour market competition and possible wage losses. In order to examine this dynamic in the Italian labour market, we analysed the characteristics of jobs in Italy in terms of skills and tasks performed. The results show that foreigners hold more manual jobs, while natives hold jobs with more communicative and quantitative content.

Keywords Foreigner workers in Italy · Specialization activities · Level of education · Abilities

1 Background and Object

In recent decades, increasing migration flows have influenced labour market dynamics in most developed countries. In 2019 the foreign population of working age accounts for 10.7% of the employed (Italian National Institute of Statistics or in Italian "Istituto Nazionale di Statistica", ISTAT hereafter, 2020) and on 1 January 2020 there were about 5 million foreign citizens residing in Italy, making up 8.4% of

---

C. Pangallo (✉) · O. Casacchia
Sapienza Rome University, Rome, Italy
e-mail: camillapangallo@uniroma1.it

O. Casacchia
e-mail: oliviero.casacchia@uniroma1.it

C. Polli
INAPP, Rome, Italy
e-mail: c.polli@inapp.gov.it

M. Bini et al. (eds.), *Advanced Methods in Statistics, Data Science and Related Applications*, Springer Proceedings in Mathematics & Statistics 467,
https://doi.org/10.1007/978-3-031-65699-6_6

the total number of residents (ISTAT, 2021). Regularly-present foreigners are therefore also a consolidated reality in Italy, albeit to a lesser extent than in many other European countries.

Several studies have investigated the relationships between immigrant flows, workers' income and educational qualifications; Peri and Sparber [6] argue that "*...when the proportion of foreign workers with a high educational level increases in the U.S., native workers choose jobs with less analytical and more communicative content*". Zavodny [8] emphasizes that the jobs held by immigrants are more physically laborious than jobs held by U.S. natives. In other research, Peri and Sparber [5] emphasize that "*... foreign-born workers specialize in occupations intensive in manual-physical labour skills while natives pursue jobs more intensive in communication-language tasks*". Other authors [2] suggest that native workers respond to immigration by specialising in occupations requiring skills in which they have a comparative advantage. Few studies however have examined the differences between the foreign and native components in terms of requirements and characteristics of workers and work in Italy.

In this paper, the Italian and foreign employed populations in 2019 in Italy were analysed by detecting both specific information regarding the characteristics and requirements of the worker, their profession, and some socio-demographic data. The analyses refer to a period before the traumatic effect of the pandemic due to SARS-COV-2; for this reason, it does not reflect the impact of Covid-19 related changes on employment flows and outcomes.

This paper is an extended version of the paper presented at the 51st Scientific Meeting of the Italian Statistical Society on June, 2022 [4].

2 Data and Methods

The data used for the analysis were those of the Survey on Italian Occupations (ISTAT and the National Institute for the Analysis of Public Policies that is denominated in Italian "Istituto Nazionale per l'Analisi delle Politiche Pubbliche", INAPP hereafter), provided by INAPP at a level of detail equal to 3 digits and the micro-data for public use of the Labour Force Survey or "*Rilevazione Continua delle Forze di Lavoro–RCFL*"[1] (ISTAT). Reference is made to data from the year 2013 for the first survey (the year 2013 is the second edition of the ICP survey, there is no more up-to-date data available at the moment) and the year 2019 for the second survey. The Survey on Italian Occupations (*Indagine Campionaria sulle Professioni*, ICP hereafter) describes, with high analytical detail, all the professions existing in the Italian labour market, with particular reference to the content of the work performed and the organizational context in which the work takes place [3]. The ICP, borrowed from the Occupational Information Network (O*NET) survey, adopts the US conceptual

[1] The RCFL involves a total of 600,000 individuals in the survey, which corresponds to more than 1% of the resident population in Italy in 2019.

model, the *content model*, which provided the basis for the creation of the contents of the questionnaire. The questionnaire is formulated by translating the one from the O*NET survey and, where necessary, adapting it to the Italian situation.

We have built an integrated database to simultaneously detect both specific information regarding the characteristics and requirements of the worker and the profession, and socio-demographic data of the workers, through statistical matching between the Sample Survey on Professions [3] and the Labour Force Survey (ISTAT, 2019).

We have also developed two methods for the construction of composite indicators, which allow us to analyze the world of professions in Italy in terms of workers' skills and work activities. The first methodology referred to the study by Peri and Sparber [5] and the eight indicators represent a weighted average value for each profession. The eight composite indicators constructed through the methodology devised by Peri and Sparber [5] are presented and described, they are:

- an indicator called *"Manual"* because it is constructed using different abilities that contain information on dexterity of limbs, hands and fingers, body coordination and flexibility, speed and strength[2];
- two different types of indicators called *"Communicative"* which were constructed using data on cognitive skills such as understanding or expressing oneself in oral and written form;
- three different types of indicators called *"Quantitative"* which summarise information concerning, for example, the ability to perform mathematical calculations or apply the correct mathematical methods to solve a problem;
- an indicator called *"Interactive"* as it summarises data such as the ability to resolve disputes and conflicts with other people;
- an indicator called *"Analytical"* constructed to capture information such as the ability to use inductive and deductive reasoning or to analyse data and information.

The second method is the one proposed by Autor et al. [1], and, in this case, the five indicators were obtained using principal component analysis. Only two of the five composite indicators are Routine indicators, which means that they summarise activities in which computer equipment such as computers are used. The five composite indicators are named as follows:

- *"Non-Routine Analytical Tasks"* is produced by examining information that captures the originality and critical thinking of the employee but also other aspects such as decision-making and problem-solving;
- *"Non-Routine Interactive Tasks"* is produced through data expressing the concept of leadership, of monitoring and controlling the work of others;

[2] For example, 19 abilities belonging to section D of the ICP questionnaire, from question number 22 to 40, were selected for the construction of this indicator. So the abilities used are: Arm-Hand Steadiness, Manual Dexterity, Finger Dexterity, Control Precision, Multilimb Coordination, Response Orientation, Rate Control, Reaction Time, Wrist-Finger Speed, Speed of Limb Movement, Static Strength, Explosive Strength, Dynamic Strength, Trunk Strength, Stamina, Extent Flexibility, Dynamic Flexibility, Gross Body Coordination and Gross Body Equilibrium.

- *"Routine Cognitive Tasks"* considers the ability to order and store information and competence in using mathematics;
- *"Routine Manual Tasks"* is built with skills concerning dexterity of the limbs, hands and fingers;
- *"Non-Routine Manual Tasks"* summarises information related to limb and movement coordination skills and activities such as operating machines and controlling machines.

The methodology developed by Autor et al. [1] used the information present in the Dictionary of Occupational Titles (DOT), subsequently, Warman and Worswick [7] incorporated the method proposed by the three authors to the data of the O*NET database as the latter has over time replaced the DOT; for this reason, in this work, the composite indicators were obtained with the variables suggested in Warman and Worswick's [7] study.

Furthermore, in order to make the indicators constructed using the two methodologies described above comparable, composite indicators designed by Autor et al. [1] were constructed, adopting the mathematical method used by Peri and Sparber [5].

3 Results

The world of professions has been analyzed in terms of both the attitudes and skills of the worker and the activities and styles of work, according to sex, level of education, and citizenship.

This paragraph presents some of the main results relating to the indicators obtained by applying the methodology of Peri and Sparber [5]. In this case, the "Manual" and "Communicative" indicators are considered because they represent the most interesting results and for which the greatest differences have emerged between Italian and foreign workers. The comparison is between the Italian and foreign employed populations from developing countries; the employed of developed countries were excluded from the study as the results obtained are similar to those of Italian employed people. The analysis was carried out separately by sex and level of education: the workers without a school certificate or in possession of an elementary or middle school certificate are included in the low qualification group, while the workers with a university degree were in the high qualification group. Graduated workers have been excluded in this part of the analysis to highlight the differences between less-educated and highly-educated workers.

As shown in Table 1 for employed men (7,117,637 individuals), the Italians with low educational qualifications engage in activities with a moderate manual content (the score is 27.7), which is higher than that observed for employed people with citizenship in a Latin American and Oceanian country (27.5), but lower than all the others.

Table 1 Employed men: manual and communicative indicators by educational level and citizenship. Years 2019 (RCFL) and 2013 (ICP)

Citizenship groups	Size of groups in weighted absolute value	Manual Indicator (M. I.)		Communicative Indicator (C. I.)		Relative difference between low and high levels of education[a]	
		Low level of educ.	High level of educ.	Low level of educ.	High level of educ.	M. I. (%)	C. I. (%)
Italy	61,96,924	27.7	10.5	57.6	68.8	90.30	17.70
Developing countries:							
EU	1,61,564	33.5	23.1	45	53.6	36.60	17.30
Non-EU Eastern Europe	1,54,857	34.3	23.7	43.8	51.7	36.80	16.60
Asia (excluding China)	2,12,087	27.8	16.7	43	58.1	50.00	29.70
China	65,000	22.8*	16.9*	50.1	61.3	29.60	20.10
North Africa	1,41,023	31.7	21.5	44.6	51.2	38.20	13.60
Sub-Saharan Africa	1,05,424	29	17.6	44.7	56.3	49.00	23.10
Latin America and Oceania[b]	59,146	27.5	20.8	43.9	54.5	27.50	21.50
Total	71,17,637						

Note Unweighted data produce estimates that are all significant at the 5% level of significance when applying an independent samples t-test (mean comparison test) with the exception of the Chinese group (*) in the manual indicator estimates for high and low degrees

Source Own elaboration on ISTAT-INAPP data

[a] Relative difference between low and high levels of education relative is calculated considering the average between the two values as denominator

[b] This group includes the employed who hold citizenship in an Oceania country (with the exception of the Australian and New Zealand employed who have been included in the foreign employed who hold citizenship in an advanced-development country) and, as this is a small group, they have been merged with the Latin American employed. In addition, these two groups present some similar characteristics both demographically (level of education) and socio-economically (economic sectors of employment, main economic activity and type of profession carried out)

Italian workers with high educational qualifications carry out activities—as expected—with a minimum manual content (10.5). Furthermore, it is observed that Italians with both low (57.6) and high (68.8) educational qualifications carry out more communicative occupations than all other groups of foreigners. The relative differences by educational qualification in exercising activities characterized by high manual skills are high in the case of Italians (90.3%), much more moderate for the other groups. Conversely, the differences in the degree of education when it comes to activities characterized by high levels of communicativeness are more limited for Italians (17.7%), and the same is true for the other groups of foreign workers, except Asians (almost 30%).

On the employed side, Italians with low qualifications, as well as foreigners, work in professions with a not particularly high manual labour content, which is, however, similar among the various groups considered (values oscillate between 21 and 24: see Table 2). For employed women (5,272,312 individuals), similar results to those of men are observed, although the differences by nationality in the case of women appear smaller. Italian employed women with high qualifications also carry out activities with a low manual labour content (11.6), showing lower values than all the other groups of foreign female workers (again, the result is very similar to that seen for men). Italians and foreigners with low educational qualifications hold professions with a much higher communicativeness content than manual work; it can be seen that Chinese women with low educational qualifications hold professions in which the communicativeness content is on average slightly higher (49.8) than those held by Italians with the same type of qualification (47.7). Looking at the level of communicativeness, highly educated Italians carry out activities with a high level of communicativeness content (67.7), the highest compared to all the other groups.

Finally, the differences by level of education in exercising professions characterised by high manual skills are high in the case of Italians (percentage difference of 62.3%) and Chinese women (51.9%), while they are much more moderate in the case of the other groups, in particular in Latin American-Oceanian (6.9%) and Asian (14.4%) countries. The differences by qualification in the use of highly communicative activities are significant, especially in the case of Italian women (34.7%), who present the highest value compared to all the other groups of foreign female employees.

The indicators were then constructed following the Autor, Levy and Murnane (henceforth: ALM) methodology described above. To make them comparable with the first type of indicators (Peri and Sparber), the indicators devised by ALM were constructed with the same variables as those proposed by Warman and Worswick [7], this time, however, calculated using the mathematical method proposed by Peri and Sparber, who suggest the construction of a sort of weighted average and not the consideration of the factorial weights obtained by a principal component analysis.

Below are the tables with the indicator values broken down by gender, educational qualification and geographical macro-area.

Table 3 on employed men with low educational qualifications shows few differences between the indicators for all eight groups. Specifically, the range of variation of the first four indicators (Non-Routine Analytical Tasks, Non-Routine Interactive Tasks, Routine Cognitive Tasks and Routine Manual Tasks) is about 5 to 7 percentage points, while for the last indicator it is wider (about 12.5 points). Italians, only for the Non-Routine Analytical Tasks, take the highest value (36.1) of the index compared to the other groups of foreigners, the Chinese excel instead in the case of Non-Routine Interactive Tasks (33.9) and Routine Cognitive Tasks (41.2). Among the foreign employed, those from Eastern European countries mainly perform manual tasks, both routine (38.1) and non-routine (33.8). In general, the profile of Italian

Table 2 Employed women: manual and communicative indicators by educational level and citizenship. Years 2019 (RCFL) and 2013 (ICP)

Citizenship groups	Size of groups in weighted absolute value	Manual Indicator (M. I.)		Communicative Indicator (C. I.)		Relative difference between low and high levels of education	
		Low level of educ	High level of educ	Low level of educ	High level of educ	M. I. (%)	C. I. (%)
Italy	4,614,536	22.0	11.6	47.7	67.7	62.3	34.7
Developing countries:							
EU	190,746	22.9	18.6	40.4	51.8	20.7	24.8
Non-EU Eastern Europe	158,276	23.5	18.7	39.8	43.8	22.7	9.7
Asia (excluding China)	83,547	21.9	18.9	35.5	44.8	14.4	23.2
China	54,842	20.6	12.1	49.8	59.8	51.9	18.1
North Africa	32,170	24.1	18.3	40.1	51.2	27.3	24.3
Sub-Saharan Africa	33,743	24.0	19.0	38.6	52.0	23.3	29.6
Latin America and Oceania	69,978	21.5	20.1	38.5	50.2	6.9	26.3
Total	5,272,312						

Note Unweighted data produce estimates that are all significant at the 5% level of significance when applying an independent samples t-test (mean comparison test)
Source Own elaboration on ISTAT-INAPP data

workers with low educational qualifications in the various activities is quite similar to that of the Chinese. With the exception of the Chinese, foreigners who perform more routine than non-routine activities.

Looking at the population of employed men with high educational qualifications (Table 4), it is quite evident that the values of the manual indicators (Routine Manual Tasks and Non-Routine Manual Tasks), regardless of the nationality taken into consideration, are much lower than those of the other three indicators. Italian employed men with university degrees perform more non-routine analytical (55.7), non-routine interactive (49.5) and routine cognitive (52.7) tasks and fewer manual tasks (16.0 and 10.6) than foreigners.

Taking into account employed women with low educational qualifications (Table 5), as for men (Table 3), it can be seen that in general Chinese women record the highest values for each indicator, Asian and South American women the lowest.

Table 3 Employed men with low educational qualifications: ALM indicators by country group. Year 2019 (RCFL) and 2013 (ICP)

	Size of groups in weighted absolute value	ALM indicators				
		Non-routine analytical tasks	Non-routine interactive tasks	Routine cognitive tasks	Routine manual tasks	Non-routine manual tasks
Italy	3,917,143	36.1	32.1	40.5	33.5	28.5
Developing countries:						
EU	146,988	33.4	29.1	39.0	37.3	33.5
Non-EU Eastern Europe	140,516	32.0	28.7	39.0	38.1	33.8
Asia (excluding China)	181,572	30.6	27.8	35.8	34.4	26.0
China	62,494	35.7	33.9	41.2	32.1	21.3
North Africa	130,166	31.6	28.1	37.8	38.0	31.4
Sub-Saharan Africa	97,273	32.5	28.7	37.4	36.6	29.8
Latin America and Oceania	46,847	29.2	27.1	35.7	33.7	26.0
Total	4,722,999					

Note Unweighted data produce estimates that are all significant at the 5% level of significance when applying an independent samples t-test (mean comparison test)
Source Own elaboration on ISTAT-INAPP data

It can be observed that all foreign women workers perform more routine than non-routine tasks, unlike Italian women. In addition, women perform few non-routine manual tasks than the other four activities.

Table 6 refers to employed women with high educational qualifications and, as in the case of men with university degrees, it can be seen that all employed women perform fewer occupations involving manual labour and devote themselves mainly to non-routine analytical and interactive activities and routine cognitive tasks. Italians, compared to foreigners, record the highest values for the first three indicators and the lowest for the last two.

Also from the analysis of the ALM indicators, it can be concluded that the most pronounced differences are to be found mainly in the distinction of tasks into manual (Routine Manual Tasks and Non-Routine Manual Tasks) and non-manual (Non-Routine Analytical Tasks, Non-Routine Interactive Tasks and Routine Cognitive Tasks) rather than into routine (Routine Cognitive Tasks, Routine Manual Tasks) and non-routine tasks (Non-Routine Analytical Tasks, Non-Routine Interactive Tasks and Non-Routine Manual Tasks); this emerges above all when considering employed men and women with high educational qualifications.

Table 4 Employed men with high educational qualifications: ALM indicators by country groups. Year 2019 (RCFL) and 2013 (ICP)

	Size of groups in weighted absolute value	ALM indicators				
		Non-routine analytical tasks	Non-routine interactive tasks	Routine cognitive tasks	Routine manual tasks	Non-routine manual tasks
Italy	2,279,781	55.8	49.5	52.7	16.0	10.6
Developing countries:						
EU	14,576	39.3	34.8	42.0	29.5	24.0
Non-EU Eastern Europe	14,341	37.8	33.7	42.2	29.2	23.6
Asia (excluding China)	30,515	44.8	40.0	45.4	22.2	15.8
China	2,506	50.0	45.3	45.0	19.6	13.8
North Africa	10,857	36.3	33.9	38.7	27.5	20.4
Sub-Saharan Africa	8,151	41.1	37.2	44.9	23.5	18.6
Latin America and Oceania	12,299	38.6	34.7	39.4	27.6	18.8
Total	2,373,026					

Note Unweighted data produce estimates that are all significant at the 5% level of significance when applying an independent samples t-test (mean comparison test)
Source Own elaboration on ISTAT-INAPP data

Italian employees with a university degree mainly perform non-manual tasks, i.e. non-routine analytical tasks, non-routine interactive tasks and cognitive tasks, compared to foreign employees for whom the latter three indicators are lower. In general, Italians and Chinese with university degrees, regardless of gender, perform occupations with a higher analytical, interactive and cognitive content and at the same time lower manual tasks than all other groups of foreign workers. Furthermore, among foreign men with high educational qualifications, those with citizenship in non-developed EU countries and Eastern European countries are employed in occupations with a slightly higher level of manual skills than the other groups. The few foreign women with high educational qualifications who are employed in manual occupations prefer routine to non-routine occupations.

The second type of indicators (Autor, Levy and Murnane) also show greater differences between Italian and foreign workers in terms of content of manual and non-manual occupations, and the study was analysed further by adopting Peri and Sparber's Manual, Communicative and Quantitative indicators.

Data from the Survey on Italian Occupations (INAPP) confirms how the use of the attitudes and skills contained in the various occupations varies according to the

Table 5 Employed women with low educational qualifications: ALM indicators by country groups. Year 2019 (RCFL) and 2013 (ICP)

	Size of groups in weighted absolute value	ALM indicators				
		Non-routine analytical tasks	Non-routine interactive tasks	Routine cognitive tasks	Routine manual tasks	Non-routine manual tasks
Italy	1,855,147	33.0	30.3	36.4	30.3	19.2
Developing countries:						
EU	134,814	27.8	24.8	30.1	27.5	17.4
Non-EU Eastern Europe	112,888	27.1	24.4	29.7	28.7	17.7
Asia (excluding China)	62,748	23.0	20.5	26.7	26.1	15.2
China	52,632	36.5	33.9	41.2	31.4	19.2
North Africa	27,843	27.0	25.2	30.3	31.0	18.8
Sub-Saharan Africa	30,251	25.2	23.3	28.9	29.5	18.7
Latin America and Oceania	55,041	25.0	22.5	27.8	25.6	14.9
Total	2,331,364					

Note Unweighted data produce estimates that are all significant at the 5% level of significance when applying an independent samples t-test (mean comparison test)
Source Own elaboration on ISTAT-INAPP data

level of education of employees; it is possible, using this database linked to the Labour Force Survey, to also test the hypothesis of a variation in this use according to citizenship.

Figure 1 shows the value of manual, communication, and quantitative skills by educational qualification for Italian and foreign workers. Communication and quantitative skills tend to increase with higher educational qualifications, while the use of manual skills decreases.

Employed people with high educational qualifications are unsurprisingly employed in jobs with high communication content. In some cases, however, there is also a high level of manual work: for example, workers in professions such as health technicians have a manual indicator value of 21.1 (6th decile) and a communication indicator value of 63.6 (7th decile) and doctors have a manual indicator value of 18.7 (5th decile) and a communication indicator value of 73.4 (10th decile). The professions with greater communication skills held by workers with university degrees are, for example, university lecturers (86.9) and researchers and technicians with university degrees (84.5). Educated workers in highly quantitative professions are Engineers and related professions (62.7) and quantitative, physical, and chemical

Table 6 Employed women with high educational qualifications: ALM indicators by country groups. Year 2019 (RCFL) and 2013 (ICP)

	Size of groups in weighted absolute value	ALM indicators				
		Non-routine analytical tasks	Non-routine interactive tasks	Routine cognitive tasks	Routine manual tasks	Non-routine manual tasks
Italy	2,759,389	53.2	46.7	49.7	17.6	10.1
Developing countries:						
EU	55,932	38.1	34.1	37.6	25.0	15.0
Non-EU Eastern Europe	45,340	30.1	26.9	32.0	23.0	13.0
Asia (excluding China)	20,799	31.9	28.6	33.2	24.0	14.0
China	2,210	41.2	36.9	44.7	19.7	10.4
North Africa	4,327	36.4	34.0	38.6	23.8	15.8
Sub-Saharan Africa	3,492	39.9	35.1	39.5	24.4	16.0
Latin America and Oceania	14,937	36.8	32.0	35.1	25.7	14.4
Total	2,906,426					

Note Unweighted data produce estimates that are all significant at the 5% level of significance when applying an independent samples t-test (mean comparison test)
Source Own elaboration on ISTAT-INAPP data

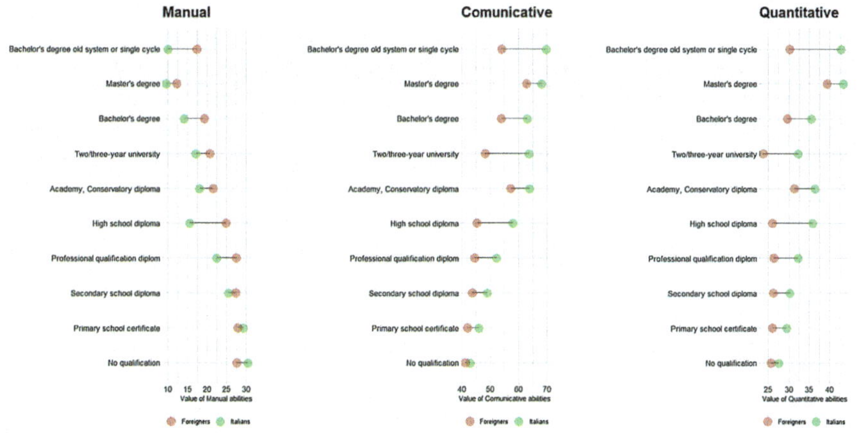

Fig. 1 Average abilities of Italian and foreign employed people in Italy by the level of education and type of ability. Years 2019 (RCFL) and 2013 (ICP). *Source* Own elaboration on ISTAT-INAPP data

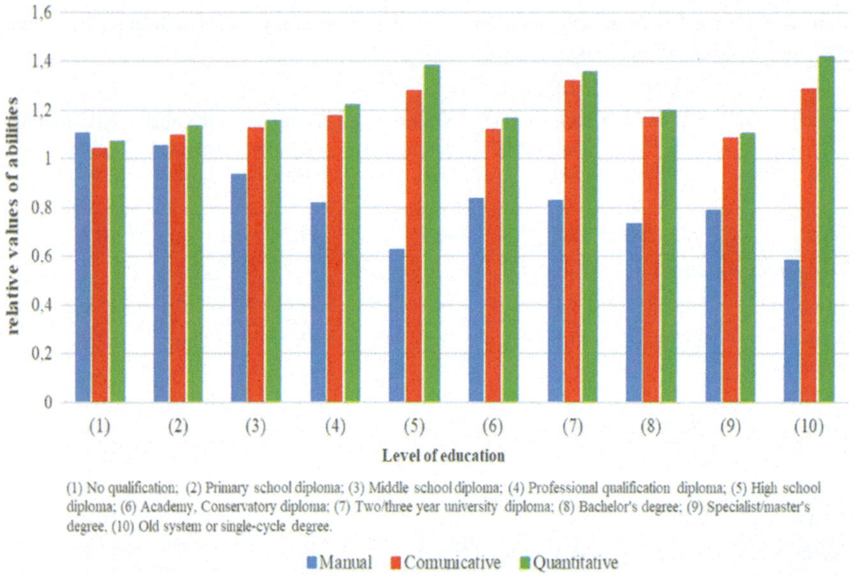

Fig. 2 Ratio between the average abilities of Italian and foreign workers by the level of education and type of ability. Years 2019 (RCFL) and 2013 (ICP). *Source* Own elaboration on ISTAT-INAPP data

science technicians (63.1). Native workers show an advantage in communication and quantitative skills, foreign workers an advantage in manual work.

Absolute skill levels have to be converted into relative values in order to obtain the comparative advantage in terms of skills of native workers over foreigners (Fig. 2). The intensity of communicative and quantitative skills is higher for Italians at practically all levels of education: Italians show a systematic comparative advantage in communicative and quantitative skills compared to manual skills. Italians also show even higher levels of quantitative skills than communicative skills, which is an unexpected result. Immigrants present a comparative advantage in manual work compared to other skills at all levels of education with the exception of the 'no qualification' category.

4 Conclusion

Observations of some significance can be made by analysing the world of professions examined in terms of both worker attitudes and skills as well as activities and work styles.

In this case, the study was conducted using alternative indicator construction methods as suggested in the literature. From our point of view, this is the first time such an analysis has been proposed. Taking the methodology suggested by Peri and Sparber as a starting point, it was observed how educational qualification and citizenship discriminate the exercise of a manual or communicative activity.

For example, considering low educational qualifications, it was noted that Italians carry out a profession with a moderate manual labour content, higher than that observed for Chinese and those employed in Latin America and Oceania, but lower than all the others; a similar situation occurs for Italians but with smaller differences by citizenship. Regardless of gender, Italian employees with low educational qualifications engage in activities with a much higher communicative content than manual labour; this is also the case for foreigners, but to a lesser extent. Furthermore, it is observed that Italians with both low and high educational qualifications perform more communicative occupations than all other groups of foreigners. Thus the differences by level of education in carrying out professions characterised by high manual skills, for both men and women, are high in the case of Italians, more moderate in the case of the other groups of foreigners. Conversely, the differences by level of education in performing activities characterised by high communicativeness are more moderate for each citizenship group. In the case of women, the value is higher for Italians than for the other groups of foreigners.

The results obtained by analysing the indicators constructed in an alternative way also confirm that the most pronounced differences by citizenship emerge mainly in the distinction of tasks into manual and non-manual, restricting the analysis to those employed with a high qualification. For this reason, in this study as well, it was appropriate to deepen the study by considering the indicators constructed using the method proposed by Peri and Sparber: the most interesting indicators were "*Manual*", "*Communicative*" and "*Quantitative*", the three sectors for which the most relevant differences emerged in the comparison between Italian and foreign employees.

Data from the INAPP Occupational Survey confirmed how the use of manual, communication and quantitative skills contained in the various occupations carried out by the respondents varies according to their level of education; through the integrated database it was possible to reveal how these aptitudes change with changes in citizenship. As people attain higher and higher educational qualifications, they tend to work in professions with more communication and quantitative content and less manual work. However, the presence of professions characterised by both a high manual and communicative content has been noted, as for example in the case of health technicians, doctors, air, sea and rail transport technicians. The professions with higher communication skills performed by workers with university degrees are, for example, university lecturers and university-educated researchers and technicians; educated workers in highly 'quantitative' professions are engineers and related professions and technicians in the quantitative, physical and chemical sciences.

Converting absolute skill levels into relative values, the intensity of communicative and quantitative skills in the professions is greater for Italians regardless of their level of education: natives have a systematic comparative advantage in communicative and quantitative skills over manual skills. Italians present even higher levels of quantitative skills than communicative skills, which appears to be an unexpected result. Foreigners, on the other hand, show a comparative advantage in manual work with respect to the other skills at all levels of education with the exception of the 'no qualification' category.

Thus, the results obtained in this work would seem to confirm part of the conclusions of Peri and Sparber's studies: in Italy, foreign workers would be complementary to native workers because they specialize in different tasks having different skills.

References

1. Autor, D.H., Levy, F., Murnane, R.J.: The skill content of recent technological change: an empirical exploration. Q. J. Econ. **118**(4), 1279–1333 (2003)
2. Gu, E., Sparber, C.: The native-born occupational skill response to immigration within education and experience cells. East. Econ. J. **43**(3), 426–450 (2017)
3. INAPP.: Nota metodologica Indagine campionaria sulle professioni (2013)
4. Pangallo, C., Casacchia, O., Polli, C.: The manual, communicative and quantitative abilities of native and foreign workers according to their level of education in Italy, In: Balzanella, A., Bini, M., Cavicchia, C., Verde, R. (eds.) SIS 2022. 51st Scientific Meeting of the Italian Statistical Society. Book of the Short Papers, pp. 1120–1126. Pearson (2022), ISBN: 9788891932310
5. Peri, G., Sparber, C.: Task specialization, immigration, and wages. Am. Econ. J. Appl. Econ. **1**(3), 135–169 (2009)
6. Peri, G., Sparber, C.: Highly educated immigrants and native occupational choice. Ind. Relat.: J. Econ. Soc. **50**(3), 385–411 (2011)
7. Warman, C., Worswick, C.: Technological change, occupational tasks and declining immigrant outcomes: implications for earnings and income inequality in Canada. Can. J. Econ./ Rev. canadienne d'économique **48**(2), 736–772 (2015)
8. Zavodny, M.: Do immigrants work in worse jobs than U.S. natives? evidence from California, IZA Discussion Papers, No. 8327, Institute for the Study of Labour (IZA), Bonn (2014)

Paid and Unpaid Work in Pandemic Times. A Study on the Division of Household Labour and the Subjective Well-Being of Working Mothers in Italy

Marina Zannella, Alessandra De Rose, Erica Aloè, and Marcella Corsi

Abstract This article draws on data from a web-based survey carried out in Italy between May and June 2020 to explore how lockdown measures affected individual well-being, family life and the use of time for paid/unpaid work. Our analysis focuses on working mothers with minor children living in couples. Descriptive results show that lockdown measures had a considerable impact on the total workload borne by women, while their male partners only slightly increased their contribution to childcare and domestic chores. The regression analysis on the relationship between mothers' satisfaction with the division of unpaid family work and their subjective well-being show that mothers who reported lower satisfaction felt more fatigued and stressed in relation to housework, childcare, and paid work during the lockdown.

Keywords Unpaid care and domestic work · COVID-19 · Time use · Subjective well-being · Gender division of work

1 Introduction

Unpaid and informal care work represented key elements in the evaluation of the social costs generated by the pandemic [9, on Italy]. Under stay-at-home and lockdown measures there has been an unprecedented increase in the demand for household production and the associated input of unpaid work, largely motivated by the impossibility of accessing paid services (e.g., laundries, restaurants, babysitters, etc.)

M. Zannella (✉)
Italian National Institute of Statistics, Rome, Italy
e-mail: marina.zannella@istat.it

A. D. Rose
Department of Methods and Models for Economics, Finance and the Territory, Sapienza University of Rome, Rome, Italy

E. Aloè · M. Corsi
MinervaLab, Sapienza University of Rome, Rome, Italy

M. Bini et al. (eds.), *Advanced Methods in Statistics, Data Science and Related Applications*, Springer Proceedings in Mathematics & Statistics 467,
https://doi.org/10.1007/978-3-031-65699-6_21

as well as benefiting from formal (e.g., school) and informal care from grandparents and other non-cohabiting relatives or friends. This additional work fell on women, adding to the already existing inequalities in the gender division of unpaid work [34]. Andrew and colleagues [2] show that in the UK, mothers who stopped working in the labour market performed significantly more domestic work than fathers in a similar situation. These findings suggest that imbalances in the gender distribution of the "extra" amount of unpaid family work caused by the pandemic cannot be explained as an exclusive effect of gender differences in employment and earnings but depend primarily on social norms governing gender roles and expectations on motherhood. Similar surveys conducted during the pandemic provide supporting results (see for example [19] for Spain; [23] for Turkey). However, studies on time use and household arrangements during the lockdowns also highlight that, even if women shouldered most of the extra unpaid workload, men's childcare time increased (e.g.: [12], on Australia; [13, 25] for Italy; [36] for Spain). These studies demonstrate that, besides the resilience of gender norms, non-normative arrangements for the distribution of unpaid care and domestic work were triggered by the lockdowns. Another aspect that has contributed to the changes in the organization of individual and family life has been the dramatic recourse to remote work. Overall, the confluence of paid and unpaid work time has produced a further intensification of work across the work/life continuum [28]. As a result, dissatisfaction with the balance between paid and unpaid work has increased significantly among women [12, on Australia]. International studies, reviewed by Seedat and Rondon [35], have documented a greater rise in psychological distress in women than in men during the lockdown, partially explained by the disproportionate burden of work that fell onto them. For many women, pre-existing work–family arrangements were already stressful, and the pandemic made the situation worse [11]. However, lockdown measures and the rise of remote work had the potential to challenge the long-standing gender differences in time use patterns and, in household where men increased their involvement in unpaid family work, women are likely to have suffered less negative consequences on their subjective well-being.

This article focuses on Italy to explore how the lockdown measures affected women's time-use, the division of unpaid work within the household, and women's degree of satisfaction toward the household division of unpaid family work. The assessment is based on an on-line questionnaire filled in by more than 1,000 individuals aged 18 or older living in Italy. Observations were collected immediately after the first lockdown and report information about paid and unpaid work before, during and after the pandemic. More specifically, the study focuses on a subsample of working mothers of minor children living in couples to test the relationship between satisfaction for the intrahousehold division of unpaid care and domestic work and levels of stress and fatigue experienced by women in association to paid and unpaid work time during the pandemic.

The paper is structured as follows. Section 2 provides the background of the article and presents the specific research questions. The data and methods are explained in Sect. 3. The main results of our analysis are shown in Sect. 4 and are discussed in Sect. 6 which concludes.

2 Background

2.1 The Impact of the Pandemic on Paid and Unpaid Work

Several studies have analysed the impact of the pandemic on the burden of work around the world. The Eurofound's *Living, working and COVID-19* survey showed that, on average in the EU-27, the pandemic has increased time devoted to unpaid care activities for both women and men, although the former continued to shoulder the greatest unpaid care burden [18]. National research leads to similar results: studies for different countries show that, during confinement periods, fathers working at home generally tended to share caregiving responsibilities more than before, especially in households where only the father stopped working while the mother performed essential work [2, 23, 37]. Similar results were also found for Italy, where men whose partners remained in their usual jobs spent more time on household tasks than before the pandemic [13]. It remains unclear whether men's increased participation in childcare and domestic work during the pandemic can lead to stable changes in the gender division of unpaid work (e.g., [4]). Andrew et al. [2] suggest that the shock of the freezing of family dynamics may have triggered new processes, generating new family arrangements, gender roles and attitudes. However, the increased involvement of fathers in childcare may reflect the pre-existing gender specialization in unpaid work, with women taking on the largest share of the routine work burden (see, among others, [10]). According to Farré et al. [19], during the first lockdown, the increase in fathers' share of care work was largely attributed to increased time spent running errands, which was the only permitted reason for leaving. An international study [20] found that during the pandemic the time spent completing household chores was linked to lower well-being especially for mothers who spent more time on necessities such as childcare and chores. In the US, a study [6] found that mothers who have greatly increased the time they spend caring for their children also disproportionately report increased stress, anxiety, and frustrations with their children. Similar findings were reported for Italy [26]. Prior time use research has shown that the gendered nature of childcare is reflected in the subjective experience of parenting time: while childcare is usually reported as a positive experience by both parents, fathers enjoy childcare time more than mothers who are more likely to experience it as less pleasant (e.g., [30] on US; [27] on UK; [40] on Itay).

As far as paid work is concerned, it was observed that during the 2020 lockdowns women were more likely to engage in paid work and caregiving simultaneously and to report more fragmented working time and reduced ability to complete a task [1]. Zoch and colleagues [41] analysed changes in care-arrangements and working conditions of employed parents in Germany during the pandemic school closures. Results from the study highlight that for mothers remote work was not associated with a higher probability of care sharing by fathers. Difficulties in combing work and family life have been particularly evident for women working in professions requiring a high degree of concentration as shown by Minello and colleagues [29] who, analysing data from in-depth interviews with academic mothers in Italy and the US, show the detrimental impacts of the pandemic on their careers. Durante et al. [17] examine self- and partner-perceived changes in contributions to household labour during the COVID-19 pandemic and how these changes positively influence women's job productivity and satisfaction.

To sum up, containment measures adopted to contrast the spread of Covid-19 entailed great changes in the organization individual and family lives. A growing body of evidence globally indicates that although men have increased their contribution to unpaid work, the additional family work generated by the pandemic has fallen disproportionately on the shoulders of women, impacting their time available for paid work, and causing psychological distress. Working mothers living in couples in which men increased their contribution to unpaid work are likely to have experienced relatively less difficulty in combining paid and unpaid work than those in which unpaid work was less equally shared. However, to our knowledge, so far, no study has analysed how changes in the amount and in the household division of unpaid family labour during the pandemic have affected women's subjective experience of time use for paid and unpaid work. Broadening the understanding of women's experience of paid and unpaid work during the pandemic is relevant for Italy since, as it will be presented in the next section, workload inequality between partners is considerably high [38, 39].

2.2 *The Italian Context*

Italy was the first European country to impose a nationwide lockdown at the beginning of March 2020. Schools have been closed nationwide for in-person activities until June, a relatively longer period compared to most OECD countries where schools began to re-open in April and May [33]. The prolonged closure of schools exposed Italian families to unprecedented pressure. Working parents, especially mothers, faced great difficulties as they were asked to take care of their children and help them study while working at home. Even before the pandemic, the gender dimension of time-use patterns was particularly evident in Italy, which presents one of the largest gender gaps in both domestic work and childcare [32]. According to the time use survey performed by the Italian National Institute of Statistics (Istat) that took place during the lockdown[1] the amount of time devoted to childcare increased for 67.2% of the people that performed it. For what concerns unpaid domestic work, the lockdown reduced the gender gap in the time devoted to this kind of activity. One third of the adult respondents, with no difference between the genders, devoted more time to cooking. Still, 82.9% of women and 42.9% of men devoted some time to meal preparation. House cleaning activities were carried out by more than one in two adult persons (54.4%). Even in this case the gender differences are very high (about 27% points), but smaller than those found in budget time surveys (46 points of difference). Unfortunately, no information on satisfaction about gender division

[1] https://www.istat.it/it/archivio/243829.

of time during lockdown or well-being indicators associated to changes in the use of time have been collected by the Istat survey.

In addition to the official data collected by Istat, a number of individual scholars made important efforts to gather timely information on family's use of time in Italy during the early stages of the pandemic, finding that the burden of unpaid work due to the lockdown overwhelmingly rested on women [3, 22]. Del Boca and colleagues [13], who used survey data collected in April 2020 on women living in dual-earner heterosexual couples, show that childcare activities were more equally shared within the couple than housework activities. Mangiavacchi et al. [25] confirm that during the lockdown Italian households experienced a greater involvement of fathers in childcare and that men whose partners continued to work at their usual workplace spent more time on housework than before. Additionally, analysis of work–life balance satisfaction shows that working women with children aged 0–5 are those who found balancing work and family more difficult during COVID-19.

Brini et al. [5] found that in Italy the pandemic has reduced the time spent in paid work (and earnings) more for fathers than for mothers, and that the number of female bread-winner households increased. De Vita and colleagues [15] analysed data collected via a web survey launched in April 2020 during the first lockdown aimed only at female workers and found high levels of tiredness and overload, greatly aggravated by the need to deal with distance learning and organizing schooling-related activities. With respect to mothers' well-being, Manzo et al. [26] found that during the lockdown mothers suffered discouragement, frustration, and exhaustion.

2.3 *Research Questions and Hypotheses*

Existing literature has demonstrated that work-life balance is more difficult for employed mothers, both before and during the pandemic, and that the considerable increases in women's work during confinement periods may have caused great psychological distress for women. Furthermore, the research is fairly consistent in demonstrating that women are happier in relationships where routine housework responsibilities are shared equally [7, 14, 21, 24]. Based on this background, this article aims at answering to the following research questions:

- How did the lockdown measures affect the gender division of family labour and women's use of time for paid and unpaid work?
- Have these changes affected the subjective experience of paid and unpaid work for mothers?

To answer these research questions, we hypothesize that men's contribution to family work during the pandemic had an impact on their female partners' subjective well-being. Our hypothesis is that women who were less satisfied with the intra-family division of unpaid work during the lockdown experienced greater levels of stress and fatigue in association to childcare, housework, and paid work.

3 Data and Methods

3.1 Survey

To carry out our survey, we used the questionnaire developed by Donehower [16] within the *Counting Women's Work* (CWW) Project as a base.[2] We translated the original questionnaire and adapted it to the Italian context and the purposes of our study by modifying some questions and asking some additional information. The final survey—structured in multiple choice questions—consisted of nine sections: household composition, health status, paid work (own and partner's), unpaid care work, unpaid domestic work, informal help to/from other households, division of unpaid care work and unpaid domestic work within the household, subjective well-being in paid and unpaid work activities, socio-demographic information. Questions were asked to the respondent in relation to three different moments: before the pandemic, during the first lockdown and in the moment when they responded to the survey (that is immediately after the end of the lockdown).[3]Responses were collected from May 22 and June 12, 2020. Data collection was conducted anonymously and participation in the survey was voluntary. The survey was promoted through the institutional website of Sapienza University of Rome and the main social networks, as well as through the mailing lists of scientific associations and professional contacts. The survey collected 1008 observations, reduced to 979 when the dataset was cleaned from missing and invalid responses.

Focusing on women, who represented the great majority of our sample, 74.6% lived with their partner (married or unmarried), and 37.9% with at least one child below 18 years old. Concerning the level of education, 24.7% of them had a college degree, 38.4% had a bachelor's degree, and 36.9% had a level of education higher than a bachelor's degree. The survey gathered detailed information about the time devoted by respondents to paid and unpaid work. To obtain the descriptive statistics, data were post-stratified to ensure consistency with the main socio-demographic characteristics of the Italian population (i.e., age, education, geographical area of residence). According to our data (Table 1), time for domestic chores increased from 2.4 h per day before the pandemic to almost 3.7 h per day during the lockdown for women with no dependent children, while it increased from almost 3 h per day to almost 5 h for women with minor children under the age of 18. Time devoted to children considerably increased for all mothers during the lockdown, although the magnitude of the increase was greater for those of younger children. In particular, childcare became similar to a full-time work for mothers of children aged less than 5 years, who devoted more than 7 h per day to this activity. Childcare time decreased only slightly after the end of the lockdown, mainly due to the prolonged school

[2] More information can be found at the CWW project webpage: https://www.countingwomenswork.org/.

[3] The lockdown in Italy ended on May 19th, 2020 and was followed by a so-called 'phase two', which still implied several restrictions, including school closures.

Table 1 Women's average daily hours of paid and unpaid work

Activity	Minor Child (ren)	N	Before		During		After	
			Mean	Std err.	Mean	Std err.	Mean	Std err.
Domestic work	Yes	308	3.07	0.14	4.88	0.14	4.15	0.15
	Not	502	2.37	0.08	3.68	0.10	3.03	0.09
Paid work	Yes	308	4.03	0.14	2.76	0.16	3.20	0.15
	Not	502	2.94	0.12	2.51	0.13	2.66	0.13
Childcare	0–2 years	75	6.19	0.23	7.26	0.20	6.87	0.19
	3-0.5 years	47	5.25	0.25	7.78	0.13	6.67	0.25
	6–10 years	90	4.45	0.26	5.77	0.27	5.66	0.27
	11–17 years	90	3.06	0.22	4.78	0.28	4.03	0.25
	Missing	6						

Note Weighted sample estimates. Sub-sample of women

closures in Italy. A similar effect was observed also for time spent on domestic chores.

Looking at paid work, on average, women without minor children reported a decrease from 2.9 h per day before the pandemic to 2.5 h per day during the lockdown followed by an increase to 2.6 h per day after the end of the lockdown. The average time devoted to paid work was equal to 4 h per day[4] for mothers of minor children before the pandemic, decreased to almost 2.7 during the confinement and increased to 3.2 h following the end of the lockdown.

3.2 Methods

In our analysis we focus on a subsample of 252 women with minor children who lived in a couple and devoted time to paid work during the pandemic. The great majority of these women (88.5%) lived in dual earners couples and only 11.5% of them reported that the male partner did not spend time on paid work during the confinement. The survey asked the respondents to report whether, compared to the pre pandemic times, they experienced changes in their levels of stress and fatigue associated with paid work, housework, and childcare. The question was asked separately for each activity. E.g., with reference to stress during childcare, the question asked: "Did you feel more or less stressed while performing childcare during the lockdown period than usual?". The respondents could answer by choosing one of the following multiple choices: much more, more, no change, less, much less. Based on this information, we built a marker of stress/fatigue in association with childcare/housework/paid work during the lockdown with 1 indicating that the respondent reported a change towards an

[4] The higher daily hours spent on paid work by mothers of young children compared with those without young children reflects their younger ages.

increase in the feeling (e.g., stress) and 0 indicating that the respondent reported no changes or a change toward less stress.

To test our hypothesis, we run six logistic regressions: one for each binary indicator of stress/fatigue associated with the three different activities analysed (i.e., more stress/fatigue in association to childcare, housework, paid work during the lockdown). Our explicative variable is the respondent level of satisfaction with the household division of family labour during the confinement compared to pre pandemic times (i.e., less satisfied, no change, more satisfied). We include the following control variables: geographic area of residence, educational level, age of the youngest child, average weekly hours spent in paid work, the male partner's share of unpaid family work.

4 Results

The majority of employed mothers in our sample reported to feel more stressed and fatigued than usual in relation to paid work, housework, and childcare during the lockdown, with the only exception of levels of stress associated with household chores for which the percentage is lower (48.5%). Looking at unpaid family work, the greatest changes are observed in the levels of fatigue experienced: 66.1 and 64% of mothers reported to feel more tired during the lockdown when carrying out housework and childcare. By contrast, levels of stress (64.1%) increased more in association with paid work than fatigue (58.6%). For all the three activities considered, among women who reported more stress/fatigue responses were concentrated on the response modality indicating more intense changes ("much more"), while the opposite is true for changes towards lower levels of stress/fatigue for which women reported to have experienced more moderate changes.

The survey asked each respondent to report the approximate share of the total household's unpaid care and domestic work performed by the partner if present. The results revealed that before the pandemic, in couples with young children where women have a paid work, the male partner's share of unpaid care and domestic work was around 27.7%. During the lockdown this share slightly increased to 30.4%. However, soon after the end of the lockdown the male partner's share of unpaid care and domestic work returned to its pre-pandemic level. About 31.5% of women in our subsample reported to be less satisfied with the household division of unpaid work during the lockdown period (including 15.5% "much less" and 15.9% "less" satisfied), 37% reported no change compared to pre pandemic times and 27.5% felt more satisfied (of which 21.5% "more" and 6% "much more" satisfied) while the information was missing for 4% of the sampled women.

Looking at the results of our models for the explicative variable in Table 2, in line with our expectations, we find that women's satisfaction with the division of unpaid family work within the household significantly affects the levels of stress and fatigue reported in association with childcare, housework and paid work during the pandemic. Mothers who reported to be less satisfied with the household's division

Table 2 Odds ratio of feeling more fatigued/stressed in association with paid work/housework/childcare during lockdown compared to usual times

Dependent variable	Effect	Odds ratio	StdErr	Wald ChiSq	Prob ChiSq
Paid work					
Fatigued	Less versus more	2.511	0.916	6.369	0.012
	Less versus same	1.963	0.650	4.158	0.041
Stressed	Less versus more	3.612	1.343	11.927	0.001
	Less versus same	2.037	0.683	4.505	0.034
Housework					
Fatigued	Less versus more	4.919	2.022	15.021	<0.001
	Less versus same	3.064	1.165	8.667	0.003
Stressed	Less versus more	3.612	1.343	11.927	0.001
	Less versus same	2.037	0.683	4.505	0.034
Childcare					
Fatigued	Less versus more	8.058	4.004	17.630	<0.001
	Less versus same	3.599	1.601	8.288	0.004
Stressed	Less versus more	6.973	3.301	16.829	<0.001
	Less versus same	2.855	1.176	6.481	0.011

Note Results of the models are controlled for: geographic area of residence, educational level, age of the youngest child, average weekly hours spent in paid work, the male partner's share of unpaid family work

of unpaid work had greater probabilities than those who reported higher satisfaction to experience more stress and fatigue while performing these activities during the lockdown period as compared to pre pandemic times. For all the activity considered, women who reported to be less satisfied with the intrahousehold division of unpaid work during the lockdown showed statistically significant higher odds of feeling more stressed and fatigued also than those who reported the same levels of satisfaction as compared to usual times. The magnitude of the effect of the explicative variable was greater when the dependent variable was "fatigue" than "stress" for both childcare and housework, the opposite is true for paid work. Our results are in line with those of prior research based on Italy finding that levels of tiredness and overload for working women during the pandemic were aggravated by the need to deal with unpaid care work especially in presence of young children [15].

5 Concluding Remarks

In this article we used data from a web survey conducted soon after the end of the first lockdown to study how measures adopted in Italy in contrast to the diffusion of COVID-19 affected women's time-use and the intra-household division of labour.

More specifically, the study focused on employed mothers of minor children and tested if lower levels of satisfaction for the division of unpaid work within the household were significantly related with higher stress and tiredness reported by mothers in association with time devoted to paid and unpaid work in the lockdown. Similar to Brini et. al [5], our results highlight that, on average, Italian women experienced a decrease of hours of paid work during the lockdown. However, women considerably increased their time for unpaid care and domestic work (in particular, those with minor children). Women reported that their partners only slightly increased their contribution to unpaid care and domestic work during the lockdown and that they returned to their pre-lockdown levels soon after its end. The extent of changes in men's use of time for unpaid work during the pandemic does not suggest that a real and lasting change in the division of unpaid work was triggered. The lack of rebalancing of gender roles shows its effects also on the subjective well-being experienced by mothers during the initial phase of the health crisis. Employed mothers reported to feel more stress and tiredness in association to paid and unpaid work during the lockdown compared to usual time. Moreover, we found the existence of a statistically significant positive relationship between the probability of experiencing higher levels of stress and fatigue in association with childcare, paid work and housework and lower levels of satisfaction with the household division of family work during the confinement period.

A strength of our analysis was the rapid-response nature of the survey and the consequent ability to gauge changes in mothers' unpaid work responsibilities and feelings during the pandemic as it was occurring. This rapidity also came at a cost, especially in terms of the need rely on a relatively small sample. In addition, because of the voluntary nature of survey participation, our sample is somewhat biased since it was mainly composed of working women with an educational level above a bachelor's degree. However, women with a lower level of education were fairly well represented in the sample. These drawbacks point to the need for further research using a more detailed time use survey, based on a more heterogeneous sample and including longitudinal information, to provide a richer account of how and why unpaid work changed and how it affected mothers' well-being during the different phases of the pandemic. Despite this caveat, results from our survey contribute to a broader understanding of the effects of the unequal division of unpaid work within the household on women's lives and their mental health and this may bring new insights on the discussion on family policies.

Acknowledgements The authors are grateful to Gretchen Donehower for developing and sharing the original version of the survey questionnaire and to the participants of the workshop 'Gender Inequalities and Economic Theory and Policies for a Post-Pandemic World' (Levy Eco-nomics Institute, Bard College, 21-23 September 2022) for valuable comments on a preliminary draft of this paper. The usual disclaimers apply.

References

1. Adams, A.: The gender wage gap on an online labour market: the cost of interruptions. CEPR Working Paper, DP14294 (2020). Available from: https://cepr.org/publications/dp14294
2. Andrew, A., Cattan, S., Costa Dias, M., Farquharson, et al.: The gendered division of paid and domestic work under lockdown. IZA Discussion Paper 13500 (2020). Available from: https://docs.iza.org/dp13500.pdf
3. Biroli, P., Bosworth, S., Della Giusta, M., Di Girolamo, et al.: Family life in lockdown. Front. Psychol. **12**, e3259 (2020). https://doi.org/10.3389/fpsyg.2021.687570
4. Boll, C., Schüller, S.: The situation is serious, but not hopeless: evidence-based considerations on the intra-couple division of childcare before, during and after the COVID-19 lockdown, SOEP papers on Multidisciplinary Panel Data Research, No 1098, German Socio-Economic Panel (SOEP). DIW Berlin, Berlin (2020)
5. Brini, E., Lenko, M., Scherer, S., Vitali, A.: Retraditionalisation? work patterns of families with children during the pandemic in Italy. Demogr. Res. **45**, 957–972 (2021). https://doi.org/10.4054/DemRes.2021.45.31
6. Calarco, J.M., Anderson, E.M., Meanwell, E.V., Knopf, A.: "Let's Not Pretend It's Fun": How COVID-19-Related School and Childcare Closures are Damaging Mothers' Well-Being, SocArXiv Papers (2020). https://doi.org/10.31235/osf.io/jyvk4
7. Charbonneau, A., Lachance-Grzela, M., Bouchard, G.: Housework allocation, negotiation strategies, and relationship satisfaction in cohabiting emerging adult heterosexual couples. Sex Roles **81**, 290–305 (2019). https://doi.org/10.1007/s11199-018-0998-1
8. Chen, J., Lu, H., Melino, G., Boccia, S., et al.: COVID-19 Infection: The China and Italy Perspectives. Cell Death Dis. 11(438) (2020). https://doi.org/10.1038/s41419-020-2603-0
9. Corsi, M., Aloè, E., Zacchia, G.: A Feminist Reading of Italy's North-South Dualism in the Wake of COVID-19. Fem. Econ. **27**(1–2), 203–216 (2021). https://doi.org/10.1080/13545701.2021.1876904
10. Craig, L.: Does father care mean fathers share? a comparison of how mothers and fathers in intact families spend time with children. Gend. Soc. **20**(2), 259–281 (2006). https://doi.org/10.1177/0891243205285212
11. Craig, L.: Coronavirus, domestic labour and care: gendered roles locked down. J. Sociol. **56**(4), 684–692 (2020). https://doi.org/10.1177/1440783320942413
12. Craig, L., Churchill, B.: Working and caring at home: gender differences in the effects of Covid-19 on paid and unpaid labor in Australia. Fem. Econ. **27**(1–2), 310–326 (2021). https://doi.org/10.1080/13545701.2020.1831039
13. Del Boca, D., Oggero, N., Profeta, P., Rossi, M.: Women's and men's work, housework and childcare, before and during COVID-19. Rev. Econ. Household **18**, 1001–1017 (2020). https://doi.org/10.1007/s11150-020-09502-1
14. DeMaris, A.: The 20-year trajectory of marital quality in enduring marriages: does equity matter? J. Soc. Pers. Relat. **27**(4), 449–471 (2010). https://doi.org/10.1177/0265407510363428
15. De Vita, L., Mazali T., Campanella, G.: The impacts of smart working on women. Lessons from the first lockdown in Italy. Int. Rev. Soc. (2022). https://doi.org/10.1080/03906701.2022.2064666
16. Donehower, G.: Counting Women's Work: Unpaid care work and Covid-19 (2020). https://www.countingwomenswork.org/news/unpaid-care-work-and-covid19-take-the-survey
17. Durante, K., Rodgers, Y., Kaplowitz, et al.: COVID-19 and Changes in the Gendered Division of Unpaid Labor, Job Productivity, and Job Satisfaction. Rutgers University Center for Women and Work Working Paper (2021)
18. Eurofound.: Living, Working and COVID-19, Publications Office of the European Union, Luxembourg (2020). Available from: https://www.eurofound.europa.eu/sites/default/files/ef_publication/field_ef_document/ef20059en.pdf
19. Farre, L., Fawaz, Y., Gonzalez, L., Graves, L.: How the Covid-19 Lockdown affected gender inequality in paid and unpaid work in Spain? IZA Discussion Paper No. 13434 (2020)

20. Giurge, L.M., Whillans, A.V., Yemiscigil, A.: A multicountry perspective on gender differences in time use during COVID-19. Psychol. Cognitive Sci. **118**(12), e2018494118 (2021). https://doi.org/10.1073/pnas.2018494118
21. Hu, Y., Yucel, D.: What fairness? gendered division of housework and family life satisfaction across 30 countries. Eur. Sociol. Rev. **34**(1), 92–105 (2018). https://doi.org/10.1093/esr/jcx085
22. Kulic, N., Dotti Sani, G.M., Strauss, S., Bellani, L.: Economic disturbances in the COVID-19 crisis and their gendered impact on unpaid activities in Germany and Italy. Eur. Soc. **23**(Supp. 1), S400–S416 (2020). https://doi.org/10.1080/14616696.2020.1828974
23. Ilkkaracan, I., Memiş, E.: Transformations in the gender gaps in paid and unpaid work During the COVID-19 pandemic: findings from Turkey. Fem. Econ. **27**(1–2), 288–309 (2021)
24. Lennon, M.C., Rosenfield, S.: Relative fairness and the division of housework: the importance of options. Am. J. Sociol. **100**(2), 506–531 (1994). https://doi.org/10.1086/230545
25. Mangiavacchi, L., Piccoli, L., Pieroni, L.: Fathers matter: Intrahousehold responsibilities and children's wellbeing during the COVID-19 lockdown in Italy. Econ. Hum. Biol **42** (2021). https://doi.org/10.1016/j.ehb.2021.101016
26. Manzo, L.K., Minello, A.: Mothers, childcare duties, and remote working under COVID-19 lockdown in Italy: cultivating communities of care. Dialogues Hum. Geogr. **10**(2), 120–123 (2020). https://doi.org/10.1177/2043820620934268
27. McDonnell, C., Luke, N., Short, S.E.: Happy moms, happier dads: gendered caregiving and parents' affect. J. Fam. Issues **40**(17), 2553–2581 (2019). https://doi.org/10.1177/0192513X19860179
28. Mezzadri, A.: Social reproduction and pandemic neoliberalism: planetary crises and the reorganisation of life, work and death. Organization **29**(3), 379–400 (2022). https://doi.org/10.1177/13505084221074042
29. Minello, A., Martucci, S., Manzo, L.K.: The pandemic and the academic mothers: present hardships and future perspectives. Eur. Soc. **23**(sup1), S82–S94 (2021). https://doi.org/10.1080/14616696.2020.1809690
30. Musick, K., Meier, A., Flood, S.: How parents fare: mothers' and fathers' subjective wellbeing in time with children. Am. Sociol. Rev. **81**(5), 1069–1095 (2016). https://doi.org/10.1177/0003122416663917
31. Nomaguchi, K., Milkie, M.A.: Parenthood and well-being: a decade in review. J. Marriage Fam. **82**(1), 198–223 (2020). https://doi.org/10.1111/jomf.12646
32. OECD.: The Pursuit of Gender Equality: An Uphill Battle, Paris (2017)
33. OECD.: OECD Employment Outlook 2020: Worker Security and the COVID-19 Crisis, Paris (2020)
34. Raile, A.N.W., Raile, E.D., Parker, D.C.W., et al.: Women and the weight of a pandemic: A survey of four Western US states early in the Coronavirus outbreak. Gend. Work Organ. Feminist Front. 1–12 (2020). https://doi.org/10.1111/gwao.12590
35. Seedat, S., & Rondon, M.: Women's wellbeing and the burden of unpaid work. bmj, 374 (2021).
36. Seiz, M.: Equality in confinement: non-normative division of labor in Spanish dual-earner families during the COVID-19 lockdown. Fem. Econ. **27**(1–2), 345–361 (2021). https://doi.org/10.1080/13545701.2020.1829674
37. Sevilla, A., Smith, S.: Baby steps: The gender division of childcare during the Covid-19 Pandemic, IZA Discussion Paper No. 13302 (2020)
38. Zannella, M., Hammer, B., Prskawetz, A., Sambt, J.: A quantitative assessment of the rush hour of life in Austria, Italy and Slovenia. Eur. J. Popul. **35**(4), 751–776 (2019). https://doi.org/10.1007/s10680-018-9502-4
39. Zannella, M., De Rose, A.: Stability and change in family time transfers and workload inequality in Italian couples. Demogr. Res. **40**, 49–60 (2019). https://doi.org/10.4054/DemRes.2019.40.3
40. Zannella, M., De Rose, A.: Fathers' and mothers' enjoyment of childcare: the role of multitasking. Vienna Yearbook Popul. Res. **19**(1), TBA-OLF (2021). https://doi.org/10.1553/populationyearbook2021.res3.1
41. Zoch, G., Bächmann, A.C., Vicari, B.: Who cares when care closes? care-arrangements and parental working conditions during the COVID-19 pandemic in Germany. Eur. Soc. **23**(sup1), S576–S588 (2021). https://doi.org/10.1080/14616696.2020.1832700

Matrix-Variate Hidden Markov Models: An Application to Employment Data

Salvatore D. Tomarchio, Antonio Punzo, and Antonello Maruotti

Abstract Hidden Markov models (HMMs) are a powerful tool for modeling heterogeneous time-dependent data structures. Recently, a family of parsimonious matrix-variate HMMs has been introduced in the literature. In this manuscript, we apply this family of models to a dataset measuring employment levels for 187 countries in the world over 26 years. Each country is evaluated using the well-known three-sector model of the economy and by taking into account the gender of the workers. By using the HMMs flexibility, different states are detected and the switches of the countries among the states over time are analyzed.

Keywords Employment · Hidden Markov models · Matrix-variate

1 Introduction

The statistical literature related to hidden Markov models (HMMs) is large and constantly evolving (for a survey, see [25]). Being dependent mixture models, HMMs provide more flexibility than single distributions and their mathematical structure makes them suitable for many real-world applications (see, e.g. [4, 6, 12, 13, 24]).

HMMs have been also used for the analysis of univariate (one-way) or multivariate (two-way) longitudinal data (see, e.g. [16, 18, 19]). In this context, they are also referred to as latent Markov models [1]. Recently, [23] have introduced a family of 98 parsimonious HMMs for matrix-variate (three-way) longitudinal data. In a

S. D. Tomarchio (✉) · A. Punzo
Department of Economics and Business, University of Catania, Catania, Italy
e-mail: daniele.tomarchio@unict.it

A. Punzo
e-mail: antonio.punzo@unict.it

A. Maruotti
Department of Law, Economics, Politics, and Modern Languages, LUMSA University, Rome, Italy
e-mail: a.maruotti@lumsa.it

M. Bini et al. (eds.), *Advanced Methods in Statistics, Data Science and Related Applications*, Springer Proceedings in Mathematics & Statistics 467,
https://doi.org/10.1007/978-3-031-65699-6_11

nutshell, these models can analyze datasets where for each statistical unit, and at each time frame, a $P \times R$ matrix of variables is observed. This kind of data can be conveniently arranged into four-way arrays having dimension $P \times R \times I \times T$, being I the sample size and T the time length.

In this manuscript, we use the models introduced by [23] to analyze the evolution in the employment levels of 187 countries in the world over 26 years. Specifically, for each country, the employment information is available in a three-sector model format. The three-sector model was proposed in the first half of the 20th century [7, 9, 10], and it established the basic division by which economic structure is largely understood to date. According to this economic model, the total production of a country can be divided into three sectors: primary, secondary, and tertiary. The primary sector commonly includes agriculture, hunting, forestry, fishing, and industries connected to the natural environment. The secondary sector typically involves various forms of manufacturing and activities that produce goods from products of the primary sector. The tertiary sector refers to any kind of service, spanning from banking and legal assistance to retail, hospitality, transportation, and entertainment.

The theory behind this categorization assumes that the development of a country's economy goes through the transition from the primary to the tertiary sector. Relatedly, with economic growth, the number of employed in the primary sector relatively decreases in favor of the secondary sector, and subsequently to the tertiary sector [15]. The discussion about whether this representation of a country's economy is debatable or not is not the purpose of this work. However, it is very resilient considering that even the World Bank (from which we downloaded the data) and other well-known institutes generally render this information in a three-sector model fashion [20].

Through the utilization of HMM characteristics, our primary objective is to explore the existence of latent states within the world's countries, from the above-discussed point of view. Thus, this exploration aims to discern distinct stages of economic development. Secondly, our objective is to trace the advancement of countries as they transition from states primarily grounded in the primary sector to those emphasizing the secondary or tertiary sectors. Additionally, we disassemble this information by gender, facilitating an analysis of gender-based employment segregation within specific sectors.

The manuscript is organized as follows. In Sect. 2, we summarize the methodological foundation of the 98 parsimonious HMMs introduced by [23]. In Sect. 3, we provide specific details about the dataset and results when the considered HMMs are fitted. Finally, some concluding remarks are given in Sect. 4.

2 Matrix-Variate HMMs: Methodology and Estimation

Generally speaking, matrix-variate HMMs (MV-HMMs) are a particular kind of time-dependent mixture consisting of two processes:

1. a hidden first-order Markov chain $\{S_{it}; i = 1, \dots, I, t = 1, \dots, T\}$ with state-space $\{1, \dots, k, \dots, K\}$, satisfying the Markov property

$$\Pr(S_{it} = s_{it}|S_{i1} = s_{i1}, \ldots, S_{it-1} = s_{it-1}) = \Pr(S_{it} = s_{it}|S_{it-1} = s_{it-1}),$$

2. a state-dependent observation process $\{\mathcal{X}_{it}; i = 1, \ldots, I, t = 1, \ldots, T\}$, with $\mathcal{X}_{it} \in \mathbb{R}^{R \times P}$, satisfying the conditional independence property

$$\begin{aligned} &f(\mathcal{X}_{it} = \mathbf{X}_{it}|\mathcal{X}_{i1} = \mathbf{X}_{i1}, \ldots, \mathcal{X}_{it-1} = \mathbf{X}_{it-1}, S_{i1} = s_{i1} \ldots, S_{it} = s_{it}) \\ =&f(\mathcal{X}_{it} = \mathbf{X}_{it}|S_{it} = s_{it}), \end{aligned}$$

   being $f(\cdot)$ a matrix-variate probability density function (pdf).

Herein, the functional form of $f(\cdot)$ is that of the matrix-variate normal distribution, i.e.

$$f(\mathbf{X}_{it}|S_{it} = k; \boldsymbol{\theta}_k) = \frac{\exp\left\{-\frac{1}{2}\operatorname{tr}\left[\Sigma_k^{-1}(\mathbf{X} - \mathbf{M}_k)\Psi_k^{-1}(\mathbf{X} - \mathbf{M}_k)'\right]\right\}}{(2\pi)^{\frac{PR}{2}}|\Sigma_k|^{\frac{R}{2}}|\Psi_k|^{\frac{P}{2}}}, \quad (1)$$

where $\mathbf{M}_k$ is the $P \times R$ matrix of means, Σ_k and Ψ_k are the $P \times P$ and $R \times R$ covariance matrices related to the P rows and R columns, respectively, tr() is the trace operator and $\boldsymbol{\theta}_k = \{\mathbf{M}_k, \Sigma_k, \Psi_k\}$.

In addition to $\boldsymbol{\theta}_k$, we also have the parameters of the hidden Markov chain. In detail, we have the initial probabilities $\pi_{ik} = \Pr(S_{i1} = k)$, $k = 1, \ldots, K$, which are collected in the K-dimensional vector $\boldsymbol{\pi}$, and the transition probabilities governing the switching among the states

$$\pi_{ik|j} = \Pr(S_{it} = k|S_{it-1} = j), \quad t = 2, \ldots, T \text{ and } j, k = 1, \ldots, K, \quad (2)$$

contained in the $K \times K$ transition probability matrix Π. In (2), k identifies the current state and j refers to the one previously visited. For easiness, we consider homogeneous MV-HMMs, that is $\pi_{ik|j} = \pi_{k|j}$ and $\pi_{ik} = \pi_k$, $i = 1, \ldots, I$.

To introduce parsimony in (1), the well-known eigen-decomposition of a covariance matrix [5] is applied to Σ_k and Ψ_k. This creates 14 parsimonious structures for Σ_k (EII, VII, EEI, VEI, EVI, VVI, EEE, VEE, EVE, VVE, EEV, VEV, EVV, VVV) and 7 for Ψ_k (II, EI, VI, EE, VE, EV, VV). The positions of the letters in the acronyms for Σ_k refer to the volume, shape, and orientation of the k-state, respectively, whereas in the acronyms for Ψ_k the letter of the volume is dropped out for identifiability reasons (for further details, see [23]). For both covariance matrices, the letter 'E' stands for equal, 'V' means varying, and 'I' denotes the identity matrix. By combining both parsimonious structures, a family of 98 MV-HMMs is obtained.

Parameter estimation is conducted by a full maximum likelihood approach based on the expectation conditional maximization (ECM) algorithm [17], and on recursions widely used in the HMM literature [2]. An iterative minorization-maximization (MM) algorithm [3] is also adopted to update some of the parameters of the EVE, VVE, and VE MV-HMMs during the ECM algorithm. For notational clarity, each MV-HMM is labeled by separating with a dash the acronyms used for the row and column covariance matrices, respectively. For further details concerning estimation and computation, see [23].

3 Real Data Analysis

As introduced in Sect. 1, the object of our analysis is a dataset concerning the employment levels of $I = 187$ countries in the world over 26 years. On a related note, the time window goes from 1994 to 2019 and consists of five-year data, i.e. $T = 6$. For each country, we measure the level of employment in the $R = 3$ sectors of the economy (primary, secondary, and tertiary) for $P = 2$ genders (males and females). Therefore, the dataset has a $2 \times 3 \times 187 \times 6$ four-way structure.

The dataset is freely accessible from the World Bank database at https://databank.worldbank.org/source/world-development-indicators. According to the data supporting information, employment is defined as people of working age engaged in any activity to produce goods or provide services for pay or profit. Data are mainly collected via nationally representative labor force surveys (e.g., household surveys and population censuses) and are expressed in percentages.

In analyzing this data set, several aspects are worth to be investigated. The first concerns the existence of areas with similar employment levels across the world. Another aspect of interest is the strength of time dependence as measured by the transition probability matrix, as well as how the countries move between the hidden states. This latter aspect can be particularly useful in identifying broad shifts in employment and stages of countries' development.

3.1 Numerical Results and Discussion

We fit the 98 parsimonious MV-HMMs for $K \in \{1, \ldots, 9\}$ using the FourWayHMM package [22] for the R statistical software. According to the Bayesian information criterion (BIC; [21]), that we use for model selection, the best-fitting model is the VEV-VV MV-HMM with $K = 7$ states (BIC=11780.05).

First of all, we can try to interpret the hidden states detected by this model by looking at its estimated mean matrices. This information is reported via the parallel coordinate plots of Fig. 1. Specifically, each subplot refers to gender and illustrates the estimated mean employment levels over the three sectors, for each of the 7 hidden states.

By starting from the subplot related to females, we note that as we move from the first to the seventh state, the average levels of employment in the primary sector roughly increase, whereas those in the tertiary sector generally decrease. A similar pattern is replicated in the subplot referring to males. Thus, the first (reasonably expected) information we gather is the presence of states having quite different levels of economic development. It is also interesting to notice that, in the first five states, females show lower average employment levels than males when the primary sector is considered. The opposite occurs in the sixth and seventh states. Similarly, with the exclusion of the fifth state, females have a lower average enrollment than males in the secondary sector. Conversely, except for the seventh state, females display higher average employment levels than males in the tertiary sector.

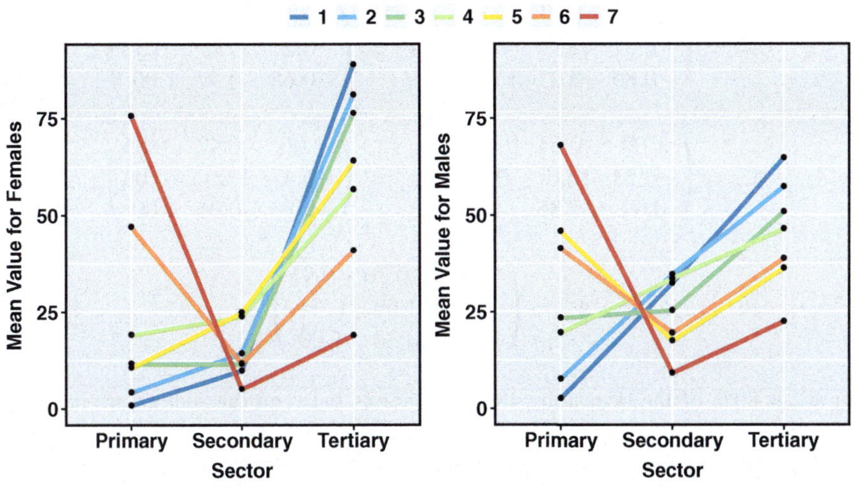

Fig. 1 Parallel coordinate plots for the mean matrices estimated by the VEV-VV MV-HMM, over the three sectors, split by gender and colored according to the estimated classification

The breakdown of employment levels by sex discloses the so-called gender segregation problem. Indeed, there is the tendency in perhaps all existing societies for some occupations to be strongly associated with females and some with males [11]. This is particularly harmful to females since their segregation in certain sectors often results from cultural attitudes that prevent them from entering industrial employment [8]. Segregating one sex in a narrow range of occupations significantly reduces economic efficiency by reducing labor market flexibility and thus the economy's ability to adapt to change. We noted how this phenomenon is practically present across all the detected states.

Useful insights can also be obtained from the analysis of the estimated correlation matrices, labeled by adding an asterisk to the corresponding covariance matrix symbols. By starting with gender-related correlation matrices, we have

$$\Sigma_1^* = \begin{pmatrix} 1.00 & 0.27 \\ 0.27 & 1.00 \end{pmatrix}, \Sigma_2^* = \begin{pmatrix} 1.00 & 0.37 \\ 0.37 & 1.00 \end{pmatrix}, \Sigma_3^* = \begin{pmatrix} 1.00 & 0.35 \\ 0.35 & 1.00 \end{pmatrix},$$

$$\Sigma_4^* = \begin{pmatrix} 1.00 & 0.19 \\ 0.19 & 1.00 \end{pmatrix}, \Sigma_5^* = \begin{pmatrix} 1.00 & 0.33 \\ 0.33 & 1.00 \end{pmatrix}, \Sigma_6^* = \begin{pmatrix} 1.00 & 0.31 \\ 0.31 & 1.00 \end{pmatrix}, \Sigma_7^* = \begin{pmatrix} 1.00 & 0.36 \\ 0.36 & 1.00 \end{pmatrix}.$$

Therefore, we observe that the correlations between genders are positive, relatively weak, and similar among the states. Only the fourth state seems to slightly deviate from the others because of its even lower values. As concerns the sector-related correlation matrices, we have

$$\Psi_1^* = \begin{pmatrix} 1.00 & 0.31 & -0.44 \\ 0.31 & 1.00 & -0.99 \\ -0.44 & -0.99 & 1.00 \end{pmatrix}, \Psi_2^* = \begin{pmatrix} 1.00 & 0.17 & -0.51 \\ 0.17 & 1.00 & -0.94 \\ -0.51 & -0.94 & 1.00 \end{pmatrix},$$

$$\Psi_3^* = \begin{pmatrix} 1.00 & -0.31 & -0.85 \\ -0.31 & 1.00 & -0.23 \\ -0.85 & -0.23 & 1.00 \end{pmatrix}, \Psi_4^* = \begin{pmatrix} 1.00 & -0.21 & -0.68 \\ -0.21 & 1.00 & -0.57 \\ -0.68 & -0.57 & 1.00 \end{pmatrix},$$

$$\Psi_5^* = \begin{pmatrix} 1.00 & -0.54 & 0.04 \\ -0.54 & 1.00 & -0.86 \\ 0.04 & -0.86 & 1.00 \end{pmatrix}, \Psi_6^* = \begin{pmatrix} 1.00 & -0.47 & -0.85 \\ -0.47 & 1.00 & -0.06 \\ -0.85 & -0.06 & 1.00 \end{pmatrix},$$

$$\Psi_7^* = \begin{pmatrix} 1.00 & -0.70 & -0.91 \\ -0.70 & 1.00 & 0.34 \\ -0.91 & 0.34 & 1.00 \end{pmatrix}.$$

In about 80% of the occasions, the correlations between the sectors are negative. Interestingly, we note that the correlations between the primary and secondary sectors are positive only for the first and second states, which are those having the majority of their employment levels in the tertiary sector. The only positive correlation between the primary and tertiary sectors is in the fifth state. Relatedly, the lowest value among these sectors is displayed in the seventh state. The latter state is also the only one showing a positive correlation between the secondary and tertiary sectors. Conversely, the first two states are substantially close to a perfect negative correlation when these two sectors are considered.

Before showing how these states cluster the world countries, it is worth analyzing the estimated transition probability matrix, i.e.

$$\Pi = \begin{pmatrix} 1.00 & 0.00 & 0.00 & 0.00 & 0.00 & 0.00 & 0.00 \\ 0.16 & 0.84 & 0.00 & 0.00 & 0.00 & 0.00 & 0.00 \\ 0.00 & 0.00 & 1.00 & 0.00 & 0.00 & 0.00 & 0.00 \\ 0.00 & 0.14 & 0.00 & 0.86 & 0.00 & 0.00 & 0.00 \\ 0.00 & 0.00 & 0.03 & 0.00 & 0.97 & 0.00 & 0.00 \\ 0.00 & 0.00 & 0.07 & 0.00 & 0.00 & 0.93 & 0.00 \\ 0.00 & 0.00 & 0.00 & 0.00 & 0.00 & 0.10 & 0.90 \end{pmatrix},$$

and the number of countries that have changed their hidden state from the second time point onwards in Fig. 2.

As we can note by the estimated Π, transitions between states are not uncommon between adjacent states, whereas are generally null among distant states. Two exceptions are the first and third states due to their absorbing nature, i.e. if entered by a country they are never left. From the analysis of the bar plot in Fig. 2, we note a growing number of switches up to 2009. From that year onwards, the switches sharply declined until 2019. A possible justification for this sudden changing behavior is the great recession that affected national economies globally around 2008. Anyway, all the countries that changed their states moved to other more tertiary-based states, supporting the theory behind the three-sector model.

To have a graphical idea of the detected states and switches among them, Fig. 3 illustrates the political world maps, for each considered year, with countries

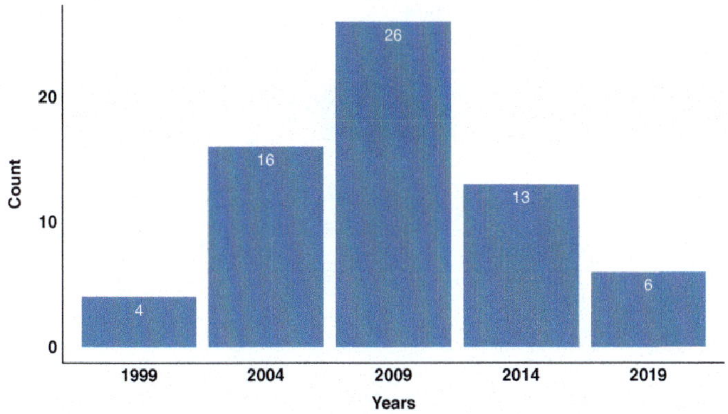

Fig. 2 Bar plot showing the number of countries that changed their hidden state from the second time point onwards

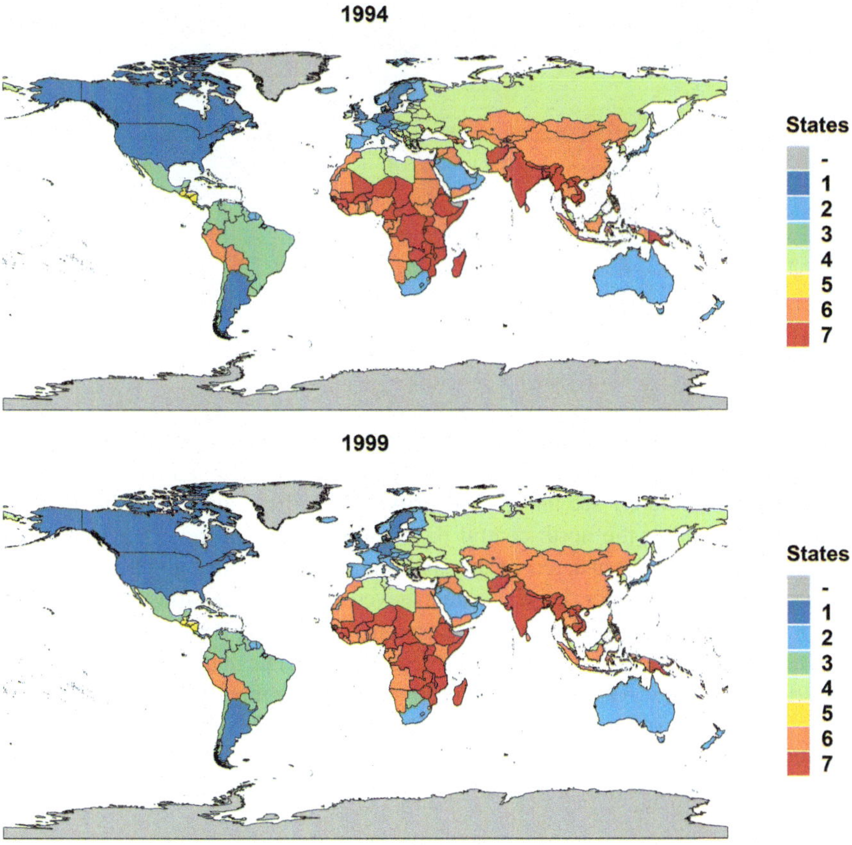

Fig. 3 Political world map with countries colored according to the estimated state memberships

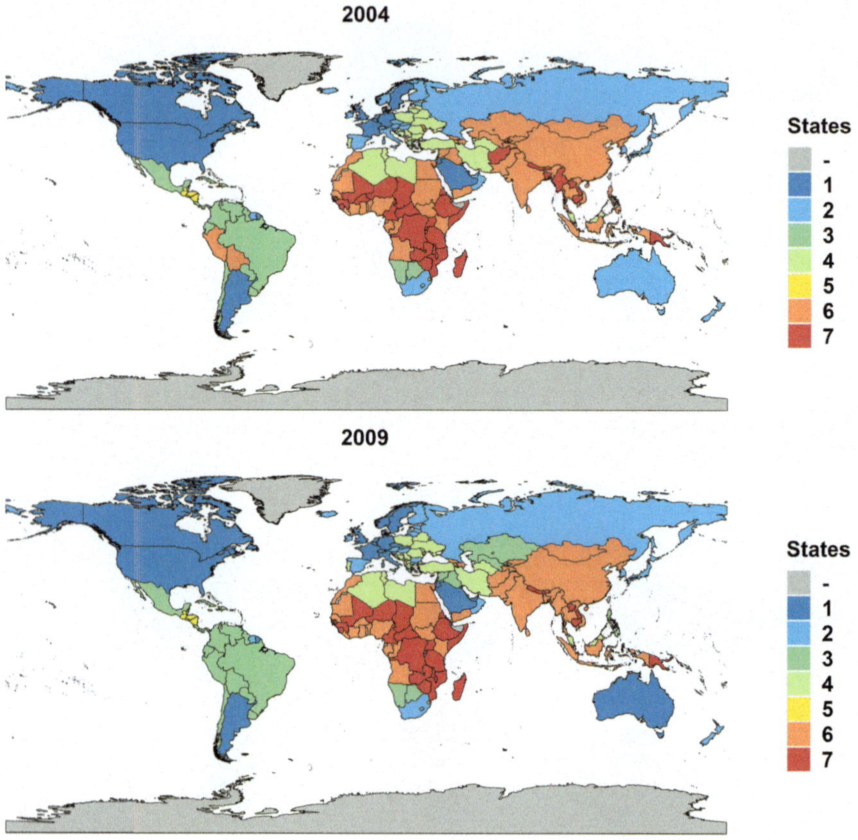

Fig. 3 (continued)

colored according to the estimated state memberships. Note that the countries not included in the analysis, because of not available data, are colored in gray.

We can immediately see that the majority of countries clustered in the first two states are located in the north-central part of the world. On the other hand, most African countries are clustered in the sixth and seventh states, although there are some exceptions that, over the years, have moved their economy towards the tertiary sector. Interestingly, the fifth state consists only of Central American countries as well as the Tonga and Samoa islands, whereas the third state mainly consists of South American countries. As concerns the third state, during the 1990s it was generally composed of Eastern European countries, Russia, and some North African countries. However, most of them have moved to more tertiary-based states over the years, relegating this state to only a handful of those who were originally there.

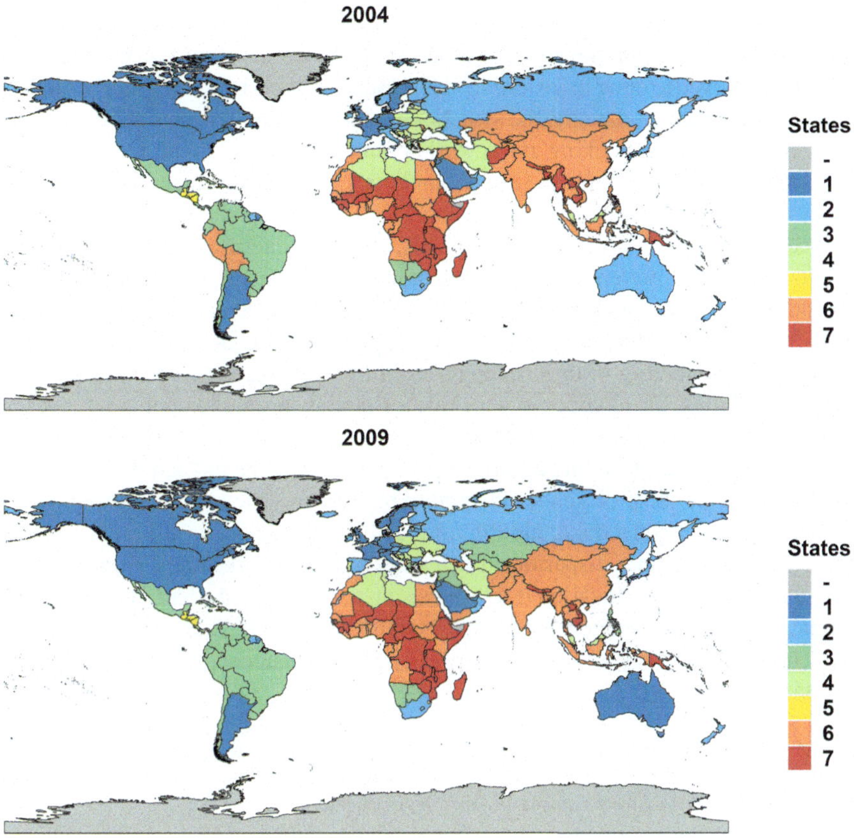

Fig. 3 (continued)

4 Concluding Remarks

Hidden Markov models (HMMs) are the state of the art in the analysis of time-dependent data. Within this framework, a family of parsimonious HHMs for the analysis of matrix-variate longitudinal has been recently introduced in the literature. These models have been herein applied to a dataset measuring the employment levels of 187 countries in the world over 26 years. This dataset disaggregates the information by gender and by using the well-known three-sector categorization of a country's economy.

Our results disclose, at first, the presence of seven hidden states in the world having quite different levels of economic development. Such information is gathered by the analysis of the estimated mean matrices, which also show evident segregation of work by gender. The estimated correlation matrices illustrated the substantially negative relationship between the three sectors and the weak correlation between the two genders. The estimated transition matrices reported the presence of two absorbing states, and that transitions between the other states mostly occur when they are adjacent. From the analysis of the switches of the countries, we noted that all of them moved to other more tertiary-based states, supporting the theory behind the three-sector model. Relatedly, the switches grew up to 2009 for then sharply decline until 2019. Lastly, by looking at the world political maps, we identified the regions in which the states are mainly clustered and tracked their movements over time.

In light of the obtained results, policies and programs should be addressed to reduce the two main critical issues. Firstly, there is a need to adopt policies that acknowledge the differentiation between the wealthiest and the poorest countries. In the latter, active involvement by the public sector is crucial to enhance the conditions and attractiveness of these regions. Emphasis should be placed on public investments to build the capabilities of these nations, enabling them to engage in a broader spectrum of activities. Key components include technology promotion policies (covering research, technical assistance, and technology transfer), skill development, training of human resources, land and agrarian reform, and adequate financing.

Secondly, the issue of gender inequality underscores the significant role played by state actions or inactions, highlighting diverse challenges in different contexts for overcoming gender disparities. Each form of state intervention, along with the associated gender stratification it reinforces, operates within a deeply ingrained ideological and cultural framework. Despite solutions cannot be easily transposed from one context to another, as discussed by [14], this does not suggest that progress is unattainable.

Acknowledgements The authors acknowledge the support of MUR, grant no. 2022XRHT8R (CUP: E53D23005950006) - The SMILE project: Statistical Modelling and Inference to Live the Environment, funded by the European Union - Next Generation EU.

References

1. Bartolucci, F., Farcomeni, A., Pennoni, F.: Latent Markov Models for Longitudinal Data. CRC Press, Boca Raton (2012)
2. Baum, L.E., Petrie, T., Soules, G., Weiss, N.: A maximization technique occurring in the statistical analysis of probabilistic functions of Markov chains. Ann. Math. Stat. **41**(1), 164–171 (1970)
3. Browne, R.P., McNicholas, P.D.: Estimating common principal components in high dimensions. Adv. Data Anal. Classif. **8**(2), 217–226 (2014)
4. Bulla, J.: Hidden Markov models with t components. Increased persistence and other aspects. Quant. Finance **11**(3), 459–475 (2011)
5. Celeux, G., Govaert, G.: Gaussian parsimonious clustering models. Pattern Recogn. **28**(5), 781–793 (1995)
6. Churchill, G.A.: Stochastic models for heterogeneous DNA sequences. Bull. Math. Biol. **51**(1), 79–94 (1989)
7. Clark, C.: The Conditions of Economic Progress. Macmillan and Co, London (1940)
8. Cohen, P.N.: The persistence of workplace gender segregation in the US. Sociol. Compass **7**(11), 889–899 (2013)
9. Fisher, A.G.: Production, primary, secondary and tertiary. Econ. Rec. **15**(1), 24–38 (1939)

10. Fourastié, J.: L'économie française dans le monde. Presses universitaires de France, Paris (1945)
11. Hesmondhalgh, D., Baker, S.: Sex, gender and work segregation in the cultural industries. Sociol. Rev. **63**(1), 23–36 (2015)
12. Lagona, F., Maruotti, A., Picone, M.: A non-homogeneous hidden Markov model for the analysis of multi-pollutant exceedances data. In: Dymarski, P. (ed.) Hidden Markov Models, Theory and Applications, pp. 207–222. InTech, Rijeka (2011)
13. Langrock, R., King, R.: Maximum likelihood estimation of mark-recapture-recovery models in the presence of continuous covariates. Ann. Appl. Stat. **7**(3), 1709–1732 (2013)
14. Mandel, H.: Configurations of gender inequality: the consequences of ideology and public policy. Br. J. Sociol. **60**(4), 693–719 (2009)
15. Marjanović, V.: Significance and influence of economic structure on economic development. Econ. Themes **48**(3), 369–382 (2010)
16. Maruotti, A., Punzo, A., Bagnato, L.: Hidden Markov and semi-Markov models with multivariate leptokurtic-normal components for robust modeling of daily returns series. J. Financial Econ. **17**(1), 91–117 (2019)
17. Meng, X.L., Rubin, D.B.: Maximum likelihood estimation via the ECM algorithm: a general framework. Biometrika **80**(2), 267–278 (1993)
18. Punzo, A., Maruotti, A.: Clustering multivariate longitudinal observations: the contaminated Gaussian hidden Markov model. J. Comput. Graph. Stat. **25**(4), 1097–1098 (2016)
19. Punzo, A., Ingrassia, S., Maruotti, A.: Multivariate generalized hidden Markov regression models with random covariates: physical exercise in an elderly population. Stat. Med. **37**(19), 2797–2808 (2018)
20. Schafran, A., McDonald, C., Lopez Morales, E., Akyelken, N., Acuto, M.: Replacing the services sector and three-sector theory: urbanization and control as economic sectors. Reg. Stud. **52**(12), 1708–1719 (2018)
21. Schwarz, G.: Estimating the dimension of a model. Ann. Stat. **6**(2), 461–464 (1978)
22. Tomarchio, S.D., Punzo, A., Maruotti, A.: FourWayHMM: Parsimonious Hidden Markov Models for Four-Way Data. R package version 1.0.0. https://CRAN.R-project.org/package=FourWayHMM (2021)
23. Tomarchio, S.D., Punzo, A., Maruotti, A.: Parsimonious hidden Markov models for matrix-variate longitudinal data. Stat. Comput. **32**(3), 1–18 (2022)
24. Zucchini, W., Guttorp, P.: A hidden Markov model for space-time precipitation. Water Resour. Res. **27**(8), 1917–1923 (1991)
25. Zucchini, W., MacDonald, I.L., Langrock, R.: Hidden Markov Models for Time Series: An Introduction Using R. CRC Press, Boca Raton (2017)

Integrating Structuralism and Diffusionism to Explain the New Italian Emigration

Francesca Bitonti

Abstract The present study presents a novel approach to understanding recent Italian emigration by integrating diffusionist and structuralist theories. Departing from traditional adaptive explanations of fertility transitions, the work delves into the innovative fusion of diffusionism and structuralism, shedding light on the dynamics of contemporary migratory movements. Drawing from the foundation laid by diffusionism in the realm of fertility studies, we extend its application to analyse the complexities of Italian emigration. Proposing an extended version of the Bass model, traditionally employed in studying the diffusion of new products and behaviours, we uncover the underlying dynamics driving the recent wave of Italian emigration. Our findings offer a fresh perspective on international migration processes, emphasizing the synergy between diffusionist and structuralist interpretations, marking a significant advancement in the study of migratory phenomena.

Keywords Diffusionism · International migration · Bass diffusion model · New Italian emigration

1 Introduction

The diffusionist theory finds its roots in the field of social sciences to explain the spread of new ideas and practices among the members of a population [26, 29]. The ground idea of diffusionism is that the interplay of social influence mechanisms, as interpersonal information exchange, social norms, and emulative processes, shape individual conduct. The diffusionist approach has found wide application also in demographic studies. From a demographic standpoint, an innovation may be intended as any idea or practice having implications on the population dynamics, namely fertility, mortality, and migrations [36]. In this sense, examples are contraceptive adoption, childbearing postponements, smoking behavior, novel migration

F. Bitonti (✉)
Department of Economics and Business, University of Catania, Catania, Italy
e-mail: francesca.bitonti@unict.it

M. Bini et al. (eds.), *Advanced Methods in Statistics, Data Science and Related Applications*, Springer Proceedings in Mathematics & Statistics 467,
https://doi.org/10.1007/978-3-031-65699-6_12

practices, and the like. Diffusionism gained momentum with the necessity to give a plausible explanation to the decline in marital fertility that occurred across Europe during the past century. This approach challenged the classical demographic transition theory, which argued that modernization and structural changes in society were the main driving forces for fertility variations. Roughly speaking, the rising cost and declining economic value of children is the key factor leading to fertility change [5, 7]. The diffusionist considerations instead argued that new ideas and norms about fertility have diffused through social interaction modifying the macro-level fertility dynamics [10]. In the 80 s, the Princeton European Fertility Project and the World Fertility Survey were the two seminal works suggesting that the fertility decline could have been influenced by the diffusion of modern contraceptive practices, new values, and new ideas of family among people [11, 12, 34]. After this introduction in the demographic debate, the diffusionist vision has been the theoretical background for several studies of fertility choices and family planning, and for the promotion of public health campaigns aimed at contrasting the spread of diseases and health-damaging behaviors [1–3, 6, 20, 24, 25, 32, 35]. Recently, modernization and diffusionist arguments have no longer been considered distinct or in opposition. Rather, they are conceptualized as strictly embedded explanations to demographic changes [27]. In adherence to this strand of thought, the objective of this work is to analyze and convey meaningful insights of international migration processes in terms of both diffusionist and structuralist approach. In particular, the analysis focuses on a novel emigration flow involving Italy in recent times. The work is organized as follows: the next section illustrates the so-called new Italian emigration; the third section introduces the methodology implemented and describes the data analyzed; the fourth one discusses the main results of our applications, while the last reports the conclusions of the work.

2 The New Italian Emigration

In recent years, Italian emigration had considerably grown compared to the 80 and 90 s, when the phenomenon was ongoing but at a smaller scale. The current emigration presents specific peculiarities, configuring a situation different from the past, which in literature is known as "the new Italian emigration".

Among the novelties, an increased share of women and well-educated individuals leaving the country has been recorded [13, 15, 30]. Furthermore, although the Southern regions continue, as in the past, to contribute to the outgoing movements, recently, the more advanced northern Italy has become the main outflow area [8, 30]. Another relevant feature is that the destinations preferred by Italian migrants mainly converge to the EU and EFTA countries [8]. Finally, the new Italian emigration appears to be paired with a change in attitude towards international migration. The globalization of the world and the availability of new communication technologies could have overcome several of the barriers hindering movements in the past, allowing Italians, especially the youngsters, to approach international mobility in a novel and more confident way [33].

Overall, the interaction among the global economic downturn, the EU integration process [14, 21, 23], and the change in attitude towards outward mobility appear to have shaped a new migration behavior that has diffused across Italy. The novel aspects characterizing the Italian emigration configure an innovative migrant behavior that diffuses among the population through particular communication channels and according to established social dynamics, but also in accordance with the economic cycle of origin and destination countries.

3 Methods and Data

The Bass model [4] is extensively employed in market research studies to analyze the diffusion process of new ideas, products, and behaviors (called "innovations") in a given population. According to the model, the pace of diffusion is driven by two forms of communication: the *external* one, such as mass media advertisement, news broadcast, or awareness campaigns, and the *internal* one, including social influence, word of mouth, and imitation. These two drivers of diffusion shape two categories of adopters: the innovators, mainly influenced by the external source, and the imitators, who adopt in response to their interaction with prior adopters. The Bass model consists of a simple first-order differential equation

$$Y'(t) = \left(p + \frac{q}{m}Y(t)\right)(m - Y(t)) = p(m - Y(t)) + q\frac{Y(t)}{m}(m - Y(t)) \quad (1)$$

where the variation of adoption $Y'(t)$ over time is proportional to the residual susceptible population $(m - Y(t))$, with m being the constant overall susceptible population, and $Y(t)$ the cumulative adoptions at time t. Rearranging the right-hand side Eq. (1) it is possible to notice that the instantaneous adoptions $Y'(t)$ is the result of the sum between two components, the external one governed by parameter p and the internal one modulated by q. Parameter p is the so-called coefficient of innovation, representing the effect of the external influence to adopt. Parameter q is the coefficient of imitation and reflects the inter-personal influence individuals can exert on each other. The basic Bass model and its extensions have found empirical applications not only in marketing studies but also in the demographic field: e.g. to analyze the diffusion of oral contraception [28], to clarify the dynamics in vaccination propensity and address public health policy [19, 22], and to study the diffusion of disease-related information during an epidemic outbreak [18]. In the present work, we propose an extended version of the traditional Bass model to capture both structural and diffusionist explanations to the new Italian emigration:

$$\frac{dY_{it}}{dt} = p_{it}(m - Y_{it}) + \frac{q_{it}}{m}Y_{it}(m - Y_{it}) \quad (2)$$

where the coefficients p and q depend on the structural characteristics of the destination countries the emigration flow is directed to:

$$p_{it} = \alpha_0 + \alpha_1 U_{it} + \gamma_2 D2_i + \gamma_3 D3_i + \cdots + \gamma_n Dn_i + \epsilon_{it} \tag{3}$$

$$q_{it} = \beta_0 + \beta_1 U_{it} + \delta_2 D2_i + \delta_3 D3_i + \cdots + \delta_n Dn_i + \epsilon_{it} \tag{4}$$

$i = 1,\ldots,n$ are the destination countries at each time interval t, U_{it} is the unemployment rate for destination country i at time t, $D2_i + \cdots + Dn_i$ are $n-1$ dummy variables identifying the $i = 2,\ldots,n$ destination countries respectively, lastly the erratic component is defined as $\epsilon_{it} \sim N\left(0, \sigma^2\right)$. The extended version of the Bass model has been fitted using an ad-hoc and original numerical optimization algorithm minimizing the mean squared errors (MSE) in the R software, which automatically estimated all the model parameters. The optimization procedure employs various search routines to estimate the parameters given a set of selected initial values. In the present application, 1500 different combinations of starting values for the model parameters have been exploited to estimate the model. The choice of the starting values was based on the results yielded by a meta-analysis of 15 different studies estimating 213 Bass models [31]. The main advantage of the proposed model is to parsimoniously provide the time-varying parameters p and q for each destination country considered. With the same interpretative rather than forecasting objective found in literature [9, 16, 17], this work intends to offer a combined diffusionist and structuralist perspective of the new Italian emigration process. The basic reasoning is that the Italian crisis, combined with the spread of mass media information about the more dynamic labor market and the better living conditions in some EU countries, might have acted as external drivers of influence. On the other end, the information conveyed by actual emigrants about the possibility to be better-off could have triggered word-of-mouth effects and emulative behaviors.

Annual counts of Italian citizens who emigrated from 2002 to 2018 towards the main EU destination countries (Austria, Belgium, France, Germany, Spain, and United Kingdom—pre-Brexit) were retrieved from the Italian National Institute of Statistics (Istat) repository (available at: http://dati.istat.it/) Data on unemployment rates come from the World Bank online database (available at: https://databank.worldbank.org/).

4 Results

The results of the extended Bass model fit are displayed in Fig. 1. During the period 2002–2018, the outgoing flows (black dots) have increased for each destination country, with Germany, France, and United Kingdom accounting for the highest portion of movements. The global model (black lines) shows, in general, a good fit to the real data, losing goodness of fit when the data present discontinuities.

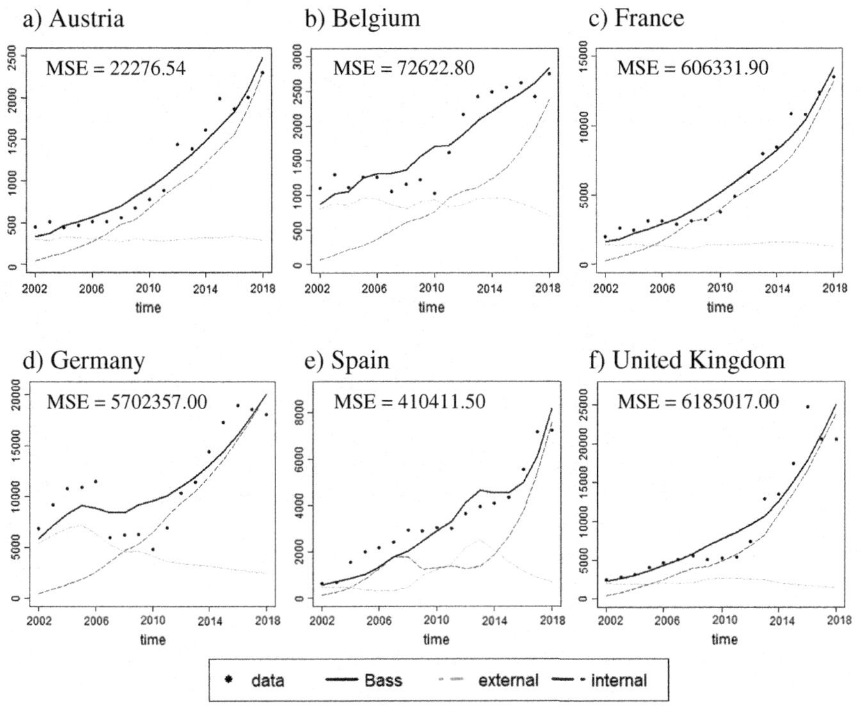

Fig. 1 Extended Bass model fitting to the Italian annual emigrations toward the main EU destination countries. Time period: 2002–2018 (note: different y-axis scales)

The external component (light grey dashed lines) accounts for a little proportion of emigrations, while the internal one (dark grey dashed lines) is the main driver of the outgoing trends. Progresses in the EU integration and the 2007 financial crisis have accelerated the spread of novel migratory behaviors in Italy, characterized by an increased attitude towards international mobility and imitation. We proposed an interpretation of the new trends in Italian out-migration flows combining diffusionist and structuralist reasoning in an ad-hoc methodological framework. The proposal is that of a "toy model", namely a simplistic model aimed to provide interpretation to complex real-world phenomena; a typical example is the Lotka–Volterra model in population ecology.

5 Conclusions

The integration of diffusionism and structuralism, as demonstrated through our extended Bass model, provides profound insights into the complexities of the new Italian emigration phenomenon. One of the key findings of our study is the paradigm shift in Italian emigration dynamics from the consideration of purely structural pushing factors to sociological ones. In particular, our model illustrates that internal factors, notably social influence and word-of-mouth effects, might play a pivotal role in driving emigration trends. This internal component emerged as the dominant force behind the surge in emigration. Besides this, our analysis uncovered the influence of broader socio-economic factors, such as the deepening integration within the European Union and the impact of the 2007 financial crisis. These events acted as catalysts, accelerating the adoption of new migratory behaviors among Italians. The intricate interplay of these external influences with internal dynamics created a synergy that significantly shaped the emigration landscape.

By integrating diffusionist and structuralist approaches, our study provides a nuanced understanding of the new Italian emigration. The diffusionist perspective, emphasizing the role of interpersonal influence and social norms, intertwines seamlessly with the structuralist lens, which considers economic indicators and destination-specific variables. This synergy, embodied in the extended Bass model, not only captures the complexities of the emigration phenomenon but also offers a theoretical framework for comprehending similar migratory patterns globally.

As we look to the future, further refinements of the proposed model, incorporating additional variables and refining the parameters, hold promise for enhancing the accuracy of predictions and understanding the evolving nature of migratory patterns. Additionally, exploring the application of our integrated approach in diverse demographic contexts and investigating the impact of digital communication channels on migration dynamics could pave the way for more comprehensive analyses in the field.

In conclusion, our study marks an advancement in migration research by seamlessly blending diffusionist and structuralist theories. By unravelling the intricate threads that weave the fabric of contemporary Italian emigration, we offer a pioneering methodological approach that holds transformative potential for the broader study of international migration.

References

1. Abraham, A.J., Roman, P.M.: Early adoption of injectable naltrexone for alcohol-use disorders: findings in the private-treatment sector. J. Stud. Alcohol Drugs **71**(3), 460–466 (2010)
2. Alvergne, A., Gibson, M.A., Gurmu, E., Mace, R.: Social transmission and the spread of modern contraception in rural Ethiopia. PLoS ONE **6**(7), 22515 (2011)
3. Backer, T.E., Rogers, E.M.: Diffusion of innovations theory and work-site aids programs. J. Health Commun. **3**(1), 17–28 (1998)
4. Bass, F.M.: A new product growth for model consumer durables. Manag. Sci. **15**(5), 215–227 (1969)
5. Becker, G.S.: A Treatise on the Family. Harvard University Press (1993)
6. Bengtsson, T., Dribe, M.: The historical fertility transition at the micro level: Southern Sweden 1815–1939. Demogr. Res. **30**(17), 493–534 (2014)
7. Bongaarts, J., Watkins, S.C.: Social interactions and contemporary fertility transitions. Popul. Dev. Rev. 639–682 (1996)

8. Bonifazi, C.: Da dove si parte, dove si va. Il Mulino **6**, 49–56 (2018)
9. Bunea, A.M., Della Posta, P., Guidolin, M., Manfredi, P.: What do adoption patterns of solar panels observed so far tell about governments' incentive? Insights from diffusion models. Technol. Forecast. Soc. Chang. 120240 (2020)
10. Carlsson, G.: The decline of fertility: innovation or adjustment process. Pop. Stud. **20**(2), 149–174 (1966)
11. Cleland, J.: Potatoes and pills: an overview of innovation diffusion contributions to explanations of fertility decline. In: Casterline, J.B. (ed.), Diffusion processes and fertility transition: selected perspectives, pp. 39–65. National Academy Press, Washington D.C (US) (2001)
12. Coale, A.J., Watkins, S.C. (eds.).: The Decline of Fertility in Europe: the Revised Proceedings of a Conference on the Princeton European Fertility Project. Princeton University Press (1986)
13. Colucci, M.: Formazione e reclutamento degli italiani che emigrano. Il Mulino **6**, 33–40 (2018)
14. De Rose, A., Strozza, S. (eds.).: Rapporto sulla popolazione. L'Italia nella crisi economica. Il Mulino (2015)
15. Fondazione Migrantes: Rapporto Italiani nel Mondo 2020 (2020)
16. Furlan, C., Guidolin, M., Guseo, R.: Has the Fukushima accident influenced short-term consumption in the evolution of nuclear energy? an analysis of the world and seven leading countries. Technol. Forecast. Soc. Chang. **107**, 37–49 (2016)
17. Guidolin, M., Mortarino, C.: Cross-country diffusion of photovoltaic systems: modelling choices and forecasts for national adoption patterns. Technol. Forecast. Soc. Chang. **77**(2), 279–296 (2010)
18. Gündüç, S.: A study on the effects of diffusion of information on epidemic spread. Int. J. Model. Simul. Sci. Comp. **10**(3), 1950015 (2019)
19. Kahana, D., Yamin, D.: Accounting for the spread of vaccination behavior to optimize influenza vaccination programs. PLoS ONE **16**(6), e0252510 (2021)
20. Lesthaeghe, R.: Long-term spatial continuities in demographic innovations: insights from the Belgian Example, 1846–2000. Population Studies Center Research Report, 10–695 (2010)
21. Livi Bacci, M.: Italiani verso la Gran Bretagna. Neodemos (2014)
22. Onofrio, A., Manfredi, P., Poletti, P.: The interplay of public intervention and private choices in determining the outcome of vaccination programmes. PLoS ONE **7**(10), 45653 (2012)
23. Pugliese, E.: Tutto il mondo è paese: la nuova emigrazione italiana. Il Mulino **6**, 8–22 (2018)
24. Ramseyer Winter, V.: Diffusion of innovations theory: a unifying framework for HIV peer education. Am. J. Sex. Educ. **8**(4), 228–245 (2013)
25. Rogers, E.M.: Diffusion of drug abuse prevention programs: spontaneous diffusion, agenda setting, and reinvention. NIDA Res. Mg. **155**, 90–105 (1995)
26. Rogers, E.M.: Diffusion of Innovations. The Free Press, New York (1962)
27. Rosero-Bixby, L., Casterline, J.B.: Modelling diffusion effects in fertility transition. Popul. Stud. **47**(1), 147–167 (1993). https://doi.org/10.1080/0032472031000146786
28. Sharif, M.N., Ramanathan, K.: Binomial innovation diffusion models with dynamic potential adopter population. Technol. Forecast. Soc. Chang. **20**(1), 63–87 (1981)
29. Strang, D., Meyer, J.W.: Institutional conditions for diffusion. Theor. Soc. **22**(4), 487–511 (1993)
30. Strozza, S., Tucci, E.: I nuovi caratteri dell'emigrazione italiana. Il Mulino **6**, 41–48 (2018)
31. Sultan, F., Farley, J.U., Lehman, D.R.: A meta-analysis of diffusion models. J. Mark. Res. **27**, 70–77 (1990)
32. Svenkerud, P.J., Singhal, A., Papa, M.J.: Diffusion of innovations theory and effective targeting of HIV/AIDS programmes in Thailand. Asian J. Commun. **8**(1), 1–30 (1998)
33. Tirabassi, M.: Migranti da sempre. Il Mulino **6**, 24–32 (2018)
34. Tolnay, S.E.: The spatial diffusion of fertility: a cross-sectional analysis of counties in the American South, 1940. Am. Sociol. Rev. **60**(2), 299–308 (1995)
35. Vitali, A., Aassve, A., Lappegård, T.: Diffusion of childbearing within cohabitation. Demogr. **52**, 355–377 (2015)
36. Vitali, A., Billari, F.C.: Changing determinants of low fertility and diffusion: a spatial analysis for Italy. Popul. Space Place **23**(2) (2017)

Scaling, Correlation and Decomposition Approaches in Social Applications

Scaling UX-AI Products: CFA and PLS-SEM Comparison

Cataldo Rosanna, Grassia Maria Gabriella, and Zavarrone Emma

Abstract Since their appearance, artificial intelligence (AI) products have been increasingly integrated into our daily life and today they are becoming an ever more pervasive reality. However, our positive expectations often exceed the current potential performance of AI products, causing disappointment and frustration. Starting from this consideration and taking smart speakers as an example, this work uses a new scale validation to evaluate the user experience with AI infused products. The work compares the results of a three-order model from the perspectives of Partial Least Squares-Structural Equation Modeling and the confirmatory approach.

Keywords Artificial intelligence · Scale validation · CFA · PLS-SEM

1 Introduction

For years, artificial intelligence (AI) has been quietly infiltrating systems and devices that we use daily. Since the mid-2010s, we have witnessed an exponential increase in the presence of AI in people's daily lives, and new products are being developed, such as smart thermostats, which adapt to user habits, and systems that enhance everyday tools, such as Netflix or Amazon recommendations [50]. This is expected to reach 640 million devices installed by 2024 [9]. The advent of these successful new products requires measuring user satisfaction with their use. Satisfaction is determined not only by the functionality and completeness of the product, but also by the user's

C. Rosanna · G. M. Gabriella
Social Sciences Department, University of Naples Federico II, Naples, Italy
e-mail: rosanna.cataldo2@unina.it

G. M. Gabriella
e-mail: mariagabriella.grassia@unina.it

Z. Emma (✉)
Humanities Department, Iulm University, Milan, Italy
e-mail: emma.zavarrone@iulm.it

M. Bini et al. (eds.), *Advanced Methods in Statistics, Data Science and Related Applications*, Springer Proceedings in Mathematics & Statistics 467,
https://doi.org/10.1007/978-3-031-65699-6_4

overall experience with it [55]. Proposing measurement scales to frame the user experience (UX) of a new generation of products infused with AI (herein after UX-PIA) has been the aim of many researchers [25, 31, 37, 55]. An exploratory literature review [48], based on 129 papers between 2000 and 2020, resulting from the query of "UX evaluation" and "UX assessment" keywords, has highlighted the overlapping ability of measurement scales to capture the UX generalist dimensions without covering aspects linked to human-computer interaction (HCI) and AI. These results are consistent with previous studies, which already tackled the issue [4, 37, 43, 52], identified the main trends and paved the way for new UX research. They highlighted, for instance, that AI should cope with ever-evolving technologies and non-human intelligence [37]. In general, all these studies focused only on usability, pleasantness of use, emotion/affect, and enjoyment. Furthermore, the literature revealed no agreement on terminology, and sometimes the same words are used without careful attention to the semantic nuances or interchangeability of terms such as dimensions, descriptors, and items. Our research is set in this context. The purpose of this work is the validation of a scale, defined as AIXE ® (AI user eXperience Evaluation), that takes into consideration, in addition to the latent dimensions already known in the literature, other aspects *linked to the subjective evaluation.* Several studies demonstrated that UX is a dynamic concept, influenced by contextual aspects, such as place, social and temporal aspects of use, but also by the subjective context; therefore, UX is linked to the specific emotional states of users [4, 33, 55]. AIXE® was developed in the research project (Meet-AI), funded by the Design Department of Politecnico di Milano. It aims to build a new evaluation method that specifically addresses the UX-PIA, by comprehending and highlighting the peculiarities they have in relation to other interactive products of common use [48]. The Meet-AI project was expected to output a dashboard where evaluations of UX-PIA can provide insights into the strategies to be implemented in order to increase market share.

The paper aims to validate the AIXE® scale under the perspective of the classical common factor model and through a Partial Least Squares—Structural Equation Modeling (PLS-SEM) *ceteris paribus* with a reflective approach, in order to compare the final results in terms of loading coefficients and goodness of fit (GoF). The implied output is strictly connected to the construct validity of AIXE® scale: the change of the method used in the constructs should be invariant. The idea is to understand if the two different methods lead to the same results. Starting from this scenario, the research question concerns the ability to validate the new AIXE® scale from the perspectives of both PLS-SEM and Confirmatory Factor Approach (CFA). In particular, our paper, therefore, intends to answer the following research question (RQ):

- RQ: Can AIXE® be validated with the CFA and PLS-SEM approaches? Do these two approaches give the same results?

In detail, the analysis is based on smart speakers, such as Amazon Echo or Google Home, which have become popular additions to homes worldwide in the past two years. These devices offer people a new way to interact with computer systems in their homes, via voice, without the need to touch a device [5]. The following

Sect. 2 is devoted to introducing the theoretical framework; Sect. 3 presents the data and compares the results from the perspectives of both PLS-SEM and confirmatory approach. Finally, some remarks are presented in Sect. 4.

2 Theoretical Framework

The evaluation process of an AI-infused product, such as smart speakers, follows any well-known model of service evaluation where we have to manage already pre-configured experiences or expectations and observed perceptions. Since the 1980s, the literature has devoted considerable space to this issue, which is now widely practiced to evaluate services using structural methodologies in which expectations and past experiences (latent variables; LVs) condition the choice of response (mode of observable variables). Our study fits into this framework: we propose the validation of an innovative measurement scale in two different ways: the confirmatory analysis and the PLS-SEM approach. The construction of a measurement scale is one of the applications that allow the contribution of LVs to be estimated. In this section, the methodological aspects related to the choice of methodology and the new proposed validation of the scale will be addressed. The section is organized as follows:

2.1 a brief overview of the possible methodologies available;
2.2 the description of the validation process of the scale;
2.3 the proposal of validation using PLS-SEM in order to have a scale that offers the characteristics of both approaches.

2.1 Methodologies

Structural equation modeling (SEM) is used in several disciplines to test empirical theories, making it a desirable method [27, 47]. There are two types of SEMs: the Covariance-based Structural Equation Modeling (CB-SEM) [30] and PLS-SEM [42, 51]. Fornell and Bookstein [17], Chin and Newsted [10], Hair et al. [23], Hair et al. [22], Hair et al. [21] and Jöreskog and Wold [28] provide recommendations for when to use CB-SEM or PLS-SEM. Concerning the CB-SEM, Jöreskog [29] suggested that the CFA had been more widely used in social sciences during the previous 50 years[1]. Synthetically, according to Hair et al. [19], CB-SEM and PLS-SEM are complementary rather than competitive. These methods cannot be considered as substitutes for each other, or in Petter's words [36] "Neither of the techniques is generally superior to the other, and neither of them is appropriate for all situations". Therefore, the choice of method is closely related to the research objectives to be achieved. According to Jöreskog and Wold [28]: "the primary purpose of the CB-SEM approach is to study the structure of the observables. The primary purpose of

[1] For a detail discussion on these methods, see Sarstedt et al. [46], Hair et al. [21], Jhantasana [27].

the PLS-SEM is to predict the indicators by means of the components expansion". In line with this notion, Hair et al. [23] recommend:

1. Covariance based Structural Equation Modeling (CB-SEM):
   - The goal is theory testing, theory confirmation, or the comparison of alternative theories;
   - Error terms require additional specification, such as the covariation;
   - The structural model has non-recursive relationships;
   - The research requires a global goodness-of-fit criterion.
2. Partial Least Squares Structural Equation Modeling (PLS-SEM):
   - The goal is prediction and theory development;
   - Formatively measured constructs are part of the structural model;
   - The structural model is complex;
   - The sample size is small and/or the data are non-normally distributed;
   - The plan is to use latent variable scores in subsequent analyses.

Once the research objective has been defined, other important factors must be considered, such as the type of data available, the minimum sample size and the complexity of the model. Model complexity has been characterized almost entirely by two aspects:

1. Type of relationship between the LV and MVs, in the measurement model, and between the LVs in the structural model;
2. Whether the higher order model is/can be nested.

About the first point, generally, in the measurement model, a LV can be related to its single indicators or MVs by a reflective or formative relationship. A block is defined as reflective if the LV is assumed to be a common factor that reflects itself in the MVs, while in the formative case, each manifest variable or each sub-block of MVs represents different dimensions of the underlying concept. The structural model specifies the relationships between the LVs; if it is supposed to depend on other LVs an LV is called endogenous, otherwise, it is exogenous.

With regards to the second point, nowadays, many constructs refer to a multidimensional reality, which cannot be synthesized with a single block. In this circumstance, a hierarchical model has been introduced. They are considered as explicit representations of multidimensional constructs, which exist at a higher level of abstraction and are related to other constructs at a different level of abstraction, completely mediating their influence from or on their underlying dimensions [54]. Typically, higher-order models are characterized by the number of levels in the model (often restricted to second-order models) [41] and the different relationships between the Higher Order Constructs (HOC) and Lower Order Constructs(LOC) [11, 15]. The higher-order construct contains a certain number of underlying sub-constructs or components.

Each of the hierarchical construct models types is characterized by different relationships between the HOC and the LOC: the reflective relationship and the formative relationship [32]. The researcher can choose among a reflective-reflective, reflective-formative, formative-reflective or formative-formative model [26, 53]. Each type represents the specific relationship between HOC and LOC, as well as the measurement model used in the LOCs [11]. For detailed reviews see Jarvis et al. [26], Edwards [15] and Crocetta et al. [11].

In hierarchical constructs—so-called higher-order constructs—as opposed to unidimensional constructs, a higher-order construct contains several layered structures of constructs and therefore, involves several dimensions. Its application allows researchers to match the level of abstraction of predictor and outcome variables in order to reduce model complexity. As a result, a model's parsimony increases because fewer parameters must be estimated [15]. Although generally, any number of levels of abstraction are conceivable, second-order constructs clearly prevail in empirical studies. The most well-explored type of second-order construct in the SEM literature is the common factor of common factors, also known as the reflective-reflective second-order construct [6, 35]. It is used to operationalize a multidimensional concept that is assumed to cause several concepts, each itself measured by a set of observable variables.

Our paper considers a third-order model starting from what was stated in Dash [13] about the similarity of results between CB-SEM and PLS-SEM. The proposed model is shown in Fig. 1, where we can see that both methods can handle LVs and their relationships.

Specifically, when the measurement models have four or more indicators and when the indicator loadings meet the common standards (≥ 0.70), there is practically no difference between CB-SEM and PLS-SEM in terms of parameter accuracy [38, 46]. Thus, the extensively discussed PLS-SEM bias is of no practical relevance for the vast majority of applications [2].

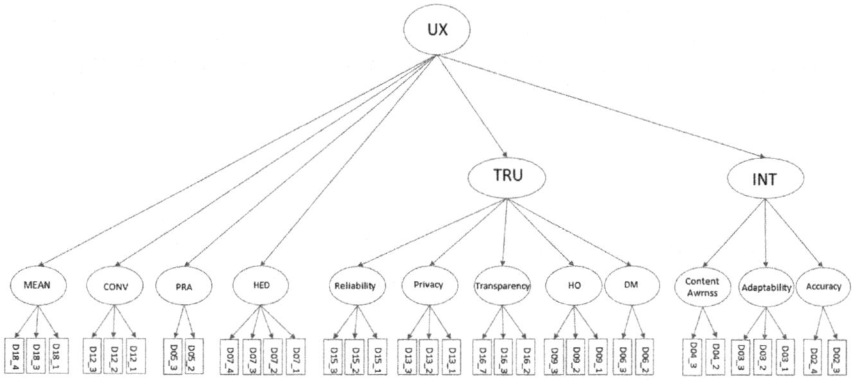

Fig. 1 Reflective hierarchical model

We would like to verify whether this similarity is also valid in the case of scale validation, as in our research. However, the test is even more complex because it takes place in the context of a model with reflective higher dimensions.

2.2 Development Scale

The AIXE® scale was developed by adopting the classical paradigm proposed by Gerbing and Anderson [18] under the reflective approach. Briefly, we report the four steps, of which the first two steps were realized conjointly with Meet-AI, a research group of Politecnico of Milan, while the last two steps have been realized in this work.

1. Literature review;
2. Focus group and item generation;
3. Item selection;
4. Model validation.

From the comprehensive work of Meet-AI group, the researchers highlighted the four most frequently assessed dimensions as the most significant for a holistic method, because these dimensions synthesize the legacy of UX evaluation. These are pragmatic, hedonic, aesthetic, and affective, which were identified by Sciannamè and Zavarrone [48] as second-order constructs. Each second-order construct can be organized in terms of the first-order constructs as described in Table 1. According to Sciannamè and Zavarrone [48], the pragmatic dimension concerns the use of an artifact not just in terms of helpfulness, efficiency, and functionality, but also from more user-friendly characteristics such as ease of use, simplicity, clearness, navigation, learnability, reliability, and convenience; the hedonic dimension concerns enjoyability and excitement, creativity, inventiveness, and innovativity; aesthetics can mainly be interpreted as appearance or attractiveness; finally, the affective component is largely influenced by psychology; in fact, valence and arousal are by far the most common descriptors in this category.

These research questions are linked to the survey of Meet-AI focus group of Milan, which introduced other overarching qualities, alongside the previous aspects that are mainly evaluated as "important". The innovative dimensions not explored in the literature identified by the focus group step are:

Table 1 Latent dimensions of UX evaluation and their descriptors

Second order constructs	First order constructs
Pragmatic	Helpfulness, Efficiency, and Functionality, easiness, simplicity, clearness, navigation, learnability, reliability, and convenience
Hedonic	Enjoyability, excitement, creativity, inventiveness, innovativity
Aesthetic	Appearance, attractiveness
Affective	Valence and arousal

Note This table represents a synthesis of Sciannamè and Zavarrone's theory [48]

- **Trustworthiness**: concerns accuracy, data management, data protection, reliability, and transparency;
- **Conversational** dimension: is mostly relatable to accuracy, context awareness, understanding, feedback quality, fluidity and naturalness;
- **Intelligence**: presents some traits reminescent of human capabilities (e.g., learning, understanding needs, companionship), and others strictly linked to the machine dimension (e.g. data elaboration, connectivity);
- **Meaningfulness**: mostly appeals to the human-computer/product relationship.

Starting from the descriptors listed above and the relative dimensions, and starting from a work of Zavarrone and Cataldo [56], the proposed model suggests the configuration of an endogenous UX third-order dimension linked through the second and first order constructs according to Fig. 1 and Table 2. Figure 1 shows an hierarchical structure, in which the a construct involves more than one dimension. In detail, herein, UX dimension is conceived as a third-order construct affecting second-order dimensions, which in turn shape first-order LVs underlying specific aspects of the second-order dimensions. This hierarchical structure allows us to analyze the relationships among the first-level LVs, defined as descriptors, and the latent dimensions representing the second factorial level, and the relationships among latent dimensions and the UX dimension, which represents the third-order construct. The conceptual model derives from the implementation of the hypotheses before they are formulated, and configured in the path diagram of Fig. 1. The path diagram highlights two parts:

1. Measurement model, which refers to all the items (and their relationships) associated with the LVs, indicated by squares; and
2. Structural model, which refers to the LVs, indicated by circles; and their relationships, indicated by segments.

The study focuses on a reflective-reflective structural and measurement model (i.e., a model resulting from the combination of reflective lower-order and reflective higher-order constructs). We considered the higher-order construct as reflective, because the general concept is manifested by several specific dimensions, themselves latent (unobserved). The same concept applies to first-order constructs, which manifest themselves through specific items (questionnaire's questions- MVs).

The process of identifying the underlying characteristics that influence the phenomenon under investigation is achieved by developing a comprehensive set of items. Typically, battery assessments encompass a range of items that effectively capture the key dimensions of latent variables. In order to achieve this outcome, more iterations of reducing the quantity are executed. Steps 2 and 3 exemplify this method. During step 2, we saw the emergence of a series of questions through the use of focus groups. These questions were derived from certain latent dimensions that were extrapolated from existing literature. This comparative analysis encompassed a range of disciplines, incorporating specialists from the field of AI-infused product development, alongside firm managers and statisticians. During this phase, a total of 129 questions were developed and subsequently assigned a standard scale consisting of four degrees. In order to conduct a preliminary analysis, we have opted to allocate these items on a conventional 4-point Likert scale. The concept that a traditional

Table 2 List of the descriptors related to their dimensions in the hierarchical model

Third order construct	Second order constructs	First order constructs	Items
	Intelligence (INT)	Accuracy	D02-3 Accuracy in anticipating needs D02-4 Accuracy in matching needs
		Adaptability	D03-1 Adaptability to habits D03-2 Adaptability to needs D03-3 Adaptability over time
		Context awareness	D04-2 The context affects its behaviour D04-3 The context affects its performance
UX	Trustworthiness (TRU)	HO	D09-1 Control of operations of the system D09-2 Control how the system behaves D09-3 Control how the system performs its tasks
		Data management	D06-2 Manage the data collected by the system D06-3 Manage the data that the system uses
		Privacy	D13-1 Protection of privacy D13-2 Protection of data D13-3 Protection of private information
		Reliability	D15-1 Reliance of system's behaviour D15-2 Reliance of system's responses D15-3 Reliance of system's over time
		Transparency	D16-2 Trasparency about system's processing D16-3 Trasparency in showing what system's decision depend on D16-7 Trasparency in the way system adapts to habits
		Pragmatic (PRA)	D05-2 Customize the system's behaviour D05-3 Customize the system's to habits
		Hedonic (HED)	D07-1 System's empathy with user D07-2 Systems behaviour according to the relationship build with user D07-3 System's empathy in anticipating user needs D07-4 System's empathy towards user needs
		Conversational dimension (CON)	D12-1 System's voice pleasant D12-2 System's voice natural D12-3 System's voice likable
		Meaningfulness (MEAN)	D18-2 System adds meaning to user life D18-3 System adds something to user life D18-4 System has value for user

score does not impede the progress of a scale from a reflective perspective has been highlighted in recent scholarly investigations. During the third step, a set of questions was presented to experts who were not part of the focus group. The answers provided by these experts were then incorporated into the analysis of the pretest. As a result, the initial pool of items was reduced to a total of 65 items (Table 2). In the fourth and final step, after the application of the Exploratory Factor Analysis (EFA) and CFA, the pool of items was reduced to 33.

2.3 CFA and PLS-SEM Statistical Validation

The proposed model was validated using both CFA and PLS-SEM, through the comparison of the composite reliability[2] of the two approaches, according to the step previously described. In the last step, PLS-SEM has been performed[3] to validate and test the assumptions of the proposed conceptual model [12]. The PLS-SEM approach is considered to be very appropriate when the study aims to develop and extend a theory [2]. The SEM approach allows the simultaneous testing of different factors and their relationships. In particular, SEM is useful for the assessment of the relationships among various latent constructs while allowing for the reduction of error in the model [21]. In order to assess the fit of the hypothesized underlying factor structure, several fit indices have been examined [24, 34, 39]:

- The internal consistency of the dimensions was assessed by Cronbach's alpha;
- The goodness of fit for CFA was evaluated by applying the comparative fit index (CFI), the Root Mean Squared Error of Approximation (RMSEA), and the Standardized Root Mean Square Residual (SRMR). For PLS-SEM GoF was calculated;
- The convergent validity was assessed through three criteria: the composite reliability (CR) value, which should be greater than 0.7, the average validity extracted (AVE), which should exceed 0.5 [19], and the factor loadings of all items, which should be greater than 0.7 [16].

When determining the model's goodness of fit, however, according to Brown [7], it is important to consider both the model's parameter estimates and the fit indices outlined above [49].

3 Data and Statistical Validation

3.1 Data

The scale construction was realized with two different samples: the first sample was used to identify the common factor structure, and the second sample to validate the

[2] To evaluate the internal consistency of each construct, composite reliability and Cronbach's α indices are used, which represent the most commonly used quality criterion for the measurement model.

[3] The analysis was performed in R using the package "plspm" [45], including non-metric quantification method [44] for the ordinal MVs on which the AIXE ® questionnaire is based.

Table 3 Socio-demographic analysis

	First Wave (W1)
Partecipants	622
N. of items	75
Gender	57% Female; 43% Male
Country	54%UK; 46% USA
Age	28% 18–30; 27% 31–40; 17% 41–50; 14% 51–60; 14% over 60
Device	62% Amazon Echo; 17% Google Nest; 8% Apple; 13% Other
Use Device	49% Frequently; 35% Sometimes; 12% Seldom; 4% Never
	Second Wave (W2)
Partecipants	705
N. of items	36
Gender	58% Male; 42% Female
Country	50% UK; 50% USA
Age	27% 18–30; 23% 31–40; 18% 41–50; 14% 51–60; 18% over 60
Device	64% Amazon Echo; 17% Google Nest; 13% Apple; 6% Other
Use Device	48% Frequently; 36% Sometimes; 12% Seldom; 4% Never

previous common factor structure. Both waves of data collection used stratified random sampling without replacement from a young American and English population (aged 18 or over) recruited via a worldwide market research company. The strata were gender and country.

In the first wave, 705 people were initially selected to participate in the study. Thirty-three interviewees refused to participate in the research and 672 people agreed to complete the survey, 40 of which, however, were eliminated in the pre-treatment phase because they had missing exceeding 5% of the responses. The total number of valid surveys to be analyzed was 622. Missing data were replaced with the closest value.

Socio-demographic variables were measured conjointly with the device type for both samples (Table 3).

Of the participants who frequently used a device, such as Amazon Echo, 58% were male, and 50% were under 40 years old.

3.2 *Results of Statistical Validation*

In order to test the RQ, it is useful to recall the analytical procedure followed during the scale development, because the final results directly depend on it. The scale development was conducted using a two step procedure, as indicated in the last two steps of Sect. 2.2:

- EFA was conducted on the first wave of data collection to reduce a large number of indicators to a more manageable set and to identify the latent dimensions that characterize the scale;
- CFA was conducted to test the reliability and validity of the constructs.

Usually, EFA is used to explore an item set associated with a construct and subsequently refine it. In our work, EFA was conducted on W1. The ordinal nature of the collected scores suggested the use of a polychoric correlation matrix composed of 75 items and the weighted least squares (WLS) method for the estimation. Further items reduction was performed and a set of 36 items out of the 75 submitted and the related 12 dimensions were selected based on higher performances (factor loadings $\geq$0.5). A reduced questionnaire, including only the 36 acceptable items resulting from the EFA, was then submitted to a new sample of 705 people in the second wave (W2). The aim is to analyze the hypothesized factor model that describes how the latent variables are measured by the manifest indicators, and to test it in a confirmatory setting with a CFA. In this work PLS-SEM approach has been applied in order to test the model, too. In detail, the validation stage was characterized by the W2 by applying CFA with a diagonally weighted least squares (WLSMV) estimator and PLS-SEM analysis. The CFA excluded three items, so the item final scale was composed of 12 latent dimensions with 33 items measured on a 4 point Likert-type scale (from “almost never” to “almost always”). The CFA and PLS-SEM goodness of fit index and Cronbach coefficient for each dimension and overall are shown in Table 5. Firstly, Table 4 presents the loadings of the first-order constructs. As can be seen, all the items presented high loadings on their items with individual values above 0.7, for both CFA and PLS-SEM. Even for the CFA approach, the item of accuracy (D02-4) and the item of hedonic dimension (D07-1) had a value of 0.6, lower than the threshold. However, these items were selected anyway because the literature suggests, according to Awang et al. [3] and Afthanorhan et al. [1], that the minimum value of factor loading is 0.60. Moreover, some studies also accepted a value of factor loading of at least 0.50. This practice is partially similar for PLS-SEM but much more convenient, because not many fitness indexes are deployed. Hair et al. [20] suggested that the acceptable item loading is 0.708, but item loading between 0.40 and 0.70 is acceptable if AVE and CR are achieved. The reliability and validity are sensitive to the number of items per construct as well as the value of factor loading.

The results presented in Table 5 show that the model presents a good fit because all measures exceeded their respective thresholds. Regarding internal consistency, all dimensions had high Cronbach coefficients for both approaches (all above 0.8). It is generally held that a Cronbach coefficient of over 0.7 is acceptable [8, 14]. Regarding goodness of fit measures, for CFA the RMSEA was 0.042, whereas the SRMS was 0.033. In general, an SRMR and RMSEA around 0.05 indicate a good fit of the model to the data. The CFI was equal to 0.986 and a conventional threshold for a good fitting model is greater than 0.90. Whereas RMSEA and SRMS are absolute measures, the CFI is a relative measure and assesses the relative improvement of fit of a model

Table 4 CFA and PLS-SEM loadings

Items	CFA	PLS-SEM
Accuracy		
D02-3	0.790	0.916
D02-4	0.612	0.909
Adaptability		
D03-1	0.739	0.914
D03-2	0.832	0.910
D03-3	0.706	0.885
Content-awareness		
D04-2	0.769	0.933
D04-3	0.779	0.927
HO accuracy		
D09-1	0.731	0.887
D09-2	0.808	0.919
D09-3	0.814	0.890
Data management		
D06-2	0.822	0.939
D06-3	0.865	0.941
Privacy		
D13-1	0.865	0.921
D13-2	0.865	0.944
D13-3	0.829	0.929
Reliability		
D15-1	0.745	0.886
D15-2	0.745	0.892
D15-3	0.753	0.884
Transparency		
D16-2	0.745	0.898
D16-3	0.734	0.912
D16-7	0.743	0.898
Pragmatic		
D05-2	0.736	0.917
D05-3	0.810	0.919
Hedonic		
D07-1	0.619	0.870
D07-2	0.753	0.885
D07-3	0.844	0.899
D07-4	0.854	0.894
Conversational dimensions		
D12-1	0.771	0.886
D12-2	0.724	0.891
D12-3	0.714	0.919
Meanfulness		
D18-2	0.779	0.900
D18-3	0.803	0.911
D18-4	0.767	0.878

Table 5 CFA and PLS-SEM goodness of fit and Cronbach coefficients

		N Items	CFA	PLS-SEM
Cronbach coefficients	UX	33	0.967	0.968
	INT	7	0.873	0.879
	PRA	2	0.800	0.807
	HED	4	0.914	0.914
	TRU	14	0.941	0.944
	CONV	3	0.866	0.872
	MEAN	3	0.873	0.877
Goodness measures			RMSEA = 0.042 SRMR = 0.033 CFI = 0.986	GoF = 0.698
Composite reliability			0.99	0.97
AVE			0.89	0.73

compared with the baseline model.[4] For PLS-SEM, the GoF can be used as an index of fit to the model. This index was equal to 0.97, which was greater than the threshold of 0.7. For the two approaches, the CR, used to measure the internal consistency of items, and AVE, used to measure how much the variations of the items scale can be explained by the constructs on average, were computed. In addition, the constructs met the needed requirements for reliability and convergent validity. Overall, CR and AVE had values higher than 0.75 and 0.5, respectively, confirming the validity and reliability of the UX-AI scale.

4 Remarks and Further Developments

It is important to consider an AI-infused product from the prospective of user involvement. Successful AI-infused products require users to be satisfied when they experience them. This is determined not only by the functionality and completeness of the software, but also by the overall UX of the software product. The work was built upon the research and the results developed within the Meet-AI research project, which led to the construction of the AIXE® scale for evaluating UX with AI-infused products. The purpose of this work was the validation of this AIXE® scale, which takes into consideration, in addition to the aspects already known in the literature, other aspects linked principally to subjectivity. The RQ was related to how a measurement scale can be validated and its use extended also for predictive purposes. This RQ was answered by the satisfactory application of the PLS-SEM methodology. The AIXE® scale has been validated from the perspectives of both PLS-SEM and the CFA

[4] In CFA approach, the baseline model is freely estimated model, which serves as a benchmark against which the fit of more restricted models is compared [39].

approaches. The analysis showed that the model presented a good fit because all measures exceeded their respective thresholds. This shows that the subjective perceptions play a relevant role in every moment of people's lives, especially in the interactions with the surrounding world. These subjective perceptions were not explored in the literature but could be related to the innovative dimensions identified by the focus group of the Meet-AI research project: trustworthiness, conversation, intelligence and meaningfulness. Therefore, further analysis in the context dimension is needed to identify more UX aspects to be included in the UX satisfaction scale. From a methodological point of view, the use of PLS-SEM leads the researcher to analyze the relationships between the different constructs and to intervene on specific elements to improve the entire model. Obviously, a generalization of this result, which offers ample opportunities in the subsequent construction of instruments that can help formulate strategies, is possible only if certain conditions are met: a reflective-reflective scheme and high factor loadings. The future objective is to use this analysis to create a UX-dashboard based on the scoring of the CB-SEM and PLS-SEM models. The dashboard would allow a comparison of the composite indicators relevant to the different latent dimensions and the quantification of the value normalized of the UX. With this dashboard, we can track, analyze, and visualize key performance indicators in order to address investment in the smart device sector.

References

1. Afthanorhan, A., Awang, Z., Rashid, N., Foziah, H., Ghazali, P.: Assessing the effects of service quality on customer satisfaction. Manag. Sci. Lett. **9**, 13–24 (2019)
2. Astrachan, C., Patel, V., Wanzenried, G.: A comparative study of CB-SEM and PLS-SEM for theory development in family firm research. J. Fam. Bus. Strat. **5**, 116–128 (2014)
3. Awang, Z., Afthanorhan, A., Asri, M.: Parametric and non parametric approach in structural equation modeling (SEM): the application of bootstrapping. Mod. Appl. Sci. **9**, 58 (2015)
4. Bargas-Avila, J., Hornbæk, K.: Old wine in new bottles or novel challenges: a critical analysis of empirical studies of user experience. In: Proceedings Of The SIGCHI Conference On Human Factors In Computing Systems, pp. 2689–2698 (2011)
5. Bentley, F., Luvogt, C., Silverman, M., Wirasinghe, R., White, B., Lottridge, D.: Understanding the long-term use of smart speaker assistants. Proc. ACM Interact. Mobile Wear. Ubiquitous Technol. **2**, 1–24 (2018)
6. Bollen, K.: Structural Equations with Latent Variables. Wiley (1989)
7. Brown, T.: Confirmatory Factor Analysis for Applied Research. Guilford Publications (2015)
8. Cataldo, R., Grassia, M., Lauro, N., Marino, M.: Developments in Higher-Order PLS-PM for the building of a system of Composite Indicators. Quality & Quantity. **51**, 657–674 (2017)
9. Ceconello, M., Others: AI-infused products so far. An analysis from a design standpoint. DESIGN INTERNATIONAL, pp. 11–28 (2022)
10. Chin, W., Newsted, P.: Structural equation modeling analysis with small samples using partial least squares. Stat. Strat. Small Sample Res. **1**, 307–341 (1999)
11. Crocetta, C., Antonucci, L., Cataldo, R., Galasso, R., Grassia, M., Lauro, C., Marino, M.: Higher-order PLS-PM approach for different types of constructs. Soc. Indic. Res. **154**, 725–754 (2021)
12. Dabbous, A., Barakat, K., Sayegh, M.: Enabling organizational use of artificial intelligence: an employee perspective. J. Asia Bus. Stud. (2021)

13. Dash, G., Paul, J.: CB-SEM vs PLS-SEM methods for research in social sciences and technology forecasting. Technol. Forecasting Soc. Change **173**, 121092 (2021)
14. DeVellis, R., Thorpe, C.: Scale Development: Theory and Applications. Sage Publications (2021)
15. Edwards, J.R.: Multidimensional constructs in organizational behavior research: an integrative analytical framework. Organ. Res. methods, Sage Publications **4**(2), 144–192 (2001)
16. Fornell, C., Larcker, D.: Evaluating structural equation models with unobservable variables and measurement error. J. Mark. Res. **18**, 39–50 (1981)
17. Fornell, C., Bookstein, F.: Two structural equation models: LISREL and PLS applied to consumer exit-voice theory. J. Mark. Res. **19**, 440–452 (1982)
18. Gerbing, D., Anderson, J.: An updated paradigm for scale development incorporating unidimensionality and its assessment. J. Mark. Res. **25**, 186–192 (1988)
19. Hair Jr, J., Hult, G., Ringle, C., Sarstedt, M.: A Primer on Partial Least Squares Structural Equation Modeling (PLS-SEM). Sage Publications (2021)
20. Hair, J.F., Jr., Risher, J.J., Sarstedt, M., Ringle, C.M.: When to use and how to report the results of PLS-SEM. Emerald Publishing Limited, European business review (2019)
21. Hair, J., Gabriel, M., Patel, V.: AMOS covariance-based structural equation modeling (CB-SEM): guidelines on its application as a marketing research tool. Brazil. J. Market. **13** (2014)
22. Hair, J., Sarstedt, M., Ringle, C., Mena, J.: An assessment of the use of partial least squares structural equation modeling in marketing research. J. Acad. Mark. Sci. **40**, 414–433 (2012)
23. Hair, J., Ringle, C., Sarstedt, M.: PLS-SEM: indeed a silver bullet. J. Market. Theory Pract. **19**, 139–152 (2011)
24. Hu, L., Bentler, P.: Cutoff criteria for fit indexes in covariance structure analysis: conventional criteria versus new alternatives. Struct. Equ. Model. **6**, 1–55 (1999)
25. Kremer, S., Schlimm, A., Lindemann, U.: The ExodUX framework: supporting comprehensive user experience design. In: 2017 Portland International Conference on Management of Engineering and Technology (PICMET), pp. 1–10 (2017)
26. Jarvis, C., MacKenzie, S., Podsakoff, P.: A critical review of construct indicators and measurement model misspecification in marketing and consumer research. J. Consum. Res. **30**, 199–218 (2003)
27. Jhantasana, C.: Using Latent Variables for Confirmatory Composite Analysis
28. Joreskog, K.: The ML and PLS techniques for modeling with latent variables: historical and comparative aspects. Systems Under Indirect Observation, Part I, pp. 263–270 (1982)
29. Jöreskog, K.: A general approach to confirmatory maximum likelihood factor analysis. Psychometrika **34**, 183–202 (1969)
30. Jöreskog, K.: Some contributions to maximum likelihood factor analysis. Psychometrika **32**, 443–482 (1967)
31. Lachner, F., Naegelein, P., Kowalski, R., Spann, M., Butz, A.: Quantified UX: Towards a common organizational understanding of user experience. In: Proceedings Of The 9th Nordic Conference On Human-computer Interaction, pp. 1–10 (2016)
32. Lauro, N., Grassia, M., Cataldo, R.: Model based composite indicators: new developments in partial least squares-path modeling for the building of different types of composite indicators. Soc. Indic. Res. **135**, 421–455 (2018)
33. Law, E., Roto, V., Hassenzahl, M., Vermeeren, A., Kort, J.: Understanding, scoping and defining user experience: a survey approach. In: Proceedings of the SIGCHI Conference on Human Factors in Computing Systems, pp. 719–728 (2009)
34. MacCallum, R., Browne, M., Sugawara, H.: Power analysis and determination of sample size for covariance structure modeling. Psychol. Methods **1**, 130 (1996)
35. Mulaik, S., Quartetti, D.: First order or higher order general factor? Struct. Equ. Model. **4**, 193–211 (1997)
36. Petter, S.: haters Gonna hate: PLS and information systems research. ACM SIGMIS Database: The DATABASE Adv. Inf. Syst. **49**, 10–13 (2018)
37. Pettersson, I., Lachner, F., Frison, A., Riener, A., Butz, A.: A Bermuda triangle? A Review of method application and triangulation in user experience evaluation. In: Proceedings Of The 2018 CHI Conference On Human Factors In Computing Systems, pp. 1–16 (2018)

38. Reinartz, W., Haenlein, M., Henseler, J.: An empirical comparison of the efficacy of covariance-based and variance-based SEM. Int. J. Res. Mark. **26**, 332–344 (2009)
39. Reise, S., Widaman, K., Pugh, R.: Confirmatory factor analysis and item response theory: two approaches for exploring measurement invariance. Psychol. Bull. **114**, 552 (1993)
40. Rhemtulla, M., Brosseau-Liard, P., Savalei, V.: When can categorical variables be treated as continuous? A comparison of robust continuous and categorical SEM estimation methods under suboptimal conditions. Psychol. Methods **17**, 354 (2012)
41. Rindskopf, D., Rose, T.: Second order factor analysis: Some theory and applications. Multivar. Behav. Res. **23**, 51–67 (1988)
42. Rigdon, E., Sarstedt, M., Ringle, C.: On comparing results from CB-SEM and PLS-SEM: Five perspectives and five recommendations. Market.: ZFP-J. Res. Manag. **39**, 4–16 (2017)
43. Rivero, L., Conte, T.: A systematic mapping study on research contributions on UX evaluation technologies. In: Proceedings of the XVI Brazilian Symposium on Human Factors in Computing Systems, pp. 1–10 (2017)
44. Russolillo, G.: Non-metric partial least squares (2021)
45. Sanchez, G., Trinchera, L., Sanchez, M., FactoMineR, S.: Package 'plspm'. (Citeseer)
46. Sarstedt, M., Hair, J., Ringle, C., Thiele, K., Gudergan, S.: Estimation issues with PLS and CBSEM: where the bias lies! J. Bus. Res. **69**, 3998–4010 (2016)
47. Schuberth, F.: Confirmatory composite analysis using partial least squares: setting the record straight. RMS **15**, 1311–1345 (2021)
48. Sciannamè, M., Zavarrone, E.: The qualities of AI-infused products. Reflections on emerging UX dimensions. In: Embedding Intelligence. Desigenerly Reflections on AI-infused Products, pp. 48–76 (2022)
49. Solheim, E., Berg-Nielsen, T., Wichstrøm, L.: The three dimensions of the Student-Teacher Relationship Scale: CFA validation in a preschool sample. J. Psychoeduc. Assess. **30**, 250–263 (2012)
50. Spallazzo, D., Sciannamè, M., Others Forewords: Embedding Intelligence. Designerly Reflections on AI-infused Products (2022)
51. Tenenhaus, M., Vinzi, V., Chatelin, Y., Lauro, C.: PLS path modeling. Comput. Stat. Data Anal. **48**, 159–205 (2005)
52. Vermeeren, A., Law, E., Roto, V., Obrist, M., Hoonhout, J., Väänänen-Vainio-Mattila, K.: User experience evaluation methods: current state and development needs. In: Proceedings of the 6th Nordic Conference on Human-Computer Interaction: Extending Boundaries, pp. 521–530 (2010)
53. Wetzels, M., Odekerken-Schröder, G., Van Oppen, C.: Using PLS path modeling for assessing hierarchical construct models: guidelines and empirical illustration. MIS Quarter. 177–195 (2009)
54. Wynne, C.: Issues and opinion on structural equation modelling. Manag. Inf. Syst. Q. **22**, 1–8 (1998)
55. Zarour, M., Alharbi, M.: User experience framework that combines aspects, dimensions, and measurement methods. Cogent Eng. **4**, 1421006 (2017)
56. Zavarrone, E., Cataldo, R.: CFA & PLS-PM for UX-AI product infusedn. SIS 2022-Book of the Short Papers, pp. 981–986 (2022)

Measures of Interrater Agreement for Quantitative Scales Based on the Standard Deviation

Giuseppe Bove

Abstract Interrater agreement for measurements on quantitative (discrete or continuous) scales is usually evaluated by overall measures across targets (subjects or objects) like the intraclass correlation coefficient. In this paper, a new index that allows to evaluate the agreement between two or more raters for each single target is presented, and a global measure for the whole group of targets is obtained as the arithmetic average of the single values. Some advantages of the proposed indices are highlighted by an application related to the measurement of tumor sizes.

Keywords Interrater agreement · Intraclass correlation · Quantitative scales

1 Introduction

Agreement among measurements provided by two or more raters (humans or devices) is considered in applications regarding education, psychometrics, biomedical sciences, and other disciplines. For instance, the agreement among clinical diagnoses provided by more physicians on a nominal scale is analysed for identifying the best treatment for the patient, or the agreement among ratings of educators who assess on a new ordinal rating scale the language proficiency of a corpus of argumentative (written or oral) texts is considered to test reliability of the new scale. Besides, agreement among quantitative measurements can be analysed, for instance, when radiologists measure the tumor size of lung cancer patients who could be considered in a clinical trial. In all these applications, the main interest is to analyse interrater absolute agreement, that is the extent raters assign the same (or very similar) values on the rating scale. Many indices have been proposed to measure the agreement between ratings or measurements of two or more raters on a whole group of targets (for a review see, for example, [4, 17, 19]). Moreover, methods to detect subsets of raters who demonstrate a high level of interobserver agreement were considered, for

G. Bove (✉)
Dipartimento di Scienze della Formazione, Via del Castro Pretorio 20, Roma, Italy
e-mail: giuseppe.bove@uniroma3.it

M. Bini et al. (eds.), *Advanced Methods in Statistics, Data Science and Related Applications*, Springer Proceedings in Mathematics & Statistics 467,
https://doi.org/10.1007/978-3-031-65699-6_22

instance, by [10]. Less frequently agreement on a single target has been considered [16], in spite of the fact that having evaluations of the agreement on the single target is particularly useful, for example, in situations where the rating scale is being tested, and it is necessary to identify any changes to improve it, or to request the raters for a specific comparison on the single case in which agreement is poor. Recently, [2, 3] proposed target-specific measures of absolute agreement for ratings provided on an ordinal and nominal scale, respectively. In the next sections, after a brief review of indices of interrater agreement for quantitative measurements, single-target measures based on the standard deviation that allow to overcome some limitations of the reviewed indices are proposed. Furthermore, global measures of agreement on the whole group of targets obtained as the arithmetic average of the single-target values of the indices will be also considered and applied to a data set concerning the agreement of radiologists measuring lung tumor size.

2 Measures of Interrater Agreement for Quantitative Scales: A Brief Review

There are several methods that are employed to reveal the patterns of agreement between scores provided for a group of targets by raters using a quantitative scale X. A common simple and recommended approach to analyse agreement is based on the graphical method proposed by [1], according to which, for each pair of raters, differences in the paired scores versus the mean of the paired scores are plotted to determine if the fitted line is flat and if the observations are within two standard deviations of the mean. For quantitative discrete scales with a limited number of levels, weighted versions of kappa-type indices are available (e.g., [7]), they are mostly formulated as

$$kappa_w = \frac{P_{a(w)} - P_{e(w)}}{1 - P_{e(w)}},$$

where $P_{a(w)}$ is the observed weighted proportion of agreement and $P_{e(w)}$ is the weighted proportion of chance agreement. In the case of only two raters and q levels of the scale, these quantities correspond to

$$P_{a(w)} = \sum_{k,l}^{q} w_{kl} p_{kl},$$

$$P_{e(w)} = \sum_{k,l}^{q} w_{kl} \pi_{k+} \pi_{+l},$$

where: w_{kl} is the weight given to the difference between levels k and l, p_{kl} is the proportion of targets assigned to the levels k and l by the first and the second rater, respectively, π_{k+} and π_{+l} are differently defined accordingly to the definition of chance agreement (e.g., for Cohen's weighted kappa, π_{k+} is the observed proportion of targets the first rater assigned to level k, and π_{+l} is the observed proportion of targets the second rater assigned to level l). Formulas for $P_{a(w)}$ and $P_{e(w)}$ for the case of three or more raters can be found in [7]. The idea of adjusting $P_{a(w)}$ for chance agreement $P_{e(w)}$ in $kappa_w$ is because in some situations raters could be unclear about the assignment of a level to a target and they could agree by pure chance. The denominator (1-$P_{e(w)}$) is the maximum value assumed by $(P_{a(w)} - P_{e(w)})$. $kappa_w$ assumes a maximum value equal to 1 in the case of perfect agreement between the ratings of the raters (in this case it is $P_{a(w)} = 1$), a value equal to 0 when the agreement found in the observed ratings is equal to that obtained as a result of the chance ($P_{a(w)} = P_{e(w)}$) and negative values if the agreement is less than that obtained as a result of chance ($P_{a(w)} < P_{e(w)}$). Most kappa-type indices share the same definition of $P_{a(w)}$, but differ on the expression to compute chance agreement, a controversial aspect of $kappa_w$. Besides, $kappa_w$ has other drawbacks: (1) weights w_{kl} can be defined in different ways and this makes difficult comparisons between different studies; (2) it is formulated in terms of agreement statistics based on all pairs of raters, but some authors argue that simultaneous agreement among three or more raters can be alternatively considered (e.g., see [20]); (3) agreement expected by chance depends on the observed proportions of targets allocated to the levels of the scale by each rater, and this implies that the measure of agreement depends on the marginal distributions of the categories of the scale observed for each rater, a source of some paradoxes (for this aspect see, e.g., [15]). For quantitative discrete scales with any number of levels or for continuous scales, the intraclass correlation coefficient (ICC) is the traditional approach, by which the agreement on the whole group of targets can be evaluated using mean squares from an analysis of variance (ANOVA). In a theoretical formulation, ICC is the between-target variance divided by the sum of the between-target variance and the error variance (this sum is the ratings total variance). The expressions of the different types of variance depend on the experimental design chosen for the particular application (see, e.g., [14, 18]). A high value of ICC indicates a good agreement among raters, because it is obtained when the between-target variance exceeds the error variance (that includes the within-target variance) by a wide margin. However, a low ICC value is not necessarily an indication of poor agreement, because a severe restriction in the range of ratings assigned in good agreement by the raters can cause low values of the between-target variance and low values of the ICC (the restriction of variance problem, [11]). The limitations of the kappa-type indices and of the ICC can be overcome defining target-specific measures of interrater agreement that work separately with each target i in the corresponding row of ratings in the *targets* $\times$ *raters* data matrix. For the case of a quantitative discrete scale X with q levels, in the research field of organizational studies, [8] proposed the index

$$r_{WG,i} = 1 - \frac{s_i^2}{\sigma_E^2},$$

where s_i^2 is the observed variance of the ratings in target i, σ_E^2 is the variance obtained from a theoretical null distribution representing a complete lack of agreement among raters (e.g., the uniform distribution). Roughly speaking, the null distribution conceptually represents no agreement, which means that to calculate $r_{WG,i}$, one makes a direct comparison between the observed variance in raters' ratings with the variance one would expect if there was no agreement among raters. Higher numbers indicate a greater agreement. For raters in perfect agreement, we have $s_i^2 = 0$, with a corresponding value $r_{WG,i} = 1$. In applications, $r_{WG,i}$ values greater than 0.7 (possibly 0.8) are considered associated with high level of interrater absolute agreement (see [12], p. 836 Table 2). What constitutes a reasonable standard for random ratings is highly debated. Often researchers define the no agreement, or the null distribution, in terms of a uniform distribution, apparently the default in most research. When the null distribution is assumed as uniform, the equation for the corresponding variance is

$$\sigma_{EU}^2 = \frac{(q^2 - 1)}{12},$$

where q refers to the number of levels of the scale X. For a total lack of agreement, the observed variance approaches the variance obtained from the theoretical null distribution. This leads $r_{WG,i}$ to approach 0. A global measure of agreement for the whole group of N targets can be defined as the arithmetic average of the $r_{WG,i}$ values ($\overline{r}_{WG} = 1/N \sum_{i=1}^{N} r_{WG,i}$). We remark that there is an ambiguity in the definition of the null distribution for $r_{WG,i}$. For instance, [8] recommended using the uniform distribution, but [13] remarked that negative values could be obtained when the ratings are strongly concentrated in the tails of the distribution. Heavy-tailed distributions arise when raters show strong disagreement. James et al. [9] recommended to set negative values to 0, this could be an undesirable heuristic, however, because it results in lost information. So, [13] recommended using maximum dissensus (i.e., maximum variance), recognizing that the ratings in a maximum variance distribution are equally distributed between the maximum $max(X)$ and the minimum $min(X)$ of the scale. In this case of maximum variance (MV) is

$$\sigma_{EMV}^2 = [0.5(max(X)^2 + min(X)^2)] - [0.5(max(X) + min(X))]^2.$$

Thus, the use of maximum dissensus essentially rescales $r_{WG,i}$ such that all values are within the range of 0 and 1. Scales with larger numbers of levels would have larger values of σ_{EMV}^2. Burke et al. [5] cautioned against the sole use of maximum dissensus because its use may lead to overestimation of interrater agreement, and recognizing that a resolution of the null response variance issue with respect to applications of $r_{WG,i}$ index remains problematic, they propose the average deviation index

$$AD_i = 1/k \sum_{j=1}^{k} \|x_{ij} - \overline{x}_i\|,$$

where k is the number of raters, x_{ij} is the level of the scale X rater j assigns to target i, $\overline{x}_i$ is the mean the ratings assigned by the k raters to target i. An advantage of AD_i is that does not require explicit modeling of null or random response variance, a disadvantage is that it is not a normalized index of agreement, and need a null response range to be determined. Thus, AD values will be difficult to compare across scales with a different numbers of levels. Other possible indices for quantitative discrete scales are reviewed, for instance, in [12].

3 Single-Target Measures of Interrater Agreement for Quantitative Discrete or Continuous Scales Based on the Standard Deviation

The single-target measures reviewed in Sect. 2 are mostly applied to ordinal scales (e.g., Likert scales), in spite of the fact they should be applied to any quantitative discrete or continuous scale X for which min(X) and max(X) are known (or can reasonably be defined). For instance, in the case of a continuous scale, the index $r_{WG,i}$ can be easily computed by assuming a uniform null distribution, for which the variance is

$$\sigma^2_{EU} = \frac{(min(X) - max(X))^2}{12}.$$

To overcome some of the problems encountered in the applications of $r_{WG,i}$ and other single-target measures of agreement, we propose a new procedure to measure absolute agreement for quantitative scales by using a geometrical orthogonal regression framework. We consider first the case of two raters and then the case of several raters.

Two raters

The scores of two raters (X_1, X_2) can be depicted in a scatter plot, the raters are in perfect agreement $(X_1 = X_2)$ if the regression goes through the origin with a slope of 1. Normally, the points in the scatter plot deviate from perfect agreement, so the level of agreement can be explored as the dispersion around the straight line $x_1 = x_2$. For target i having the scores x_{i1}, x_{i2} $(i = 1, 2, \ldots, N)$, the squared length of the orthogonal projection of the corresponding vector on the straight line $x_1 = x_2$ is

$$\frac{(\sum_{j=1}^{2} x_{ij})^2}{2},$$

given by the ratio between the squared scalar product $(\sum_{j=1}^{2} x_{ij})^2$ between the two vectors from the origin to the points (x_{i1}, x_{i2}), $(1, 1)$, respectively, and the squared length of the unit vector $(1, 1)$. So, for the Pythagorean theorem, the squared distance l_i^2 of the point (x_{i1}, x_{i2}) from the straight line $x_1 = x_2$ is the difference between the squared length of the vector from the origin to the point (x_{i1},x_{i2}) and the squared length of the orthogonal projection of the corresponding vector on the straight line $x_1 = x_2$

$$l_i^2 = \left(\sum_{j=1}^{2} x_{ij}^2 - \frac{(\sum_{j=1}^{2} x_{ij})^2}{2}\right).$$

An equivalent way to write the previous math expression is

$$l_i^2 = 2\left(\frac{\sum_{j=1}^{2} x_{ij}^2}{2} - \left(\frac{\sum_{j=1}^{2} x_{ij}}{2}\right)^2\right) = 2s_i^2$$

with s_i^2 the observed variance of the scores x_{i1}, x_{i2}. Finally, we have that the distance l_i of the point (x_{i1}, x_{i2}) from the straight line $x_1 = x_2$ is proportional to the standard deviation of the scores x_{i1}, x_{i2}, because

$$l_i = \sqrt{2}s_i.$$

When $min(X)$ and $max(X)$ are known for the quantitative scale X, a target-specific measure of absolute agreement g_i can be defined as

$$g_i = 1 - \frac{l_i}{max(l_i)} = 1 - \frac{s_i}{((max(X) - min(X))/2)}$$

with $g_i = 0$ for maximum disagreement and $g_i = 1$ for perfect agreement. Frequently, in the applications the minimum and maximum of a continuous scale X cannot be defined, in this case an alternative relative measure of agreement can be obtained (disregarding the constant $\sqrt{2}$) by the coefficient of variation

$$CV_i = \frac{s_i}{\overline{x}_i},$$

with $\overline{x}_i$ the arithmetic average of the scores x_{i1}, x_{i2}, (the denominator is the absolute value of $\overline{x}_i$ if $\overline{x}_i < 0$). Global measures of agreement are obtained by averaging the target-specific values

$$g_{ave} = \frac{1}{N}\sum_{i=1}^{N} g_i, \quad CV_{ave} = \frac{1}{N}\sum_{i=1}^{N} CV_i$$

for the two cases, respectively.

Two or more raters

The scores of k raters $(X_1, X_2, \ldots, X_k)$ can be represented in k dimensions, so the level of agreement can be explored as the dispersion of the points $(x_{i1}, x_{i2}, \ldots, x_{ik})$, $(i = 1, 2, \ldots, N)$ around the straight line $x_1 = x_2 = \cdots = x_k$. For target i having the scores $x_{i1}, x_{i2}, \ldots, x_{ik}$, the expression of the distance l_i of point $(x_{i1}, x_{i2}, \ldots, x_{ik})$ from the straight line $x_1 = x_2 = \cdots = x_k$ is

$$l_i = \sqrt{\left(\sum_{j=1}^{k} x_{ij}^2\right) - \frac{\left(\sum_{j=1}^{k} x_{ij}\right)^2}{k}} = \sqrt{ks_i^2},$$

with s_i^2 the observed variance of the scores $x_{i1}, x_{i2}, \ldots, x_{ik}$ (the expression can be easily obtained as a generalization of the case of two raters). Formulas for g_i and g_{ave} remain the same as the ones obtained for two raters case. Generalizations of the coefficient of variation CV_i and the average CV_{ave} are immediate for any k. Some experiences with real applications suggest the following indications of thresholds for the interpretation of CV_i and CV_{ave} expressed in percentage: values less than or equal to 10% excellent agreement, values between 10% and 20% good agreement, values greater than 20% fair to poor agreement.

4 Application

A discussion of disagreement between radiologists is considered in [6]. The purpose of the study was to assess the consistency of measurements performed by readers evaluating lung tumors, this is a critical component of many cancer trials because measurements can be used, for instance, to warrant additional testing of an agent or to determine whether therapy should be continued. Patients were selected with non-small-cell lung cancer and with 40 lung lesions whose size exceeded at least 1.5 cm. Measurements were performed independently by five thoracic radiologists using printed film by computed tomography. Each radiologist read each of 40 images measuring (a) the longest diameter (unidimensional measurement) and (b) the longest diameter and the perpendicular longest diameter (bidimensional measurement) of each lesion. Measurement was repeated on the same images 5–7 days later, that means each radiologist was looking at the same image (and the same tumor size) twice (Table 6.18 in [4] contains the data of the two replications of the unidimensional measurements). Variation between and within (the two replications of) the five radiologists measuring tumor size is examined in [6] with different methods, in order to ascertain how to improve measurement consistency in terms of reduction of incorrect interpretation of tumor response. In the following, we are only interested in analysing agreement applying the proposed indices CV_i and CV_{ave} to the unidimensional measurements of the five radiologists in the first and second replication. Some descriptive statistics regarding the first replication of the unidimensional measure-

Table 1 Descriptive statistics of tumor measurements (centimeters) of five radiologists on N = 40 lung lesions (first replication)

Radiologist	Mean	Median	Range	SD
1	3.92	3.80	1.5 to 8	1.59
2	3.71	3.80	1.2 to 7.8	1.50
3	4.42	4.20	1.5 to 9	1.54
4	4.37	4.10	1.5 to 9	1.60
5	4.14	3.95	1.7 to 9	1.54
Total	4.11	4.00	1.2 to 9	1.58

Table 2 Intraclass correlation values and descriptive statistics of the distributions of the coefficient of variation (percentage) in repetitions 1 and 2

Repetition	ICC	CV_{ave}	CV_{min}	CV_{max}	$CV_i \leq 10\%$	$10\% < CV_i \leq 20\%$	$CV_i > 20\%$
1	0.837	12.89	1.27	39.24	20	13	7
2	0.837	12.89	1.27	39.24	20	13	7

ment are provided in Table 1. The similarity of the means in Table 1 reflects a pretty good level of agreement, with radiologist 2 reporting the smallest mean tumor size and the smallest standard deviation. The good level of agreement is confirmed by the values of the intraclass correlation coefficient (random factorial design assumed) and the CV_{ave} provided in Table 2, where it can also be noted that at least half of the forty CV_i values are less or equal to 10% in both the repetitions.

On the other hand, some high CV_i values (not reported here) are present in each repetition (e.g., images 16, 22, 30, 34), these images could be selected for a comparison between radiologists and to detect particular types of lesions (irregular edge and/or irregular contour) difficult to measure. This aspect cannot be analysed by global measures of absolute agreement like $kappa_w$ or ICC.

In [6] the measurements referring to the same radiologist in the two replications (intraobserver) and the measurements referring to two different radiologists in the same repetition (interobsever) were compared for each tumor to detect the presence of misclassifications (differences in the measured size of tumors higher than a predefined cutoff). The level of intraobserver misclassifications was lower than the level of interobserver misclassifications, so it was adviced that the same reader should perform measurements for any one patient for the duration of the clinical cancer trial. The indices CV_i and CV_{ave} were computed for each radiologist in the two replications and for each pair of radiologists in each replication. CV_{ave} values are provided in Table 3. It can be noted that off-diagonal values are higher than diagonal values in Table 3, and this confirms that intraobsever measurements have less dispersion than interobserver measurements.

Table 3 Pairwise intraobserver and interobserver CV_{ave} values (percentage) of five radiologists

Radiologist	1	2	3	4	5
1	4.26				
2	8.80	4.54			
3	9.30	13.03	3.47		
4	6.84	10.55	7.39	4.36	
5	6.06	9.69	7.84	5.26	4.20

A comparison was performed between the results of CV_i and those obtained with $r_{WG,i}$. We remark that the application of the $r_{WG,i}$ index required the assumption of a null range for the possible values of the measurements of the tumor sizes. We know that patients were selected with lung lesions whose size exceeded at least 1.5 cm. So, for each patient, we assumed a uniform null distribution in the range of 1.5 to the maximum measurement observed for the same patient among the five radiologists. In spite of the fact that the correlation coefficient between the two indices was not low (in the first repetition was −0.78 and in the second repetition −0.7), the $r_{WG,i}$ index gave results only partially similar to those obtained by the CV_i index. The value of the global measure of agreement was low ($\overline{r}_{WG} = 0.5$) for the presence of several negative values for $r_{WG,i}$ (if we fix the negative values to 0, we obtain $\overline{r}_{WG} = 0.61$, still an indication of moderate agreement). Negative values for $r_{WG,i}$ were obtained not only in correspondence of high values of CV_i but also in two cases of good agreement in which CV_i was low. We have also compared the results of CV_i with those obtained by setting the expected random variance σ_E^2 in $r_{WG,i}$ to the maximum dissensus value σ_{EMV}^2. The value of the global measure of agreement increases to $\overline{r}_{WG} = 0.83$ for the absence of negative values for $r_{WG,i}$, and as expected the patient values of the index showed overestimation of interrater agreement in several cases.

5 Conclusions

This contribution provides new target-specific measures of interrater agreement, extending to scores on quantitative scales the proposals for categorical data presented in [2, 3]. Besides, a measure of agreement for the whole group of targets is obtained as the average of the target-specific values. The index presents advantages respect to the intraclass correlation coefficient, kappa-type and the $r_{WG,i}$ indices. Preliminary results obtained in the application to the measurement of tumor sizes seem encouraging. The approach considered is mainly descriptive and it is based on a geometrical formalization. Future developments may concern: (1) an accurate definition of reliable thresholds useful for the interpretation of the level of agreement in the applications; (2) the study of the sampling properties of the index.

References

1. Bland, J.M., Altman, D.G.: Statistical methods for assessing agreement between two methods of clinical measurements. Lancet **327**(8476), 307–310 (1986)
2. Bove, G., Conti, P.L., Marella, D.: A measure of interrater absolute agreement for ordinal categorical data. Stat. Methods Appl. **30**(3), 927–945 (2021)
3. Bove, G.: A subject-specific measure of interrater agreement based on the homogeneity index. In: Porzio, G.C., Rampichini, C., Bocci, C. (eds.) CLADAG 2021 Book of Short Papers, pp. 272–275. Firenze University Press, Firenze (2021)
4. Broemeling, L.D.: Bayesian Methods for Measures of Agreement. Chapman & Hall/CRC, London (2009)
5. Burke, M.J., Finkelstein, L.M., Dusig, M.S.: On average deviation indices for estimating interrater agreement. Organ. Res. Methods **2**, 49–68 (1999)
6. Erasmus, J.J., Gladish, G.W., Broemeling, L., Sabloff, B.S., Truong, M.T., Herbst, R.S., Munden, R.F.: Interobserver and intraobserver variability in measurement of non-small-cell carcinoma lung lesions: Implications for assessment of tumor response. J. Clin. Oncol. **21**(13), 2574–2582 (2003)
7. Gwet, K.L.: Handbook of inter-rater reliability, 4th edn. Advanced Analytics LLC, Gaithersburg, MD, USA (2014)
8. James, L.R., Demaree, R.G., Wolf, G.: Estimating within-group interrater reliability with and without response bias. J. Appl. Psychol. **69**, 85–98 (1984)
9. James, L.R., Demaree, R.G., Wolf, G.: r_{WG} an assessment of within group interrater agreement. J. Appl. Psychol. **78**, 306–309 (1993)
10. Landis, J.R., Koch, G.G.: An application of hierarchical kappa-type statistics in the assessment of majority agreement among multiple observers. Biometrics **33**, 363–374 (1977)
11. LeBreton, J.M., Burgess, J.R.D., Kaiser, R.B., Atchley, E.K., James, L.R.: The restriction of variance hypothesis and interrater reliability and agreement: are ratings from multiple sources really dissimilar? Organ. Res. Methods **6**, 80–128 (2003)
12. LeBreton, J.M., Senter, J.L.: Answers to 20 questions about interrater reliability and interrater agreement. Organ. Res. Methods **11**(4), 815–852 (2008)
13. Lindell, M.K., Brandt, C.J.: Measuring interrater agreement for ratings of a single target. Appl. Psychol. Meas. **21**, 271–278 (1997)
14. McGraw, K.O., Wong, S.P.: Forming inferences about some intraclass correlation coefficients. Psychol. Methods **1**(1), 30–46 (1996)
15. Marasini, D., Quatto, P., Ripamonti, E.: Assessing the interrater agreement for ordinal data through weighted indexes. Stat. Methods Med. Res. **25**, 2611–2633 (2016)
16. O'Connell, D.L., Dobson, A.J.: General observer-agreement measures on individual subjects and groups of subjects. Biometrics **40**, 973–983 (1984)
17. Shoukri, M.M.: Measures of Interobserver Agreement and Reliability. Taylor and Francis Group, Boca Raton Florida (2011)
18. Shrout, P.E., Fleiss, J.L.: Intraclass correlations: use in assessing rater reliability. Psychol. Bull. **86**(2), 420–428 (1979)
19. von Eye, A., Mun, E.Y.: Analyzing Rater Agreement. Manifest Variable Methods. Lawrence Erlbaum Associates, Mahwah, New Jersey (2005)
20. Warrens, M.J.: Equivalences of weighted kappas for multiple raters. Stat. Methodol. **9**, 407–422 (2012)

Decomposing Inequality After Asymmetric Shocks: An Analysis of Italian Household Consumption

Federico Attili and Michele Costa

Abstract The massive shocks that have plagued the world economy over the past 15 years have altered the distribution of income, wealth and opportunities. The economic literature is replete with research testifying to the asymmetric effects of the financial crisis, COVID-19 and inflation. Although their impact on overall inequality has been widely explored, we believe there is a gap in the study of inequality factors and divides between socio-economic subgroups. Indeed, the common methodology for measuring inequality between subgroups, usually based on a comparison of averages, fails to fully account for asymmetric changes in distributions. We propose to employ a novel decomposition of the Gini index to overcome this issue. Our analysis focuses on COVID-19 and makes use of Italian household consumption micro-data to explore main inequality factors such as age, gender and geography. What emerges is that the increase in the age and gender gap is significantly smaller than it would be using the conventional methodology.

Keywords Inequality decomposition · Asymmetric shocks · Gini index · COVID-19 · Monte Carlo experiment

1 Introduction

In little more than a decade, most of the world's economies have suffered three of the largest shocks since the second world war. The economic repercussions of the financial crisis and the COVID-19 restrictions have been studied extensively. Many efforts are now devoted to the study of high inflation, mainly driven by rising energy prices. All three shocks affected asymmetrically economic agents, business and personal income [4, 7, 13, 16]. As a result, the debate on the implications of

F. Attili · M. Costa (✉)
Department of Economics, University of Bologna, Piazza Scaravilli, 2, Bologna, Italy
e-mail: michele.costa@unibo.it

F. Attili
e-mail: federico.attili2@unibo.it

M. Bini et al. (eds.), *Advanced Methods in Statistics, Data Science and Related Applications*, Springer Proceedings in Mathematics & Statistics 467,
https://doi.org/10.1007/978-3-031-65699-6_18

asymmetric shocks for monetary policy, prices and risk sharing flourished, receiving adequate attention and support. On the contrary, methods to study inequality have barely evolved, while we argue that the usual assessment of inequality is not robust to asymmetric shocks when the focus is on comparisons between population subgroups. Indeed, traditional methods generally compare distributions of subgroups by looking at the difference between their means. However, the comparison of means is not representative of the distance between distributions when the distributions are heterogeneously and asymmetrically affected by shocks.

In this paper, we focus on Italian household consumption after COVID-19, which represented a very strong shock with serious health, psychological, social and economic consequences. We show that usual methods to assess inequality between subgroups do not fully capture the heterogeneous and asymmetric modifications of the consumption distribution of subgroups. At the same time, we propose a complementary methodology helping to deliver a more complete picture of inequality between groups.

The need to evaluate the effects of COVID on inequality has immediately been recognized. It led to a thriving literature (see e.g. [19] for a review), to which we contribute by including the methods related to inequality decomposition in the methodological toolbox of researchers and policymakers. Indeed, the between-inequality component represents a strategic tool to explore the effects of asymmetric shocks on divides.

COVID-19 impacted Italian households consumption in 2020 through different channels. The principal ones are income reduction [8], lockdown, restrictions, and behavioral changes [12, 15]. Except for the lockdown, which flattened household consumption to unprecedented levels in all the social strata, the other channels shaped the distribution of consumption (and income) with heterogeneous intensity along its deciles and subgroups. The presence of substantial asymmetries which characterize the effects of COVID has been thoroughly investigated and highlighted by [4]. To fully understand the heterogeneous changes that have occurred, it is therefore important to measure the inequality between subgroups at all distribution points.

We exploit the Italian household budget survey (HBS) to investigate consumption inequality of household residing in Italy. With respect to three different partitions, which define subgroups based on age, gender and geographical factor, we compare how different can be the picture of household consumption inequality when considering decompositions that follow different paradigms. In particular, we compare the summer of 2019 and 2020, which are the two closest comparable periods before and after the lockdown and the main restrictions. We show that the increase in age and gender divide is much smaller if measured by comparing the whole subgroup distributions. The impact of COVID-19 on household consumption is instead homogeneous among the Italian regions and the different points of their distributions, therefore decompositions of inequality agree on a modest reduction of the geographical divide in consumption.

This paper is the extension of a previous version presented at the 51st Scientific Meeting of the Italian Statistical Society [3] and proceeds as follows. Section 2 builds a common framework on the methodologies for studying inequality factors.

Section 3 theoretically discusses the impact of shocks on between-inequality indicators. Section 4 presents the data employed and illustrates their main features. Section 5 discusses the response of different measures after simulated shocks and after COVID-19, and exploits a bootstrap algorithm to infer their confidence intervals. Section 6 concludes.

2 Inequality Factors and Inequality Decomposition

2.1 Assessment of Inequality Factors

Inequality factors such as gender, age and geographical area play extremely relevant and widely recognized roles. By taking into account two or more reference values of the inequality factor, we classify the overall population into two or more subgroups. The differences between the subgroups can be exploited in order to assess the relevance of the related inequality factor and its effects on overall inequality.

Traditionally, given a variable of interest y, such as income or consumption, with mean $\bar{y}$, and two subgroups j and h, with means $\bar{y}_j$ and $\bar{y}_h$, the most direct and immediate way to measure the relevance of an inequality factor is based on the gap between subgroups j and h, measured with the (relative) difference between the averages of the subgroups:

$$\Delta\bar{y}_{jh} = |\bar{y}_j - \bar{y}_h|/\bar{y} \tag{1}$$

which provides a classic aggregate indicator. There are also other aggregated indicators, but averages represent by far the most commonly used tool, so we can consider relation (1) as a benchmark.

A more detailed framework than relation (1) can be developed by extending the comparison to a set of points of the distributions of subgroups j and h. Let be y_{jq}, $q = 1, \ldots, Q$, the Q quantiles of subgroup j, and $y_{hq}, q = 1, \ldots, Q$, the Q quantiles of subgroup h, then we can define a sequence of $\Delta y_{jhq}, q = 1, \ldots, Q$, such as

$$\Delta y_{jhq} = |y_{jq} - y_{hq}|/\bar{y}$$

and compare the two subgroups based on the synthesis of the Δy_{jhq} given by

$$d_{jhQ} = \sum_{q=1}^{Q} \Delta y_{jhq}/Q \tag{2}$$

Compared to $\Delta\bar{y}_{jh}$, quantity d_{jhQ} allows a comparison developed on a wider set of information, which grows with Q and can be useful when the distributions of the two subgroups have different characteristics. In the following we use $Q = 10$, thus adopting the traditional inequality framework based on deciles.

The extension to the case of more than two subgroups can be easily obtained since it only involves a series of binary comparisons and their synthesis.

In this paper, we want to assess if the relevance of inequality factors and their dynamic is sensitive to the employed methodology. In particular, we compare the results of Eqs. (1) and (2), arguing that they differ in the case of asymmetric shocks because the dynamic of the difference between means is not representative of the correspondent evolution of the distributions. We extend this argument by considering the more refined methods of inequality decomposition, where the inequality between plays a key role in evaluating the importance of the underlying inequality factors and the gap between the subgroups.

Our discussion is valid for most of the inequality measures, from the generalized entropy indexes such as the Theil index, to the Zenga [18] and the Gini indexes. In the following, we refer to the Gini index G, one of the most widespread and well-known inequality measure (for a detailed discussion see, e.g., [11]), whose use has expanded from the original area of economic inequality to poverty, well-being and many other fields of economics.

In the framework of the Gini index, we take into account three of its decompositions, focusing on the inequality between component, which aims to evaluate the effects on overall inequality attributable to differences between subgroups [6, 10].

2.2 Decompositions of the Gini Index

Our first reference is the decomposition proposed by Bhattacharya and Mahalanobis [5], whose contribution is really pioneering and has paved the way for numerous further decompositions, for which it has always represented an essential benchmark.

Following Bhattacharya and Mahalanobis, the inequality between k subgroups is based on the comparison of the subgroup means. It reads as

$$G_b^{BM} = \sum_{j=1}^{k} \sum_{h=1}^{k} s_j s_h |\bar{y}_j - \bar{y}_h| / 2\bar{y} \tag{3}$$

where s_j and s_h are the share of population in subgroups j and h.

In the framework of the Bhattacharya and Mahalanobis's contribution, a residual component is present when the subgroups overlap, making their proposal a three-term decomposition.

A further relevant contribution to the Gini index decomposition literature is proposed by Yitzhaki and Lerman [20], who suggest evaluating the inequality between subgroups as

$$G_b^{YL} = 2cov(\bar{y}_j, \bar{F}_j)/\bar{y} \tag{4}$$

where F indicates the cumulative distribution.

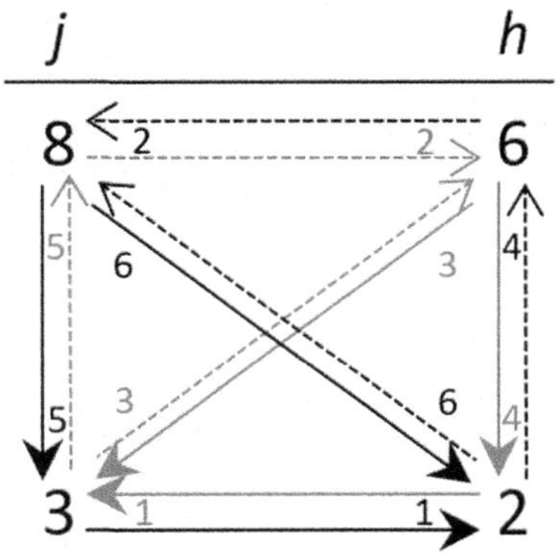

Fig. 1 Illustration of the Gini index in the case of four units divided in two groups, j and h. The arrows represent absolute pairwise differences between the four values. Their sum constitutes the numerator of the Gini index

Yitzhaky and Lerman characterize their proposal by introducing an overlapping component, clearly distinguishing it from the between-inequality component. We note that even this between-inequality measure primarily relies on the means of subgroups and their comparison.

Alongside these two well-known decompositions of the Gini index, our third reference is represented by a recent proposal by [2], which, not being as well known as the previous ones, deserves some introductory notes. The main intuition of Attili's decomposition is to measure between-inequality by comparing the values of individuals belonging to different subgroups and equally ranked in their subgroups, instead of comparing subgroup means. This between-inequality measure, taking advantage of a broader information set, goes in the direction of Eq. 2.

We present the paradigm of the decomposition by a simple scheme involving four units divided into two groups, j and h (Fig. 1). The Gini index of the four values normalises the sum of all the absolute differences indicated by the arrows.
The vertical and horizontal arrows refer to the inequality within and between subgroups, respectively.

The decomposition is finalised disentangling contributions to within and between inequality from the diagonal arrows. E.g., for the solid black diagonal difference $(8 - 2) = 6$, we split it in $(8 - 3) + (3 - 2) = 6$, attributing the first (vertical) difference to inequality within and the second (horizontal) difference to inequality between.

In the case of grey diagonals, the two involved vertical and horizontal differences are rescaled so that their sum equates the diagonal difference. This is achieved by means of a rescaling factor, calculated dividing the diagonal difference by the sum of the vertical and the horizontal one. Likewise, black arrows are rescaled by 1.
As in the example, the decomposition of G in within and between components singles out the sum of vertical from the sum of horizontal differences, where each difference is weighted by the sum of the related rescaling factors.

Generalising to the case of subgroups with size n, the between component reads

$$G_b^A = \sum_{j=1}^{k}\sum_{h=1}^{k}\sum_{i=1}^{n} |y_{ji} - y_{hi}| w_{jhi}/(2\bar{y}n^2k^2) \tag{5}$$

For each value $i = 1, \ldots, n$, it compares observation ranking in the $i-th$ position of subgroup j (y_{ji}) with the corresponding observation of the other subgroups (y_{hi}), while the weights w_{jhi} ensure correspondence with the Gini index in a fully informative within-between decomposition scheme. In this presentation, we restricted the attention to the case of equal-sized groups, but the methodology can be easily extended to the case of groups with different sizes. A comprehensive presentation is given in [2].

3 Shocks and Assessment of Inequality Factors

In this section, we study the effect of symmetric and asymmetric shocks on the indicators discussed in the previous section.

We consider a population of size N and we define the income of individual i as y_i, with $i = 1, \ldots, N$. The population is divided into two groups, j and h, of size n_j and n_h $(n_j + n_h = N)$. Let income of individual i of group j to be y_{ji}, with $i = 1, \ldots, n_j$.

We start by discussing symmetric shocks. We can model symmetric shocks as multiplicative or additive shocks. As for multiplicative shocks, we define:

$$x_i = y_i \cdot (1 + c) \tag{6}$$

where c is the variation induced by the shock.

It is simple to show that the values of both Eqs. (1) and (2) stay unchanged after substituting Eq. (6). Indeed, such a shock does not even affect inequality measures, according to the principle of scale invariance [1]. We model additive symmetric shocks as:

$$x_i = y_i + c \tag{7}$$

In this case, we can show that:

$$\Delta\bar{y}_{jh} = |\bar{y}_j - \bar{y}_h|/(\bar{y} + c)$$

and

$$\Delta y_{jhq} = |y_{jq} - y_{hq}|/(\bar{y} + c)$$

so that

$$d_{jhQ} = \frac{1}{Q \cdot (\bar{y} + c)} \sum_{q=1}^{Q} |y_{jq} - y_{hq}|$$

Compared to Eqs. (1) and (2), both Eqs. (3) and (3) vary in the same direction and proportion. If c is positive they decrease by a factor $\bar{y}/(\bar{y} + c)$, while they increase for negative shocks.

This concordance could weaken introducing stochastic shocks. We model them both as additive and multiplicative:

Multiplicative Additive

$$x_i = \begin{cases} y_i \cdot (1 + c) & \text{with prob. } p \\ y_i & \text{with prob. } 1 - p \end{cases} \qquad x_i = \begin{cases} y_i + c & \text{with prob. } p \\ y_i & \text{with prob. } 1 - p \end{cases} \tag{8}$$

It is not simple to treat Eq. (2) analytically in the stochastic case, due to the potential change in decile membership of some units after the shock. The study of shock sensitivity is also challenging for the between-inequality indicators.

In general, the analytical study of indicators is even harder when moving toward asymmetric shocks, which, as we argue from the beginning, can significantly reduce the grounds for the concordance of the indicators.

We can generalize the analysis of the effects of shocks by resorting to a mixture distribution

$$X^m = (1 - t)Y + tX$$

where Y and X are the income distribution before and after the shock, respectively, while t ranges between 0 and 1. When $t = 0$ the shock does not play any role and $X^m = Y$. When $t = 1$ it produces its maximum effect and $X^m = X$.

The mean of X^m is

$$\bar{X}^m = \bar{Y} + t(\bar{X} - \bar{Y})$$

and we can evaluate $\delta \bar{X}^m / \delta t$ in order to assess the effects of the shock on the mean.

The linear structure of the mean greatly simplifies the evaluation also in case of asymmetric shocks. For example, if the shock adds a quantity c only to the incomes below y^*, we obtain

$$X = Y + c\mathbb{1}_{Y<y*}$$

where $\mathbb{1}$ is the indicator function. The mean of the after shock distribution is

$$\bar{X} = \bar{Y} + cF(y^*)$$

At this point, we can rewrite the mean of the mixture as

$$\bar{X}^m = \bar{Y} + t(cF(y^*))$$

Table 1 Additive and multiplicative shock to individual i of group h as modelled and studied by the Monte Carlo algorithm

Asymmetry	Additive process	Multiplicative process	Probability
Shock asymmetrically affect both groups and deciles	$x_{ji} = \begin{cases} y_{ji} + c \\ y_{ji} \end{cases}$	$x_{ji} = \begin{cases} y_{ji} \cdot (1 + c) \\ y_{ji} \end{cases}$	$\begin{cases} p_{jq} \\ 1 - p_{jq} \end{cases}$

Looking at its derivative

$$\delta \bar{X}^m / \delta t = cF(y^*)$$

we can study the behaviour of the mean while the shock produces its effects. In this case $\delta \bar{X}^m$ increases linearly with t and its slope $cF(y^*)$ increases with y^*. This approach can be easily generalized to analyse the relative evolution between the averages of multiple groups.

Unfortunately, the extension to inequality measures, from the Gini index to the components of its decompositions, is far more complex [9, 14] especially in the case of an empirical (discrete) distribution, where the Gini index is based on the absolute value.

In the following, in order to overcome the difficulties related to the analytical evaluation of the impact of asymmetric shocks on the inequality measures, we exploit a Monte Carlo algorithm. It starts from the data on Italian households' consumption and aims at studying how indicators of between-inequality react to different kinds of shocks. We maintain the distinction between additive and multiplicative shocks and we model the shocks as described in the scheme of Table 1. The table encapsulates the conceptual framework used in the Monte Carlo algorithm to study the impacts of additive and multiplicative asymmetric shocks on individuals within various groups and deciles. Mathematical representations are provided for the two distinct types of shocks. The additive process adds a constant c to y_{ji}, the original value of individual i in group j, with probability p_{jq}, or does not affect y_{ji} with probability $1 - p_{jq}$. The multiplicative process, with the same probabilities, involves an increase by a factor of $(1 + c)$, indicating a proportional change rather than a fixed increment, or no change at all. The probability p_{jq} is a crucial element of the model, and denotes the heterogeneous probability of a shock occurring. This probability is not only specific to each group (j), but also sensitive to the rank of the unit i within its group (q). The different scenarios implemented, expressed in terms of the choice of p_{jq}, and the details of the Monte Carlo study, as well as its results, are illustrated in Sect. 5.

4 Data

Our analysis makes use of the household budget survey (HBS), run by Istat, which allows to study consumption expenditure behaviours of households residing in Italy according to their main socio-economic and territorial characteristics. We focus on

Table 2 Deciles and means of the Italian households' monthly consumption in 2019 and 2020, detailed by subgroups defined by age, gender and geographical factors. Age subgroups are based on the partition "Young (aged less than 65) versus Old (aged more than 64)"; the "Gender Divide" panel compares "Men versus Women"; the "Geographical divide" panel compares "North and center versus South and islands"

	2019		2020	
	Young	Old	Young	Old
d_1	591	**592**	**584**	572
d_2	**879**	844	**853**	823
d_5	**1470**	1322	**1404**	1293
d_9	**2700**	2548	**2639**	2455
d_{10}	4035	**4273**	**3953**	3723
Mean	**1784**	1705	**1731**	1604
	2019		2020	
	Men	Women	Men	Women
d_1	590	**593**	**608**	533
d_2	**873**	852	**867**	789
d_5	**1438**	1368	**1386**	1310
d_9	**2688**	2581	**2635**	2469
d_{10}	4088	**4171**	**3948**	3721
Mean	**1772**	1727	**1722**	1614
	2019		2020	
	North and Center	South and Islands	North and Center	South and Islands
d_1	**676**	501	**663**	489
d_2	**981**	716	**947**	695
d_5	**1585**	1122	**1499**	1083
d_9	**2868**	2083	**2779**	2061
d_{10}	**4397**	3253	**4117**	3101
Mean	**1926**	1392	**1838**	1356

surveys carried out in July, August and September of 2019 and 2020 (4650 and 6964 households in 2019 and in 2020, respectively), because from July to September 2020 the government removed the shutdown of most activities and many of the restrictions on social distancing and travels. By focusing on summer consumption, we improve the identification of the (potentially structural) effects of income losses and behavioral changes from 2019 and 2020.

Household consumption is equivalised for the number of household members by using the square root equivalence scale. We investigate three different bi-partitions based on age, gender, and geographical factor. They are well-recognised inequality factors, giving rise to age, gender, and geographical divides. Consistently with the theoretical development in Sect. 2, it is possible to extend the analysis to the case of k subgroups. The rationale behind the choice of two subgroups is to achieve a simpler

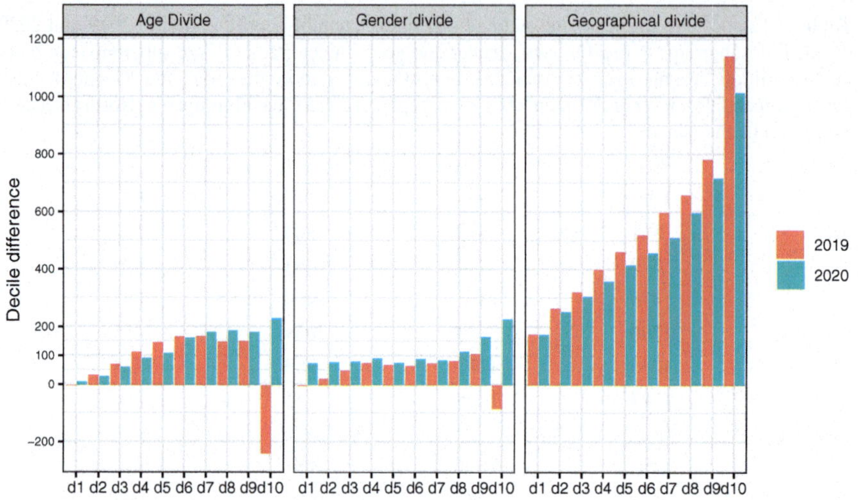

Fig. 2 The difference between deciles of different subgroups. The three panels report the comparison of subgroup deciles (in 2019 and 2020) for three partitions, based on household head characteristics. The "Age Divide" panel compares "Young (aged less than 65) versus Old (aged more than 64)"; the "Gender Divide" panel compares "Men versus Women"; the "Geographical divide" panel compares "North and center versus South and islands"

and clearer assessment of the difference between the subgroups and, consequently, of the effect of the underlying inequality factor.

In Table 2 we report deciles and means of the Italian households' monthly consumption in 2019 and 2020, detailed by subgroups defined by age, gender and geographical factors. In Fig. 2 we highlight the difference between group deciles.

The first panel compare Young-(< 65) and Old-headed(≥ 65) households. In 2019 the former have higher consumption than the latter but in the extreme deciles. Consumption of all groups and deciles decreases in 2020, but COVID-19 more evidently reduces rich Old-headed households' consumption, so that Young-headed households show higher consumption at all deciles. Higher values notoriously attract the means, so the difference between subgroup means ($\Delta \bar{y}_{jh}$) increases by 67.7% from 2019 to 2020. However, COVID-19 affected the consumption distributions of the two subgroups asymmetrically, such that, in contrast with the means, the deciles at the bottom of the distribution get closer in 2020 (Fig. 2). This is captured by a much smaller increase of d_{jh10} (+5.3%). Are the means representative of the relative dynamic of the distributions?

The second panel explores the gender divide in households' consumption. In 2019 Male-headed households have higher consumption than Female-headed households; again, extreme deciles are an exception . For the same deciles, the gender divide increases a lot in 2020, while it slightly increases in the middle of the distribution. As before, $\Delta \bar{y}_{jh}$ is more responsive (+149.2) than d_{jh10} (+77.7) to fluctuations of richer households.

In the third panel, we investigate the territorial divide. As it is well known, Italian households in the Center and North of Italy are richer than those living elsewhere. The mean and the deciles of consumption of the two subgroups corroborate this fact and suggest that, at least for consumption, the geographical divide is the strongest among the three analysed. The impact of COVID-19 on household consumption is stronger for households in the Centre and the North and, differently from the other partition, it is quite symmetric across deciles. Therefore, the reduction of the distance between subgroups is measured at around 6% by both $\Delta\bar{y}_{jh}$ and d_{jh10}.

5 Results

5.1 Simulated Shocks: A Monte Carlo Experiment

As discussed in Sect. 3, alternative indicators of inequality between groups are likely to react in different ways to asymmetric (and stochastic) shocks. Starting from the 2019 data, we use a Monte Carlo experiment to study the behavior of Eqs. (1)–(5) after different shocks. According to the probabilities p_{jq} as defined in Table 1, asymmetries produce shocks of heterogeneous intensity both between groups and between quantiles. For the sake of interpretability, we control the settings by modeling the probabilities as follows.

Regarding the source of asymmetry resulting from belonging to different groups, we define three scenarios (we refer to the partition between Group j and Group h for explanatory purposes). In the first one, Group j have a higher probability ($p_j = 75\%$) of being hit by the shock than Group h ($p_h = 25\%$). In the second one, the two groups have the same probability of 50%. In the last one, Group j has a lower probability ($p_j = 25\%$) of being hit than Group h ($p_h = 75\%$). Furthermore we introduce the asymmetry of the shock within the groups by defining $\rho_q = 0$ or 1 as the probability of being hit by a shock being in the decile q of the group.

Having p_j and ρ_q we are able to define $p_{jq} = p_j \cdot \rho_q$.

The original consumptions of 2019 correspond to $\rho_q = 0$ for all $q = 1, \ldots, 10$. We start experimenting with $\rho_q = 1$ only for the first decile; then for $q = 1, 2$; until $\rho_q = 1$ for all $q = 1, \ldots, 10$. Then we start excluding the possibility of being hit ($\rho_q = 0$) for $q = 1$; then for $q = 1, 2$; until $\rho_q = 1$ only for $q = 10$.

We simulated all the settings both for additive and multiplicative shocks, setting $c = 200$ for the additive and $c = 20\%$ for the multiplicative case. We run each experiment 100 times and take the average of the 100 replicates of the five indicators.

We summarise the results for the additive shock in Fig. 3, where, for each setting and indicator, we express the average of the simulated results as relative to the indicator before the shock. If an indicator takes a value higher (lower) than 1, simulated inequality between groups is higher (lower) than the observed value.

First, we notice that, as expected, the behaviour of $\Delta\bar{y}_{jh}$ is the same of G_b^{BM}. It is also similar to that of G_b^{YL}, but less sensitive to changes in the scenario. On the other

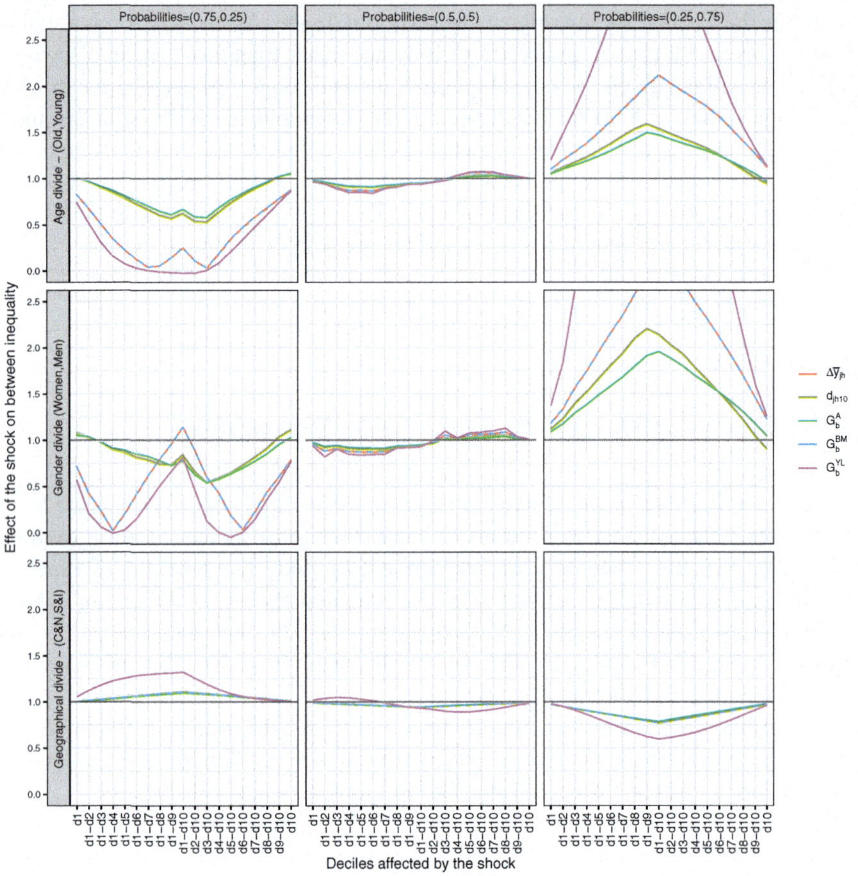

Fig. 3 Summary of simulation results, additive shock. The effect of the shock on the inequality between groups, as expressed by the ratio between the mean of the simulated indicators and their values before the shock, by inequality factor and shock probability

hand, d_{jh10} and G_b^A accordingly plot a different picture. Where do these differences come from? What information do they provide?

To deal with these questions, we link the trajectories of Fig. 3 to the consumption gaps the groups (and their deciles) had in 2019 (Table 2 and Fig. 2). Let's focus for a while only on the gender and age divides, which show interesting similarities. In the first scenario (Probabilities = (0.75,0.25)), the probabilities of a positive shock are higher for Old and Women. In 2019, Old and Women are, respectively, poorer than Young and Men in all the deciles, but in the first and the tenth. All five indicators suggest a decreasing between-group inequality while the shock propagates along the first deciles. However, when the shock only hits the first decile, d_{jh10} and G_b^A increase with respect to their values before the shock. This is because the two first deciles of the two groups are very close before the shock, while they stray after it.

More precisely, before the shock, the average distance at the first decile is less than 20 for both partitions. After the shock, with the probabilities of the first scenario, c = 200 produces an expected average increase of 150 for Old and Women, and of 50 for Young and Men. The richest groups in the first decile are now Old and Women, and the absolute difference between the two first deciles strays from less than 20 to more than 80.

Measuring the gender divide, a similar argument explains why $\Delta\bar{y}_{jh}$, G_b^{BM} and G_b^{YL} start to increase from $d1 - d5$, while d_{jh10} and G_b^A continue to decrease. Let us focus on the grey bars in Fig. 2. After the simulated shock, most of the bars become shorter, and the mean of decile differences decreases. On the contrary, the three indicators based on the mean stop decreasing at d1-d4, because the grey bars become negative after the shock and compensate for the positive ones. At d1-d4 the means are approximately equal. However, involving more deciles in the shock shortens more bars and should imply a further decrease in between-group inequality, which is not captured by the three mean-based indicators.

All the trajectories of other scenarios can be interpreted with the same approach, with the case of symmetric shock between groups leading to no drastic changes in the indicators.

As for the geographical divide, it is always quite stable, because the shock is low relative to the divide. However, the direction is the one expected, and the indicators are well correlated because, given the distributions before the shock, the shortening of the bars corresponds to the approach of the two averages.

As for the multiplicative case, we inform that the findings follow similar arguments. As natural, the divides are more reactive when extending the shocks to top deciles.

5.2 A Real Shock: The COVID-19 Impact

To compare the five indicators of between-inequality under analysis ($\Delta\bar{y}_{jh}$, d_{jh10}, G_b^A, G_b^{BM} and G_b^{YL}), Table 3 reports their values for the three bi-partitions under study, in 2019 and 2020 (G_b^A, G_b^{BM} and G_b^{YL} are reported as share of G to take into account inequality changes between the two years).

We can observe really different patterns for the three divides analyzed. Geography stands out as the main inequality factors in both years and all indicators agree in providing this suggestion. The same agreement is observed with reference to the gender divide, which is the smallest of the three in 2019, but which increases strongly in 2020, reaching the importance of the age divide.

Comparison on a time basis is illustrated in Fig. 4, which shows the percentage variation, from 2019 to 2020, of the five indicators. They agree on the direction of the dynamic of the three divides. However, they differ on the intensity of the divides evolution but for the geographical divide. It is interesting to compare the pictures of the two baseline indicators, $\Delta\bar{y}_{jh}$ and d_{jh10}, with the evolution described by the between components from the alternative Gini index decompositions. As for the age

Table 3 Observed value of $\Delta\bar{y}_{jh}$, d_{jh10}, G_b^A, G_b^{BM} and G_b^{YL} as measured on partitions—defined by age, gender and geographical factors—of the Italian households' monthly consumption in 2019 and 2020

	2019					2020				
	$\Delta\bar{y}_{jh}$	d_{jh10}	G_b^A	G_b^{BM}	G_b^{YL}	$\Delta\bar{y}_{jh}$	d_{jh10}	G_b^A	G_b^{BM}	G_b^{YL}
Age divide	0.045	0.071	0.044	0.033	0.003	0.075	0.075	0.047	0.058	0.005
Gender divide	0.026	0.036	0.027	0.019	0.001	0.064	0.064	0.041	0.048	0.003
Geogr. divide	0.304	0.304	0.148	0.218	0.077	0.286	0.285	0.142	0.208	0.069

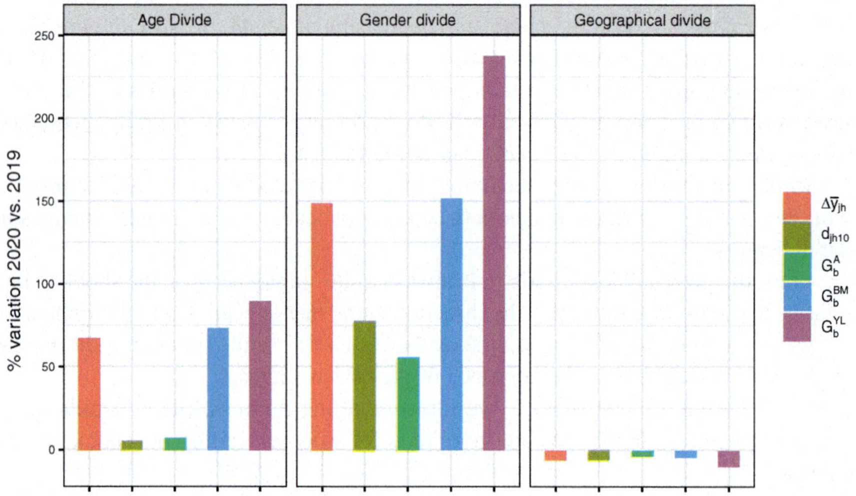

Fig. 4 The three panels show the percentage variation (from 2019 to 2020) of the measures of between-group inequality discussed in Sect. 2 for three partitions, based on household head characteristics. The "Age Divide" panel compares "Young (aged less than 65) versus Old (aged more than 64)"; the "Gender Divide" panel compares "Men versus Women"; the "Geographical divide" panel compares "North and center versus South and islands"

divide, from 2019 to 2020, it increases by 67.7% according to the means difference, while just by 5.3% according to d_{jh10}. While G_b^{BM} and G_b^{YL} are in agreement with $\Delta\bar{y}_{jh}$ for a strong increase in the age divide, G_b^A increases slightly according to d_{jh10}. To understand this difference we propose to look at Fig. 4 while taking Fig. 2 into account. The negative bar for the tenth decile in 2019 lowers the difference between the means of Young and Old, thus leading to an underestimation of the age divide by using the three mean-based indicators. The bar for the tenth decile becomes positive in 2020, overcoming the previous underestimation. This is why in 2020 G_b^{BM}, G_b^{YL} and $\Delta\bar{y}_{jh}$ show (Fig. 4) a substantial increase. Only G_b^A and d_{jh10} properly represent the information in Fig. 2.

The picture does not depart when considering the gender partition. All the indicators signal a stronger increase in the divide, ranging from +56% to +238%. Again,

Table 4 Bootstrap mean, 90% confidence interval and coefficient of variation of $\Delta\bar{y}_{jh}$, d_{jh10}, G_b^A, G_b^{BM} and G_b^{YL} as measured on partitions—defined by age, gender and geographical factors—of the Italian households' monthly consumption in 2019 and 2020. The bootstrap algorithm consists of 999 replicates. For each replicate, observations are sampled with replacement according to an i.i.d. bootstrap scheme accounting for survey weights

		2019					2020				
		$\Delta\bar{y}_{jh}$	d_{jh10}	G_b^A	G_b^{BM}	G_b^{YL}	$\Delta\bar{y}_{jh}$	d_{jh10}	G_b^A	G_b^{BM}	G_b^{YL}
Age divide	5%	0.015	0.054	0.037	0.012	0.001	0.050	0.051	0.035	0.038	0.002
	Mean	0.046	0.073	0.046	0.034	0.003	0.075	0.075	0.048	0.058	0.005
	95%	0.074	0.092	0.057	0.056	0.006	0.099	0.099	0.061	0.076	0.008
	C.V.	0.386	0.156	0.132	0.387	0.544	0.193	0.189	0.158	0.193	0.341
Gender divide	5%	0.003	0.022	0.017	0.002	0.000	0.041	0.042	0.030	0.031	0.002
	Mean	0.027	0.040	0.029	0.020	0.001	0.064	0.064	0.042	0.048	0.004
	95%	0.056	0.060	0.041	0.041	0.003	0.089	0.088	0.055	0.067	0.006
	C.V.	0.616	0.302	0.254	0.617	0.951	0.228	0.224	0.186	0.227	0.418
Geogr. divide	5%	0.278	0.278	0.138	0.201	0.065	0.264	0.264	0.133	0.192	0.059
	Mean	0.305	0.304	0.149	0.219	0.077	0.285	0.285	0.142	0.208	0.069
	95%	0.331	0.331	0.160	0.238	0.090	0.305	0.305	0.151	0.222	0.078
	C.V.	0.053	0.053	0.046	0.052	0.097	0.045	0.046	0.038	0.044	0.084

the two measures not relying on the mean capture the asymmetric change of the distributions and report the weakest increase in the gender divide.

As for the last partition, the geographical one, the agreement between $\Delta\bar{y}_{jh}$ and d_{jh10} on modest reduction of the geographical divide is confirmed by the between components of the Gini index decompositions. In this case, all the indicators do agree due to a symmetric change in the consumption distribution, leading to a quite homogeneous decrease of the divide across all the deciles.

To sum up, when deciles vary heterogeneously, the change of the difference between subgroup means does not fully account for the evolution of the distance between the subgroup distributions. The latter is better captured by the average change of the differences between quantiles (d_{jh10}), which is well represented by the evolution of Attili's between component.

With the aim of assessing the relevance of our results, and referring to statistical inference for inequality measures (see, e.g., [17]), we develop a bootstrap procedure in order to obtain confidence intervals and variability measures for all relevant indicators. Each of the 999 replicates of the bootstrap is obtained by resampling observations from the original datasets according to an i.i.d. sampling with replacement, accounting for survey weights.

Table 4 reports the observed value, the 90% bootstrap confidence intervals and the coefficient of variation for $\Delta\bar{y}_{jh}$ and d_{jh10}, as well as for the between-inequality component of the three Gini index decompositions analyzed throughout the paper. The averages of the 999 replicates of all the indicators are close to the observed values both in 2019 and 2020, meaning that the bootstrap procedure is unbiased.

The confidence intervals are quite symmetric around the observed value for all five indicators, so we can directly rely on variability measures to analyse the dispersion of their distributions. Our choice is to use the coefficient of variation to ensure comparability between five indicators ranging on different scales.

Comparing the partitions, both in 2019 and 2020 the dispersion of the indicators varies across partitions in inverse proportion to the intensity of the divide: the geographical divide is measured with the lowest uncertainty by all the indicators, while the dispersion of the indicators is the highest measuring the gender divide. This is because, when the divide is lower, the resampling is more likely to produce larger variations.

Comparing the two years, with few exceptions concerning d_{jh10} and G_b^A, the indicators have higher variability in 2019. As for the comparison of the indicators, for all partitions and both years, G_b^{YL} is the between-inequality indicator with the highest variability over the 999 replicates, while G_b^A is the stablest. In the middle, G_b^{BM}, $\Delta\bar{y}_{jh}$ and d_{jh10} have intermediate values of variability, with G_b^{BM} and $\Delta\bar{y}_{jh}$ reasonably showing similar dispersion.

Both the comparison of the two years and that of the indicators suggest that d_{jh10} and G_b^A are more robust to extreme values, which are more frequent in 2019 and affect in a more decisive manner the indicators based on the means.

6 Conclusions

We analyze the effects of asymmetric shocks on the measurement of inequality, with a specific focus on the Gini index and on the components of its decompositions.

As a case study, we evaluate the COVID-19 impact on inequality with respect to three factors: age, gender and geography. Age and gender gaps increase, while territorial divide (which is confirmed as the most important inequality factor in Italy) slightly decreases.

The assessment of these dynamics strongly depends on the employed measure. If it is based on averages, the evaluation reflects an aggregate effect, which may be attributable to some aspects of the distribution only. This is the case of age and gender gaps in Italian household consumption: the average change is driven by the group of households consuming more.
When, on the other hand, the measure of divides refers to the whole distribution, we have a more complete representation of the intervening dynamics.

Our findings are supported by a Monte Carlo study, which allows us to assess the effects of a wide range of asymmetric shocks on different inequality measures, and by a bootstrap procedure, which allows us to infer the confidence intervals and evaluate the responsiveness of indicators to extreme values.

Among the future lines of research, we intend to assess the role of other aggregated comparison measures (useful references being Wasserstein distance and Kullback-Leibler divergence) and analyze other crises as well as the effects of the post-COVID recovery in 2021.

To conclude, inequality decomposition methods, which include both mean-based measures, such as [5, 20], and measures referring to the whole distribution, such as [2], represent an excellent choice for measuring the effects on inequality of symmetric and asymmetric shocks.

References

1. Allison, P.D.: Measures of inequality. Am. Sociol. Rev. 865–880 (1978)
2. Attili, F.: Uncovering complexities in horizontal inequality: a novel decomposition of the Gini index. Soc. Ind. Res. **173**(2), 351–376 (2024)
3. Attili, F., Costa, M.: Covid-19 impact assessment and inequality decomposition. In: Book of Short Papers SIS 2022, vol. 1, pp. 1189–1094. Pearson (2022)
4. Bell, B., Bloom, N., Blundell, J.: Income dynamics in the United Kingdom and the impact of the Covid-19 recession. Quant. Econ. **13**(4), 1849–1878 (2022)
5. Bhattacharya, N., Mahalanobis, B.: Regional disparities in household consumption in India. J. Am. Stat. Assoc. **62**(317), 143–161 (1967)
6. Elbers, C., Lanjouw, P., Mistiaen, J.A., Özler, B.: Reinterpreting between-group inequality. J. Econ. Inequal. **6**(3), 231–245 (2008)
7. Fana, M., Torrejón Pérez, S., Fernández-Macías, E.: Employment impact of Covid-19 crisis: from short term effects to long terms prospects. J. Ind. Bus. Econ. **47**(3), 391–410 (2020)
8. Gallo, G., Raitano, M.: Sos incomes: Simulated effects of covid-19 and emergency benefits on individual and household income distribution in Italy. ECINEQ, Society for the Study of Economic Inequality, Tech. rep. (2020)
9. Gastwirth, J.L.: Is the Gini index of inequality overly sensitive to changes in the middle of the income distribution? Stat. Public Policy **4**(1), 1–11 (2017)
10. Giorgi, G.M.: The Gini inequality index decomposition. An evolutionary study. In: The Measurement of Individual well-being and Group Inequalities: Essays in Memory of ZM Berrebi, pp. 185–218 (2011)
11. Giorgi, G.M., et al.: Gini's scientific work: an evergreen. Metron **63**(3), 299–315 (2005)
12. Hasan, S., Islam, M.A., Bodrud-Doza, M.: Crisis perception and consumption pattern during Covid-19: do demographic factors make differences? Heliyon **7**(5), e07141 (2021)
13. Jaravel, X.: Inflation inequality: measurement, causes, and policy implications. Annu. Rev. Econ. **13**, 599–629 (2021)
14. Lambert, P.J., Decoster, A.: The Gini coefficient reveals more. Technical Report, Social Science Research Network (2005)
15. Marques Santos, A., Madrid, C., Haegeman, K., Rainoldi, A.: Behavioural changes in tourism in times of Covid-19. JRC121262 **22**(3), 121–147 (2020)
16. Milanovic, B.: After the financial crisis: the evolution of the global income distribution between 2008 and 2013. Rev. Income Wealth **68**(1), 43–73 (2022)
17. Mills, J.A., Zandvakili, S.: Statistical inference via bootstrapping for measures of inequality. J. Appl. Econ. **12**(2), 133–150 (1997)
18. Porro, F., Zenga, M.: Decomposition by subpopulations of the Zenga-84 inequality curve and the related index ζ: an application to 2014 bank of Italy survey. Stat. Methods & Appl. **29**(1), 187–207 (2020)
19. Stantcheva, S.: Inequalities in the times of a pandemic. Technical Report, National Bureau of Economic Research (2022)
20. Yitzhaki, S., Lerman, R.I.: Income stratification and income inequality. Rev. Income Wealth **37**(3), 313–329 (1991)

Exploring Factors Affecting the Evaluation of Online Learning Services. Evidence from a Social Science Bachelor's Degree

Ilaria Primerano, Giuseppe Giordano, and Maria Prosperina Vitale

Abstract The sudden shift from face-to-face to online learning during Covid-19 pandemic had important consequences for students' experiences. The present contribution aims to assess the factors affecting the student services' evaluation in terms of prior digital skills, their experience on online platforms and the perception of didactic tools. These relations are explored by adopting a Partial Least Squares Path Modeling approach. A questionnaire was administered to students' cohorts enrolled in a social science bachelor's degree program at a University in Southern Italy. The main findings reveal that the perception of online experiences is the most important factor influencing students' evaluation as well as having prior digital skills favor the usage of digital platforms and a positive assessment of distance learning tools.

Keywords Digital skills · Higher education · Online learning · PLS-PM · Service quality evaluation

1 Introduction

The adoption of online learning resources during Covid-19 pandemic has affected educational processes in the university context. The crisis generated by pandemic has, indeed, encouraged institutions to enforce the use of ICTs favoring a wider use of digital platforms with the acquisition of skills and new competences for students and teachers during lockdown restrictions [1]. The digitalization of higher education

G. Giordano · M. P. Vitale
Department of Social and Political Studies, University of Salerno, Fisciano, Italy
e-mail: ggiordano@unisa.it

M. P. Vitale
e-mail: mvitale@unisa.it

I. Primerano (✉)
CNR IRPPS - Institute for Research on Population and Social Policies, National Research Council, Rome, Italy
e-mail: ilaria.primerano@cnr.it

M. Bini et al. (eds.), *Advanced Methods in Statistics, Data Science and Related Applications*, Springer Proceedings in Mathematics & Statistics 467,
https://doi.org/10.1007/978-3-031-65699-6_20

has also opened an in-depth debate on the possibilities offered by technologies for learning scenarios going beyond the weaknesses registered for the development of digital literacy [2].

Within the analysis of the effect of pandemic on different aspects of the university life, several studies have focused on psychological consequences of lockdown on mental health [3], students' performance [4, 5], and the perception of online teaching experiences [6]. With respect to this latter aspect, the redefinition of training processes, from face-to-face to online spaces, has stimulated the assessment of factors affecting didactic experiences supported by digital tools and online resources [7], as well as technological, pedagogical, and social issues [8].

The quality of online teaching and university services' satisfaction has been examined by exploring different aspects [9], among others the effect of the type of resources used to support learning processes [10], the tools implemented in the digital platforms [1], the access to digital devices and software [11], and the role played by previous individual digital skills [12], a comprehensive review can be found in [13]. Albeit, according to the Italian digital skills national strategy [14], it is strategic a *review and strengthening of the fundamental and cross-cutting information technology culture necessary to the digital transformation of university education in all classes*, there is no doubt that the digital skills acquired in previous educational terms, have had an impact on the ability of university students to adapt to the new services offered by University.

Moving from this theoretical framework, the present contribution focusing on a Bachelor of Arts degree, aims to address the following research questions:

- *How do previous digital skills affect the use of digital platforms during distance learning, student perception and evaluation on online learning services?*
- *How do the use of digital platforms and the perception of online experiences affect the evaluation of online learning services?*

To investigate the relationships among these factors, we adopt the Partial Least Squares Path Modeling (PLS-PM) approach [15, 16]. The PLS-PM combines two sub-models in the path diagram: the *measurement model* or *outer* model, including the relations between each manifest variable (MV) and the corresponding latent variable (LV), and the *structural model*, also called *inner* or *path* model, including the relations among LVs. The direction of the relations defines different model specifications. The outer model can be *reflective*, when MVs are a reflection of the LV, or *formative*, when MVs affect a LV. The inner model includes *exogenous* and *endogenous* latent variables in case of they depend on other LVs. PLS-PM approach is defined to deal with metric data for the dependent MVs for different types of constructs [17]. However, quasi-metric data stemming from Likert scales, such as discrete ratings or rankings, are also acceptable as long as five or more scale points can be assumed to be equidistant [18]. It is also possible to include categorical variables with two levels as construct indicator [19]. In presence of more than two levels, they have to be transformed into as many dummy variables as levels minus one. Preferably, categorical variables should only play the role of exogenous variables in a structural model. A non-metric approach to PLS-PM has been addressed to handle both metric and non-metric variables at once, based on optimal scaling features [20].

The working hypotheses along with this methodological framework are investigated through real-data gathered by administering a questionnaire to students' cohorts in a university located in Southern Italy. We collect answers of 188 students enrolled in a Social Science bachelor degree program during the a.y. 2020–2021.

The contribution is structured as follows. Section 2 presents data and descriptive statistics of the variables under analysis. Section 3 reports the conceptual model with details of research hypotheses and the main findings of the estimated PLS-PM. Some concluding remarks are reported in Sect. 4.

2 Data and Descriptive Statistics

The whole data collected during the survey refers to a wider study as described in [21]. The survey has reached 188 students covering around 70% of the population enrolled in the second and third year of a social science bachelor's degree program in the academic year 2020–2021.

At our aims, we focus on a subset of items describing each dimension as follows, (see also Table 1):

- SKILLS—The prior digital skills dimension measures student competences according to some of the key components developed for the European Digital Competence Framework for Citizens (EU DigComp 2.1). The seven items are measured on a 7-points Likert scale, from *not at all* to *completely* category. The items regard the ability to use software systems for security and protection of personal data, and to avoid online dangers and fraud.
- USAGE—The usage of digital tools dimension mainly refers to students adoption of the tools implemented by universities to support distance learning. Specifically, the items refer to the use of collaborative files, sharing files, and chat conversations, measured on a 5-points Likert scale (*never, rarely, sometimes, often, always*). These three items have been recoded as binary indicators for each modality (according to the non-metric PLS-PM procedure) and thus, only five significant effects out of the fifteen have been retained in the final model specification.
- PERCEPTION—The student perception dimension is defined on a metric scale about four items measuring the distance learning activities, the use of digital platform, the self-study attitude, and its mood during online activities.
- EVALUATION—The student service evaluation dimension is measured by eight items on a 7-points Likert scale, from *not at all satisfied* to *completely satisfied.* The items refer to the effectiveness of distance learning, the organization of classrooms, exams and their own learning ability to organize study time, the quality of the relationships among peers and with teachers, the distance learning modality adopted, and the distance learning platform tools.

Table 1 Latent variables (LVs) blocks and their items (MVs) included in the PLS-PM

Dimensions	Description	N. items	Label
SKILLS	Skills measured according to the key components developed by the European Digital Competence Framework for Citizens (EU DigComp 2.1)	7	ds_cloud
			ds_collaborativetools
			ds_socialnetworks
			ds_programming
			ds_office
			ds_security
			ds_fraud
USAGE	Frequency in the use of some interactive tools available on digital platform to support distance learning activities	5	always_chat
			often_collfile
			always_collfile
			often_filesharing
			always_filesharing
PERCEPTION	Students' perception of different distance learning aspects: distance learning activities, ICT tools usage, self-study ability, and personal mood	4	item_dad
			item_ict
			item_sel
			item_mood
EVALUATION	Students' opinions on distance learning experience, platform's functionality, organization of activities, effectiveness of distance learning, quality of the relationship between peers and teachers	8	dad_experience
			organization_class
			organization_exam
			organization_studytime
			stud_relation
			tea_relation
			tea_modality
			tools_use

Figures 1 and 2 show some descriptive statistics of all the items included in each dimension. It appears that students rarely used chat and more often collaborative files and file sharing tools (Fig. 1, left panel). With respect to prior digital skills, students present a high experience in using social networks, cloud and fraud detection (Fig. 2, right panel). Furthermore, students' perceptions show moderate values for almost all items (Fig. 2, left panel), while the evaluation of relations among peers and teachers presents the lowest median values with respect to the other characteristics (Fig. 2, right panel).

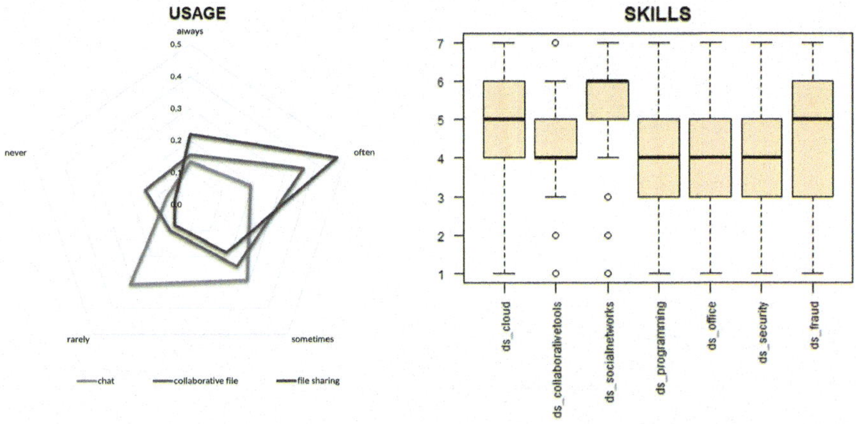

Fig. 1 Distributions of items in USAGE and SKILL dimensions

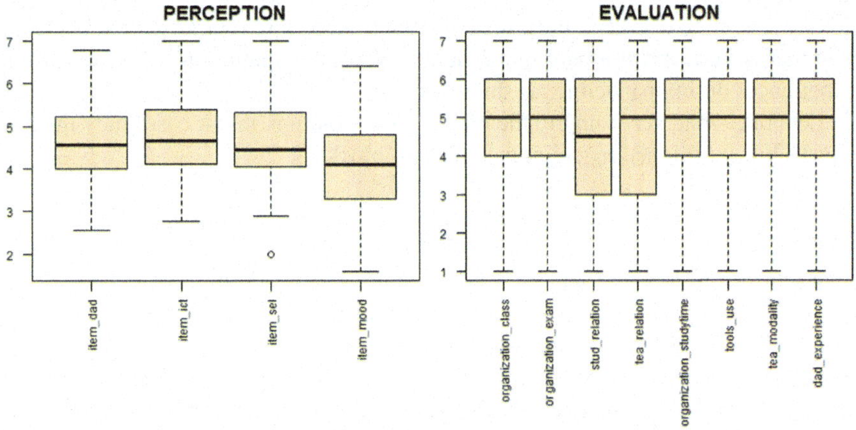

Fig. 2 Distributions of items in PERCEPTION and EVALUATION dimensions

3 The PLS-PM Conceptual Model

According to the PLS-PM notation, let **X** be a data matrix partitioned by columns in H blocks:

$$\mathbf{X} = [\mathbf{X}_1 | \ldots | \mathbf{X}_h | \ldots | \mathbf{X}_H] \quad (1)$$

of order $n \times p$, each block $\mathbf{X}_h$ $(h = 1, \ldots, H)$ has dimensions $n \times p_h$ with $\sum_{h=1}^{H} p_h = p$; where n is the number of observations and p is the number of all items.

A *path diagram* is the typical representation of a causal model where each block $\mathbf{X}_h$ is a set of MVs and is conceptually connected to a LV ξ_h. In such a diagram, rectangles

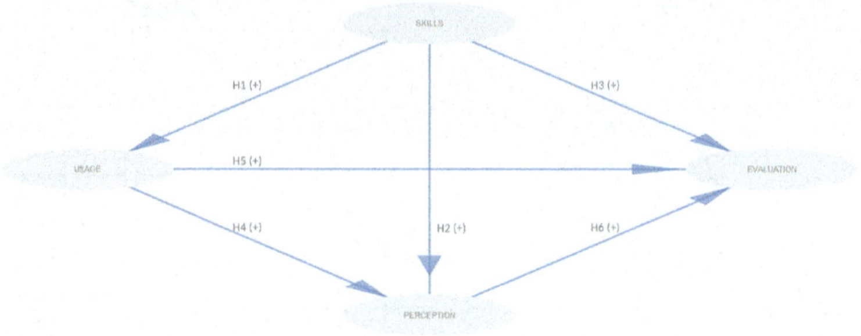

Fig. 3 The PLS-PM conceptual model

represent manifest variables ($\mathbf{x}_{hj}$, $j = 1, \ldots, p_h$), ellipses are the latent variables (ξ_h), and arrows describe the direction of the relations between them, supposed to be linear.

Going back to our case study, the 24 MVs grouped into four LVs are included in a formative conceptual model to measure students' evaluation of distance learning experiences during pandemic, as showed in Fig. 3.

Let notice that, according to the the PLS-PM notation, in our case study the data matrix $\mathbf{X}$ is partitioned by columns in $H = 4$ blocks:

$$\mathbf{X} = \left[\mathbf{X}_{1=skills}|\mathbf{X}_{2=usage}|\mathbf{X}_{3=perception}|\mathbf{X}_{4=evaluation}|\right] \tag{2}$$

with dimensions $n = 188$ and $p = 24$, i.e., $p_1 = 7$, $p_2 = 5$, $p_3 = 4$, $p_4 = 8$.

We remark that for the USAGE block, as indicated in Sect. 2, a preliminary study was conducted on all 15 items and after a deep analysis, we decided to retain only five of that. Specifically, we retrieve one item for "chat" (always_chat); two items for "collaborative files" (often_collfile; always_collfile), and two items for "file sharing" (often_filesharing; always file_sharing).

Following the research questions in Sect. 1, we assume that:

H1 Prior digital skills positively impact on student frequency of using distance learning tools [SKILLS → USAGE]
H2 Prior digital skills positively impact on student perception [SKILLS → PERCEPTION]
H3 Prior digital skills positively impact on student evaluation [SKILLS → EVALUATION]
H4 Frequency of using distance learning tools positively impacts on student perception [USAGE → PERCEPTION]
H5 Frequency of using distance learning tools positively impacts on student evaluation [USAGE → EVALUATION]
H6 Perception on distance learning aspects positively impacts on student evaluation [PERCEPTION → EVALUATION].

3.1 Results

The results of the formative PLS-PM are based on a two-step assessment procedure. In the first step, the outer loadings and weights of each item are compared to identify those that most effectively contribute to the latent construct in the outer model; while the second step concerns the assessment of the inner model.

Table 2 displays weights, loadings (see also Fig. 4), communalities, and redundancy indices of the outer model. Almost all loadings are close to or greater than the threshold value of 0.7. They also show acceptable communality values except for

Table 2 PLS-PM outer model results

	Weight	Loading	Communality	Redundancy
SKILLS				
ds_cloud	0.120	0.634	0.402	0.000
ds_collaborativetools	0.207	0.724	0.524	0.000
ds_socialnetworks	0.260	0.558	0.312	0.000
ds_programming	0.241	0.747	0.558	0.000
ds_office	0.289	0.782	0.611	0.000
ds_security	0.230	0.754	0.568	0.000
ds_fraud	0.075	0.656	0.431	0.000
USAGE				
always_chat	0.568	0.640	0.410	0.016
often_collfile	0.318	0.364	0.132	0.005
always_collfile	0.329	0.427	0.182	0.007
often_fileshar	0.532	0.398	0.158	0.006
always_fileshar	0.460	0.367	0.135	0.005
PERCEPTION				
item_dad	−0.070	0.670	0.449	0.049
item_ict	0.296	0.718	0.515	0.056
item_sel	0.316	0.775	0.600	0.065
item_mood	0.656	0.900	0.810	0.088
EVALUATION				
dad_experience	0.237	0.847	0.717	0.407
organization_class	0.269	0.828	0.685	0.389
organization_exam	0.026	0.692	0.479	0.272
organization_studytime	0.172	0.802	0.643	0.365
stud_relation	0.077	0.695	0.483	0.274
tea_relation	0.229	0.835	0.696	0.396
tea_modality	0.215	0.844	0.712	0.404
tools_use	−0.007	0.791	0.626	0.356

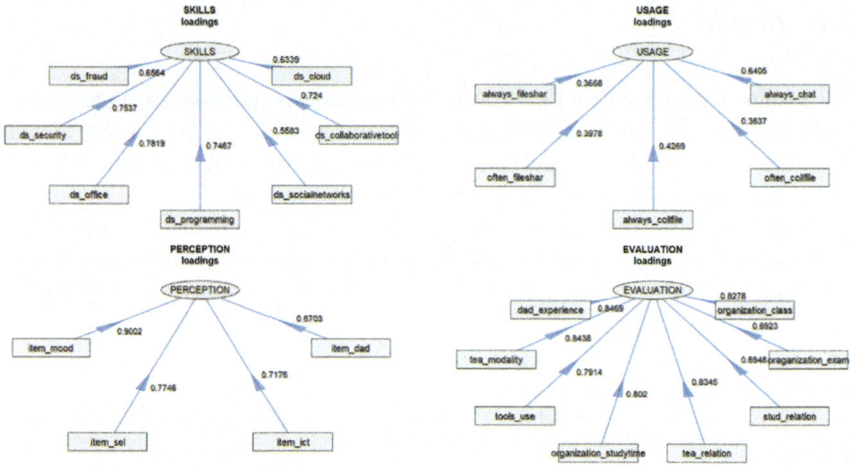

Fig. 4 Path diagram of outer model

the variables considered in the USAGE block. At the same time, low and negative weights are observed for item_dad in the PERCEPTION block, and tools_use in the EVALUATION one.

By looking at the inner model, it seems that the research hypotheses in Sect. 3 are supported by the results (see Table 3 and Fig. 5). The regression coefficients are almost all positive and significant except for the USAGE. Prior digital skills of students have a significant and positive direct effect to usage of digital tools available on digital platforms (0.200), perception (0.302), and evaluation of distance learning activities (0.117), supporting H1, H2 and H3. The usage of distance learning tools has a not

Table 3 PLS-PM inner model results

	Estimate	Std. error	t value	Pr(>|t|)
USAGE				
Intercept	0.000	0.072	0.000	1.000
Skills	0.200	0.072	2.785	0.006
PERCEPTION				
Intercept	0.000	0.069	0.000	1.000
Skills	0.302	0.071	4.259	0.000
Usage	0.084	0.071	1.188	0.236
EVALUATION				
Intercept	0.000	0.048	0.000	1.000
Skills	0.117	0.052	2.252	0.026
Usage	0.232	0.050	4.663	0.000
Perception	0.633	0.051	12.331	0.000

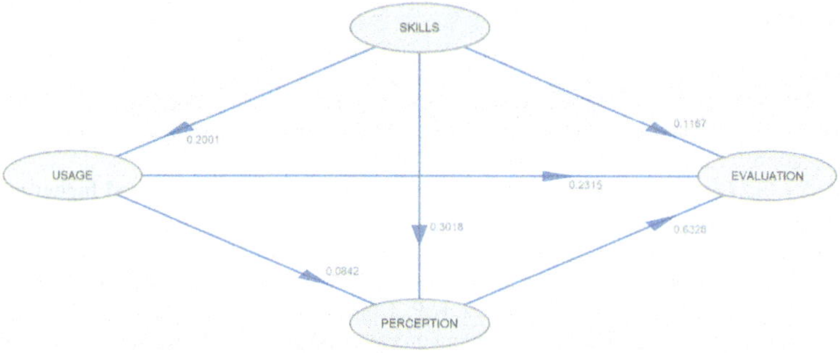

Fig. 5 Path diagram of PLS-PM inner model

significant effect on student perception (0.084), and a significant positive effect on evaluation (0.232), supporting H5. Finally, student perception is the most important factor affecting student evaluation with the highest significant and positive direct effect (0.633), supporting H6. Indeed, looking at the coefficient of determination R^2 of the endogenous latent variables, it emerges that the highest value is observed for the EVALUATION block ($R^2 = 0.568$), which also shows a communality of 0.630 and a redundancy index of 0.358.

4 Conclusion

Covid-19 pandemic has strongly affected the world of education and training, disrupting the habits and behaviors of students. Universities have been asked to face a new process of digitalization of education and services offered to their students in a rapid and emergency time frame, of which distance learning services are one of the leading results. The sudden shift, from face-to-face to online learning, has had important consequences for students.

The PLS-PM results provide useful insights to answer our research questions supporting that the perception of online experiences is the most important factor influencing students' evaluation of distance learning services. Although, having prior digital skills favor the usage and the perception as well as the positive evaluation of these services. Even if the study here presented deals with data gathered on a small sample of students enrolled in a social science bachelor's degree program, it provides useful suggestions regarding issues and benefits of virtual educational practices. Institutions need to redefine educational scenarios in which online academic environments can favor the improvement of skills and new competences useful for pedagogically renewed learning processes in university scenarios. These policy recommendations could be more effective in presence of students lacking prior digital skills or roughly speaking for not STEM disciplines.

References

1. Tejedor, S., Cervi, L., Pérez-Escoda, A., Tusa, F., Parola, A.: Higher education response in the time of coronavirus: perceptions of teachers and students, and open innovation. J. Open Innovation: Technol. Mark. Complex. **7**(1), 43 (2021)
2. Tejedor, S., Cervi, L., Pérez-Escoda, A., Jumbo, F.T.: Digital literacy and higher education during COVID-19 lockdown: Spain, Italy, and Ecuador. Publications **8**(4), 48 (2020)
3. Cao, W., Fang, Z., Hou, G., Han, M., Xu, X., Dong, J., Zheng, J.: The psychological impact of the COVID-19 epidemic on college students in China. Psychiatr. Res. **287**, 112934 (2020)
4. Gonzalez, T., De La Rubia, M.A., Hincz, K.P., Comas-Lopez, M., Subirats, L., Fort, S., Sacha, G.M.: Influence of COVID-19 confinement on students' performance in higher education. PloS one **15**(10) (2020)
5. Leorato, S.: Did transition to online teaching during Covid-19 pandemic affect students' performance? Evidence from a statistics course. Statistica Applicata-Italian J. Appl. Stat. **1**, 75–95 (2022)
6. Murphy, L., Eduljee, N.B., Croteau, K.: College student transition to synchronous virtual classes during the COVID-19 pandemic in Northeastern United States. Pedag. Res. **5**(4) (2020)
7. Cicha, K., Rizun, M., Rutecka, P., Strzelecki, A.: COVID-19 and higher education: First-year students' expectations toward distance learning. Sustainability, **13**(4), 1889 (2021)
8. Giovannella, C., Marcello, P., Donatella, P.: The effects of the Covid-19 pandemic on Italian learning ecosystems: the school teachers' Perspective at the steady state. Interact. Des. Arch.(s) J. **45**, 264–286 (2020)
9. Li, L., Yin, T.S., Mohamad, M.N.: Factors relating to student satisfaction with service quality: a systematic review. Global Bus. Manag. Res. **14** (2022)
10. Schijns, J.: Measuring service quality at an online university: using PLS-SEM with archival data. Tert. Educ. Manag. **27**(2), 161–185 (2021)
11. Weldon, A., Ma, W.W., Ho, I.M., Li, E.: Online learning during a global pandemic: perceived benefits and issues in higher education. Knowl. Manag. E-Learn.: Int. J. **13**(2), 161–181 (2021)
12. Almaiah, M.A., Al-Khasawneh, A., Althunibat, A.: Exploring the critical challenges and factors influencing the E-learning system usage during COVID-19 pandemic. Educ. Inf. Technol. **25**(6), 5261–5280 (2020)
13. Su, C.Y., Guo, Y.: Factors impacting university students' online learning experiences during the COVID-19 epidemic. J. Comput. Assist. Learn. **37**, 1578–1590 (2021)
14. Ministero per l'innovazione tecnologica e la digitalizzazione, D.M. 21 luglio 2020, Decreto e Strategia nazionale per le competenze digitali, https://assets.innovazione.gov.it/1610029655-dtd-1277-a-all1.pdf (2020)
15. Lamberti, G., Aluja Banet, T., Trinchera, L.: University image, hard skills or soft skills: which matters most for which graduate students?. Quality & Quantity 1–22 (2021)
16. Vinzi, V.E., Trinchera, L., Amato, S.: PLS path modeling: from foundations to recent developments and open issues for model assessment and improvement. Concepts, Methods and Applications, Handbook of partial least squares, pp. 47–82 (2010)
17. Crocetta, C., Antonucci, L., Cataldo, R., Galasso, R., Grassia, M.G., Lauro, C.N., Marino, M.: Higher-order PLS-PM approach for different types of constructs. Soc. Indic. Res. **154**, 725–754 (2021)
18. Rhemtulla, M., Brosseau-Liard, P.E., Savalei, V.: When can categorical variables be treated as continuous? A comparison of robust continuous and categorical SEM estimation methods under suboptimal conditions. Psychol. Methods **17**(3), 354–373 (2012)
19. Henseler, J.: Modeling, partial least squares path. In: Leeflang, P., Wieringa, J., Bijmolt, T., Pauwels, K. (eds.) Advanced Methods for Modeling Markets. International Series in Quantitative Marketing. Springer, Cham (2017)

20. Russolillo, G.: Non-Metric Partial Least Squares. Electron. J. Stat. **6**, 164–1669 (2012)
21. Primerano, I., Catone, M.C., Giordano, G., Vitale, M.P.: Assessing undergraduate students' perceptions of distance learning during the COVID-19 pandemic. In: Lombardo, R., Camminatiello, I., Simonacci, V. (eds.) Book of Short papers 10th International Conference IES 2022 Innovation and Society 5.0: Statistical and Economic Methodologies for Quality Assessment, 263-268. PKE s.r.l., Sesto San Giovanni (2022)

Discrimination via Principal Components

Discriminazione Tramite Componenti Principali

Nickolay Trendafilov, Michele Gallo, Violetta Simonacci, and Valentin Todorov

Abstract In many modern data, the number of variables is much higher than the number of observations and the within-group scatter matrix is singular. This work proposes a way to circumvent this problem by doing LDA in a low-dimensional space formed by the first few principal components (PCs) of the original data. Two approaches are considered to improve their discrimination abilities in this low-dimensional space. Specifically, the original PCs are rotated to maximize the LDA criterion, or penalized PCs are produced to achieve simultaneous dimension reduction and maximization of the LDA criterion. Both approaches are illustrated and compared on some well known data set.

Astratto *Sempre più spesso, i dati si presentano con un numero di variabili molto più alto di quello delle osservazioni, ne consegue che la matrice di dispersione intra-gruppi è singolare. Questo lavoro propone un modo per superare questo problema mediante la LDA in un sottospazio formato dalle componenti principali (PC) dei dati originali. In particolare, per migliorare la capacità di discriminazione nello spazio a dimensioni ridotte, una prima opzione si basa sulla rotazione delle PC originali, mentre una seconda consiste nell'individuare delle PC penalizzate che sono date dalla contemporanea riduzione dello spazio dimensionale e dalla massimizzazione del criterio LDA. Entrambi gli approcci sono illustrati e confrontati usando dataset noti in letteratura.*

Keywords Dimension reduction · Orthogonal rotations · Penalized PCA

N. Trendafilov (✉) · M. Gallo
University of Naples "L'Orientale", Naples, Italy
e-mail: ntrendafilov@unior.it

M. Gallo
e-mail: mgallo@unior.it

V. Simonacci
University of Naples Federico II, Naples, Italy
e-mail: violetta.simonacci@unina.it

V. Todorov
United Nations Industrial Development Organization (UNIDO), VIC, Vienna, Austria
e-mail: valentin@todorov.at

M. Bini et al. (eds.), *Advanced Methods in Statistics, Data Science and Related Applications*, Springer Proceedings in Mathematics & Statistics 467,
https://doi.org/10.1007/978-3-031-65699-6_15

1 Introduction

Many modern data $X \in \mathbb{R}^{n \times p}$ have much more variables than observations, $p \gg n$. Then, the Fisher's linear discriminant analysis (LDA) cannot be applied because the within-group scatter matrix S_W is singular. There exists a great number of approaches to circumvent this problem [6, Chap. 7.4]. This work proposes to do LDA in a low-dimensional space formed by the first few principal components (PCs) of X. Two approaches are considered to improve their discrimination abilities in this low-dimensional space. Specifically, the original PCs are rotated to maximize the LDA criterion, or penalized PCs are produced to achieve simultaneous dimension reduction and maximization of the LDA criterion. Both approaches are illustrated and compared on a well known data set. It is shown how they can be extended to clustering.

2 Revisiting PCA

Let X be an $n \times p$ data matrix which columns are centered, i.e. $1_n^\top X = 0_p^\top$, and have unit lengths, i.e. $\text{diag}(X^\top X) = I_p$. For short, such X is called whitened. Principal component analysis (PCA) of X is performed by its singular value decomposition (SVD). Assuming that the rank of X is $r \leq \min\{n, p\}$, the SVD can take the form $X = FDA^\top$, where $F^\top F = A^\top A = I_r$ and $D \in \mathbb{R}^{r \times r}$ is a positive definite diagonal matrix which diagonal entries are arranged in decreasing order. Note that $1_n^\top F = 0_r^\top$.

In PCA applications, we are looking for some $s \leq r (\leq \min\{n, p\})$ and replace the original data matrix X by its truncated SVD of the form $X_s = F_s D_s A_s^\top$, where F_s and A_s denote the first s columns of F and A respectively, and D_s is a diagonal matrix with the first (largest) s singular values of X. The matrices F_s and A_s contain the component scores and loadings respectively, and D_s^2 contains the variances of the first s PCs. Usually, we are interested in considerable dimension reduction, i.e. $s \ll r$ and in many cases we even set $s = 2$.

The PCA interpretation is based on the component loadings A_s, which reveal the importance of the original variables. The component scores F_s are used to visualize the n observations into a s-dimensional space. It is worth noting that for any orthogonal matrix $Q \in \mathbb{R}^{s \times s}$ we have:

$$X_s = F_s Q Q^\top D_s A_s^\top = (F_s Q)(A_s D_s Q)^\top = F_s D_s Q Q^\top A_s^\top = (F_s D_s Q)(A_s Q)^\top . \quad (1)$$

Now, right multiplication of X by (the projector) $F_s F_s^\top$ shows that PCA can be expressed as a least-squares (LS) projection of the data onto the s-dimensional subspace in $\mathbb{R}^n$ spanned by the columns of F_s, i.e. by the component scores. Then, PCA of X can be rewritten as

$$\min_{F^\top F = I_s} \|X - FF^\top X\|_E , \quad (2)$$

where $\|A\|_E^2 = \text{trace}(A^\top A)$ is the Euclidean (Frobenius) norm of $A \in \mathbb{R}^{n\times p}$. Thus, $X_s = F_s F_s^\top X$ is the best rank s (LS) approximation to X in $\mathbb{R}^n$.

If, in addition, the observations of X are divided into g groups, then the component scores F_s visualize also the g groups in a s-dimensional space. However, such a projection does not take into account the group structure of the data and thus, is not optimal. Linear discriminant analysis (LDA) overcomes this weakness by finding a low-dimensional space where the groups are best separated.

3 LDA of F_s

As above, let X be an $n \times p$ whitened data matrix with observations divided into g groups and membership defined by an $n \times g$ indicator matrix G with $\{0, 1\}$ elements, such that the matrix of group means is given by $\bar{X} = (G^\top G)^{-1}G^\top X$. Then, we have:

$$S_W = S_T - S_B = X^\top X - \bar{X}^\top (G^\top G)\bar{X} = X^\top X - X^\top H X \ , \tag{3}$$

where $H = G(G^\top G)^{-1}G^\top$ and S_B and S_W are the between- and within-groups scatter matrices of X [6, Ch 7.1]. The purpose of the Fisher's LDA is to find a transformation matrix $A \in \mathbb{R}^{p\times s}$, such that the *a-priori* groups are better separated in the transformed data $Y = XA$ than with respect to any of the original variables [3]. This is achieved by solving the following generalized eigenvalue problem

$$S_B A = S_W A\Lambda \ , \tag{4}$$

where Λ is the $s \times s$ diagonal matrix of the s largest eigenvalues of $S_W^{-1} S_B$ ordered in decreasing order. This is possible if S_W^{-1} exists. We stress that there are *at most* $\min\{p, g - 1\}$ non-zero eigenvalues in Λ, which is the rank of S_B.

3.1 LDA with Rotated Component Scores

Our purpose is to improve the discrimination features of the component scores. We assume that PCA is already performed and F_s is available, keeping in mind that $s \le \min\{p, g - 1\}$. For short we simply write F. As we want to do LDA on F, our data matrix X in (3) becomes $F \in \mathbb{R}^{n\times s}$, which is a centered orthonormal matrix. This simplifies (3) to:

$$S_W = I_s - F^\top H F = F^\top (I_n - H)F \ . \tag{5}$$

Then, the Fisher's LDA of F requires the solution of:

$$F^\top H F Q = Q\Lambda \ , \tag{6}$$

which is a symmetric eigenvalue problem for $F^\top HF$ or a singular value problem for HF. Equivalently, we can express (6) as the optimization problem:

$$\max_{Q^\top Q=I_s} \|HFQ\|_E \ , \tag{7}$$

meaning that LDA of F, in fact, finds an orthogonal rotation Q, such that the rotated component scores FQ maximize trace(S_B), the between-group sum of squares.

The problem with this solution is that it "adjusts" the group memberships of the objects with respect to group means obtained by the initial PCA of X. It will be seen, that this procedure is more suitable for clustering, when both the group memberships and the centroids are adjusted at every step.

3.2 LDA with Penalized Component Scores

In the previous Sect. 3.1 the dimension reduction and the discrimination are performed one after another. Here, we want to put them together into a single procedure. For this reason, we enhance PCA with discriminatory features by considering a joint minimization of the PCA objective function (2) and maximization of trace(S_B), the between-groups sum of squares of the component scores F. This results in the following problem:

$$\min_{F^\top F=I_s} \|(I_n - FF^\top)X\|_E - \|HF\|_E \ , \tag{8}$$

which is equivalent to:

$$\max_{F^\top F=I_s} \text{trace } F^\top(XX^\top + H)F \ , \tag{9}$$

and is solved as truncated EVD of $XX^\top + H$.

4 Examples

4.1 Examples with n > p (non-singular S_W)

4.1.1 Fisher's Irises

The famous Fisher's Iris data [3] contains observation of three ($g = 3$) Iris species on $n = 150$ flowers by measuring $p = 4$ variables. Traditionally, the Iris data are analyzed without standardization. Nevertheless, their basic statistics in Table 1 suggest that this is a plausible option and we analyze the sphered data instead.

Table 1 Iris data: Basic statistics

	Sepal length	Sepal width	Petal length	Petal width
Mean	5.843	3.057	3.758	1.199
Var	0.686	0.190	3.116	0.581

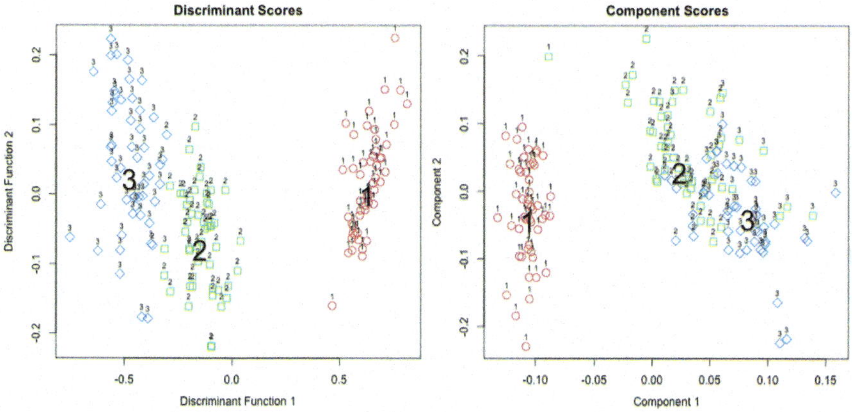

Fig. 1 Iris data: LDA (left) and PCA of sphered data

It is well known that one of the species is very well separated, but the other two are very close and difficult to split. This is exactly what is depicted in Fig. 1. The classical Fisher's LDA admits two discriminant functions, and the corresponding discriminant scores of the flowers are depicted on the left panel of Fig. 1. The number of misclassified flowers by the classical LDA is 3 (2%). There are 27 (18%) misclassified flowers with the first two CSs (right panel of Fig. 1). In the space spanned by all four CSs they reduce to 20 (13.3%).

Now, we apply the methods proposed in Sects. 3.1 and 3.2. Figure 2 depicts the projections of the flowers and the three groups on the rotated (left) and on the penalized CSs. As $\text{rank}(F^\top HF) = 2$, two rotated CSs are obtained for the Iris data. They achieve 20 (13.3%) misclassified observations, rather than four classical CSs. The penalized CSs (right) achieve quite clear separation of the groups. As $\text{rank}(XX^\top + H) = 7$, seven penalized CSs are necessary to separate the groups completely. In fact, the first four penalized CSs already achieve complete separation of the flowers, i.e. for $s = 4$, the penalized CSs $F_s \in \mathbb{R}^{150\times4}$ achieve complete separation. The right panel of Fig. 2 depicts the fist two columns of F_7. They produce only 3 (2%) misclassified flowers, which is the best achievement by the classical LDA.

The initial value of $\text{trace}(S_B) = F^\top HF$ (F from the SVD of X) is 1.19 and is maximized by the two methods to 1.41 and 3 respectively.

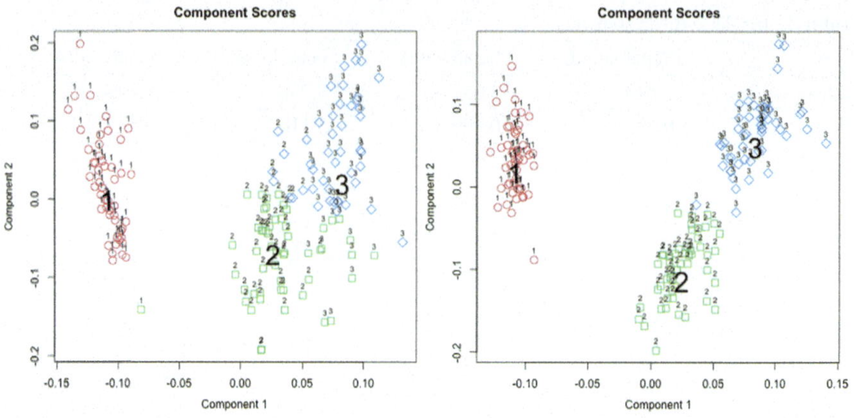

Fig. 2 Iris data: rotated CSs (left) and penalized CSs

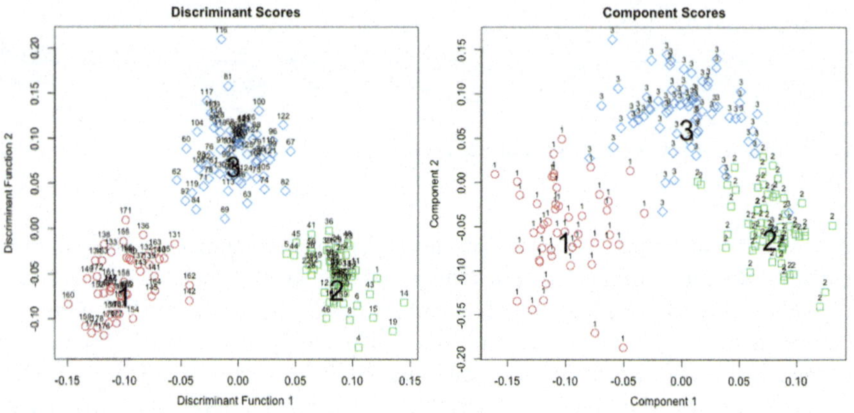

Fig. 3 Wines data: LDA (left), and PCA of sphered data

4.1.2 Three Italian Wines

Here we consider another popular but more complicated data set involving $p = 27$ physical and chemical measurements on $n = 178$ wine samples of three $g = 3$ types Italian wine, Barbera (48), Barolo (59), and Grignolino (71), coded as 1, 2 and 3. The data set is available in the `R` package `MBCbook` and is studied in detail in [2].

The variables are very different and need standardization. First, we apply LDA and PCA to the sphered data. LDA separates the wines very well, while PCA provides less clear picture as depicted on Fig. 3. There are no misclassified wines with LDA. PCA also achieves complete separation in the 27-dimensional space of all PCs. In fact, complete separation is achieved starting with 17 PCs and above. There are 8 (4.5%) wrongly classified wines if working with the first two PCs, as depicted on the right panel of Fig. 3.

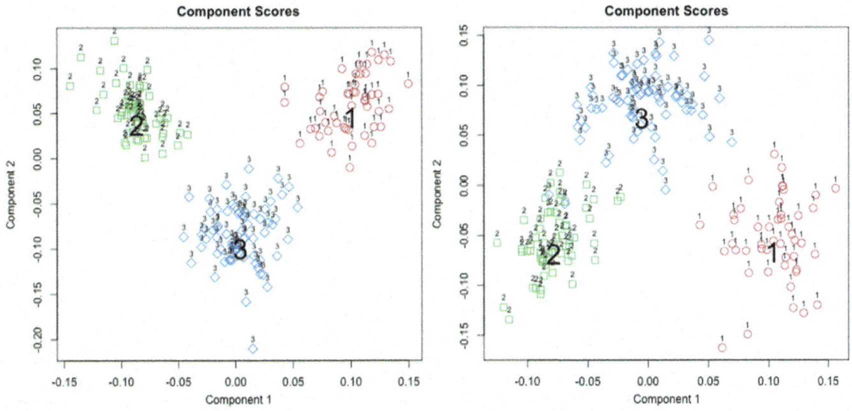

Fig. 4 Sphered wines data: rotated CSs (left) and penalized CSs

Then, we apply the methods proposed in Sects. 3.1 and 3.2. Figure 4 depicts the projections of the wines (and the three groups) on the rotated (left) and on the penalized CSs.

As rank$(F^\top HF) = 2$, two rotated CSs are used and they achieve complete separation of the groups, rather than 17 as in PCA. As rank$(XX^\top + H) = 30$, complete separation the three groups of wines is achieved by 30 penalized CSs. It turns out, that 11 penalized CSs (and above) are already enough for the complete separation of the groups. The first two penalized CSs (i.e. the first two columns of $F \in \mathbb{R}^{178\times 30}$) depicted in the right panel of Fig. 4 achieve 2 (1.1%) misclassified observations.

4.1.3 Diabetes Diagnosis

This data set consists of observations made on $n = 145$ subjects, clinically classified into $g = 3$ groups: chemical (Type 1) diabetes (36), overt (Type 2) diabetes (33), and non-diabetic (76). The measurements are made on three variables: plasma glucose response, plasma insulin response, and steady-state plasma glucose response (SSPG). The three variables are considered comparable in [5], however their basic statistics in Table 2 give us a good reason to sphere the data before analysis.

Table 2 Diabetes diagnosis: Basic statistics

	Glucose	Insulin	sspg
Mean	121.99	540.79	186.12
Std	63.93	319.56	120.94

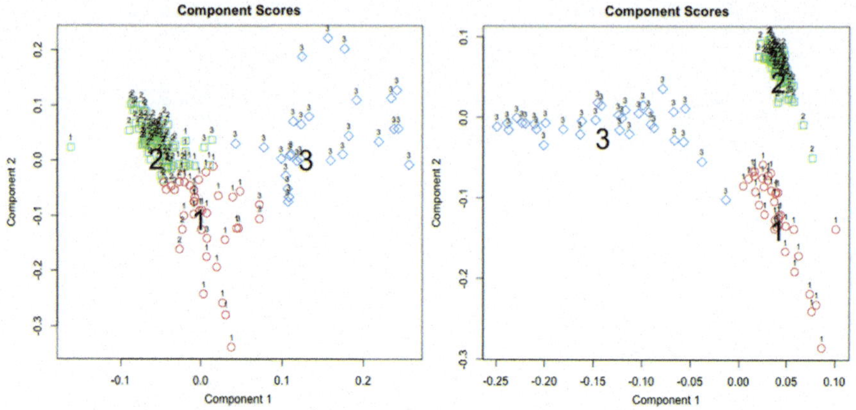

Fig. 5 Diabetes diagnosis sphered: rotated CSs (left) and penalized CSs

The classical LDA and PCA solutions of the sphered data are not quite satisfying and not depicted. The misclassified observations are respectively 20 (13.8%) and 22 (15.2%).

Then, we apply the methods from Sects. 3.1 and 3.2. We find that $\text{rank}(F^\top HF) = 2$, and that two rotated CSs also achieve 22 (15.2%) misclassified observations, instead of the original three. As $\text{rank}(XX^\top + H) = 6$, six penalized CSs provide clear separation of the groups (right panel of Fig. 5), with no misclassified observations. In fact, the first four penalized CSs ($s = 4$) already provide clear separation. The first two penalized CSs (the first two columns of $F \in \mathbb{R}^{145\times 6}$), as plotted on the right panel of Fig. 5, produce 3 (2.1%) misclassified observations.

4.2 Examples with $p \gg n$ (singular S_W)

4.2.1 MLL Data

The data set contains 72 leukemia samples divided into three groups (24 ALL, 20 MLL and 28 AML) and is studied in [1].

PCA of the centered 72×12582 data matrix projects the observations on the first two CSs with 9 (12.5%) misclassified, see Fig. 6 (left). The first seven CSs already achieve correct allocation (no misclassified observations) and explain 58.6% of the total variance. The rotated CSs achieve correct allocation (no misclassified observations) with $s = 7$ ($F \in \mathbb{R}^{72\times 7}$), as well as the restriction to the first two of them, see Fig. 6 (left).

4.2.2 Gastrointestinal (Gastro) Data

This dataset contains features extracted from colonoscopy videos used to detect gastrointestinal lesions It contains $n = 76$ lesions of $g = 3$ types: 15 serrated adenomas,

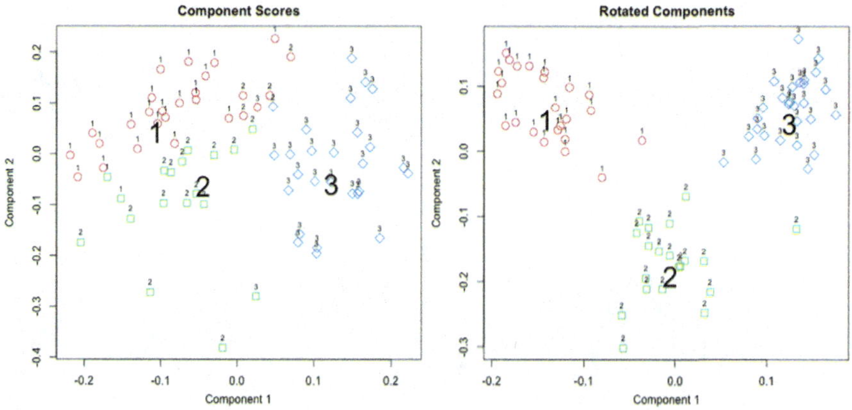

Fig. 6 MLL centered data: CSs (left) and rotated CSs

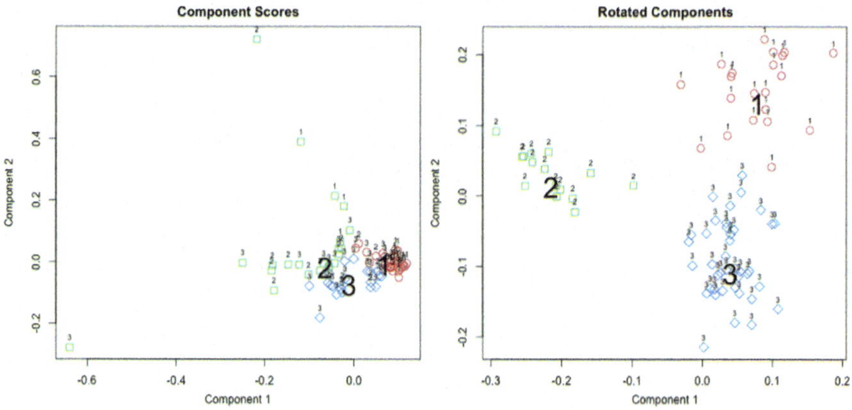

Fig. 7 Gastro centered data: CSs (left) and rotated CSs

21 hyperplastic lesions and 40 adenoma. The measurements are made on 422 textural features, 76 color features, and 200 shape features, i.e. $p = 698$ variables in total. For further details, see [4].

Every lesion appears twice because it has been recorded using two types of lights: white light (WL) and narrow band imaging (NBI). The data authors' suggest concatenating the information from the two types of light for each lesion, i.e. create a single vector of 1396 elements per lesion.

The variables/features contain zeros and thus, we analyze centered 76×1396 data matrix. PCA produces 38 (50%) misclassified observations with its first two component scores, Fig. 7 (right). PCA achieves separation of the groups with 59 CSs. In contrast, the rotated CSs achieve correct allocation with the first two columns of $F \in \mathbb{R}^{76 \times 59}$, Fig. 7 (right).

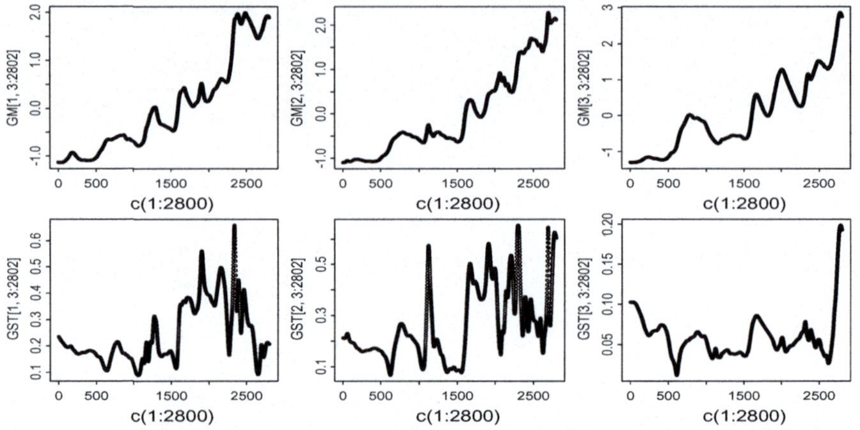

Fig. 8 NIR data: means (top row) and standard deviations within groups

4.2.3 NIR Data

We consider the NIR data set containing $n = 202$ textile samples divided into $g = 3$ groups and analyzed with a near-infrared spectrometer that produces spectra with $p = 2800$ wavelengths (variables) [2, 8.3].

The group means and standard deviations (from first to third) are given in Fig. 8. The means in all groups have the same range and pattern of a strongly increasing linear trend. The variability jumps considerably in the groups, and it is weaker in the third group than in the other two.

For this reason, we prefer to analyze the sphered data. PCA produces 56 (27.7%) misclassified observations, Fig. 9 (left). Rotated CSs achieve correct allocation with the first two columns of $F_s \in \mathbb{R}^{202\times 100}$ $(s = 100)$, Fig. 9 (right).

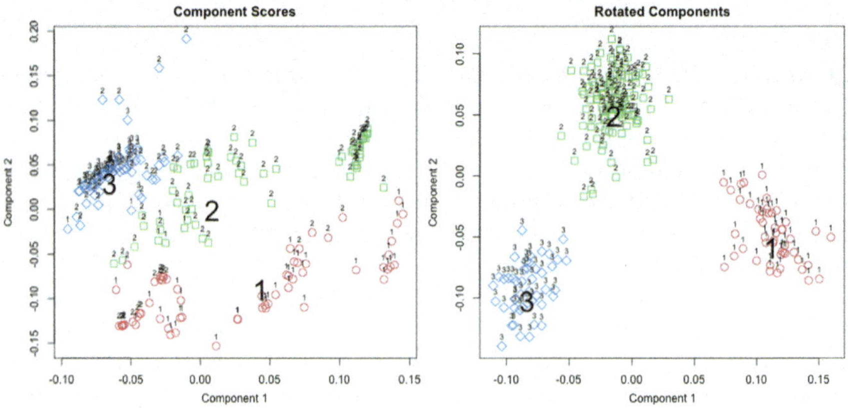

Fig. 9 NIR sphered data: CSs (left) and rotated CSs $(s = 100)$

5 Conclusion

It is demonstrated that the proposed new methods for linear discrimination exhib it quite satisfactory performance. They can be a reasonable alternative to the classical Fisher's LDA and an useful tool for discrimination andclassiâŁŁcation of grouped data with more variables than observations.

The rotated CSs work well with both types of data: $n > p$ or $p >> n$, and are particularly useful for analyzing data with $p >> n$. The penalized CSs can be a reasonable and interesting alternative to the Fisher's LDA of datawith $n > p$. However when $p >> n$, they do not seem to be as useful as the rotated CSs.

References

1. Armstrong, S.A., Staunton, J.E., Silverman, L.B., Pieters, R., den Boer, M.L., Minden, M.D., Sallan, S.E., Lander, E.S., Golub, T.R., Korsmeyer, S.J.: MLL translocations specify a distinct gene expression profile that distinguishes a unique leukemia. Nat. Genetics **30**, 41–47 (2002)
2. Bouveyron, C., Celeux, G., Murphy, B., Raftery, A.: Model-based Clustering and Classification for Data Science with Applications in R. Cambridge University Press, Cambridge, UK (2019)
3. Fisher, R.A.: The use of multiple measurements in taxonomic problems. Ann. Eugenics **7**, 179–188 (1936)
4. Mesejo, P., Pizarro, D., Abergel, A., Rouquette, O., Beorchia, S., Poincloux, L., Bartoli, A.: Computer-aided classification of gastrointestinal lesions in regular colonoscopy. IEEE Trans. Med. Imaging **35**, 2051–2063 (2016)
5. Reaven, G.M., Miller, R.G.: An attempt to define the nature of chemical diabetes using a multidimensional analysis. Diabetologia **16**, 17–24 (1979)
6. Trendafilov, N., Gallo, M.: Multivariate Data Analysis on Matrix Manifolds (with Manopt). Springer, New York, NY (2021)

Non-linear Models and Spatial Analysis

Nonlinear HAR Models and Nonlinear Least Squares: Asymptotic Properties

Emilija Dzuverovic and Edoardo Otranto

Abstract We empirically the asymptotic properties of the nonlinear least squares estimator for a nonlinear extension of the class of Heterogeneous Auto-Regressive (HAR) models for realized covariance matrices, the Hadamard Exponential HAR (HE–HAR). First, we confirm both consistent and normally distributed OLS estimates for a classical multivariate HAR specification in vectorial form, used as a benchmark. Then, we replicate the Monte Carlo experiment under different specifications and distributional hypotheses for HE–HAR extensions as a sensitivity check to verify the corresponding results' robustness. The results establish the convergence of regular HAR coefficients in most cases, while the asymptotic normality of the NLS estimates is uniquely confirmed for the HE–vech-HAR specification with log-transformed realized covariance series. The only persistent asymptotic bias is evident for a HE parameter estimate.

Keywords Hadamard exponential operator · Monte Carlo · Realized covariance matrix · Nonlinearity · Linear approximation

1 Introduction

Forecasting the covariance matrix of asset returns has paramount importance in asset pricing, portfolio allocation, and risk management. As such, the majority of introduced econometric models assume that variances and covariances are either assessable conditional upon past daily information, i.e., multivariate GARCH (MGARCH) models [12], or latent, i.e., multivariate stochastic volatility (MSV) models [5].

E. Dzuverovic
Department of Economics and Management, University of Pisa, Via Cosimo Ridolfi 10, 56124 Pisa, Italy
e-mail: emilija.dzuverovic@ec.unipi.it

E. Otranto (✉)
Department of Economics, University of Messina, Via dei Verdi 75, 98122 Messina, Italy
e-mail: eotranto@unime.it

M. Bini et al. (eds.), *Advanced Methods in Statistics, Data Science and Related Applications*, Springer Proceedings in Mathematics & Statistics 467,
https://doi.org/10.1007/978-3-031-65699-6_7

In contrast, a recent preferable realized covariance (RC) method allows treating the volatility as an "observable" quantity via the high-frequency (HF) data (e.g., [2, 6–8]). However, RC models generally face several challenges, and the most prominent include ensuring the positive definite(PD)-ness of forecasts coupled with the dimensionality issue. The latter implies the rapid increase in the number of parameters with the cross-sectional dimension that has a potential pronounced impact on the statistical efficiency of the estimation and often even renders the model estimation unfeasible. Hence, the RC modelling framework calls for parsimonious yet flexible parameterizations, capable of capturing complex serial dependencies observed in realized variances and covariances.

Among the HF data-based models, the multivariate extensions of the Heterogeneous Auto-Regressive (HAR) model [23] are widely employed, i.e., vech-HAR model [21], HAR-DRD model [44]. This class of models, originally inspired by the Heterogenous Market Hypothesis of [43] and asymmetric propagation of volatility between long and short horizons, appears capable of capturing a commonly recognized long memory of RC via simple linear regression structures.

Nevertheless, in face of the emphasized "curse of dimensionality" issue, an easy-to-implement framework contains the adoption of scalar versions of HAR-based models. This implies that all variances and covariances, or correlations, obey the same dynamics, an assumption that is obviously highly restrictive. In this regard, using the element-by-element Hadamard exponential (HE) function of the matrix, the asset pair-specific and time-varying impact coefficients of lagged realized variances and covariances, or correlations, could be defined, adding great flexibility in dynamics via a single parameter [10, 11].

A practical advantage of the HAR family of models is the straightforward estimation via OLS. By incorporating the HE extension, the nonlinear LS (NLS) could be adopted. Thus, great computational advantages in terms of simplicity, stability, and cost potentially arise on behalf of both flexible and parsimonious HE–HAR models compared to all other maximum likelihood (ML)-based RC models. It has therefore been natural to ask if the convincing NLS method would generate, at least asymptotically, unbiased, efficient, and normally distributed estimates. The approach taken to provide a reliable and empirically valid answer to this question contains the extensive Monte Carlo (MC) simulation experiment.

As the LS method has been extensively adopted within the nonlinear regression framework, the asymptotic properties of the estimator have been subject to notable research interest. In general, the consistency is theoretically established by [40, 53], among others, whereas, e.g., [46, 52] prove the asymptotic normality under certain regularity conditions.

Recently, [51] extrapolates analogous theoretical results to nonstationary and heteroscedastic models, which obviously contain a broad class of commonly used volatility models such as GARCH [14], time-varying GARCH [48], and nonlinear GARCH [41]. Correspondingly, [26] provides the empirical evidence via MC study that the proposed nonlinear iterative LS (NL-ILS) estimator is consistent and even outperforms the ML benchmark for many GARCH-type models. Not to mention, [33] derive the proof of both consistent and asymptotically normal LS estimates of the power-transformed ARCH model [37] under analogous assumptions as for standard ARCH models [17].

Earlier, the rate of the NLS convergence and asymptotic distribution have been proved and empirically supported by [22] for the long memory auto-regressive fractionally integrated moving-average (ARFIMA) model [31, 35], widely advocated within the volatility modelling framework. In addition, [19] prove the consistency of the NLS estimator for smooth transition regressions, often useful to capture the asymmetry in volatility dynamics. Nevertheless, some components have a non-Gaussian distribution that causes usual statistical methods to yield inefficient estimates and/or invalid tests.

Andreou [3] compares the common LS regression models with corresponding mixed data sampling (MIDAS) regressions [28, 29], assuming the underlying MIDAS data generating process (DGP). The analytical and numerical results are presented for the asymptotic bias and relative efficiency of the slope estimator in these two classes for a number of alternative HF volatilities. They show that the LS estimator is asymptotically biased for all cases considered. In contrast, the MIDAS-NLS slope estimator turns out to be not only unbiased but also relatively more efficient. The analogous conclusions are achieved earlier by [4], where asymptotic properties of the FLAT-LS and MIDAS-NLS are tested in models that contain either the i.i.d., ARCH(1), or AR(1) HF regressors.

Chen et al. [20] show that NLS estimates for the logarithmic auto-regressive conditional duration (log-ACD) models utilized for the HF data analyses are consistent and asymptotically normal, as well as strongly competitive with respect to the ones generated via the quasi-ML method (QMLE).

In this work, we investigate asymptotic properties of the NLS for the multivariate HE–HAR family of models [11] via MC experiments. A decision to examine NLS estimates for the HE–HAR class can be justified by the possibility of having a very flexible but parsimonious model while maintaining the computational advantages of the HAR family. Initially, we adopt the vech-HAR model of [21] in order to verify whether our exercise confirms the theoretically established asymptotic properties of linear OLS estimates (e.g., [36, 42]). Then, we consider the two HE-based specifications introduced by [11], extending the vech-HAR model of [21], and HAR-DRD model of [44], under different distributional assumptions, i.e., multivariate Normal (applied to the log transformation of RC matrices), Wishart, and Matrix F distribution.

Our results show that the convergence is achieved for conventional HAR parameters, while there is a small-medium bias for the HE coefficient. Asymptotic normality seems satisfied only for the HE–vech-HAR specification with log-transformed RC matrices under Normal distribution.

The rest of the paper is organized as follows. Section 2 introduces the HAR-type framework adopted for MC experiments; Sect. 3 presents the design of implemented simulations and provides a discussion on empirical results; Sect. 4 gives some concluding comments.

2 Modelling Framework

One of the most widespread models in the econometric literature for modelling the realized volatility is the HAR model of [23]; its linear auto-regressive form has a great advantage to be estimated via OLS. To account for not only serial but also cross-correlation dynamics across the elements of covariance matrices, due to the common movements among financial markets and the increasing degree of globalization in the economy, among others, recent literature has attempted to extend the most utilized univariate volatility models to the multivariate framework.

Let us denote the $m \times m$ RC matrix of day t as $\boldsymbol{C}_t$, the information set available at time $t-1$ as $\mathcal{F}_{t-1} = \{\boldsymbol{C}_{t-1}, \boldsymbol{C}_{t-2}, ...\}$, and the conditional expectation of $\boldsymbol{C}_t$, i.e., $\mathbf{E}(\boldsymbol{C}_t|\mathcal{F}_{t-1}))$, as $\boldsymbol{S}_t$.

2.1 Linear Vech-HAR Model

A natural extension of the HAR model was proposed by [21], introducing the vech-HAR model, i.e.,

$$\begin{aligned}\text{vech}(\boldsymbol{C}_t) = (1-\alpha_D-\alpha_W-\alpha_M)\overline{\boldsymbol{C}} + \alpha_D\,\text{vech}(\boldsymbol{C}_{t-1}) + \alpha_W\,\text{vech}(\overline{\boldsymbol{C}}_{t-1:t-5}) \\ +\alpha_M\,\text{vech}(\overline{\boldsymbol{C}}_{t-1:t-22}) + \boldsymbol{\varepsilon}_t,\end{aligned} \tag{1}$$

where vech$(\cdot)$ is the operator that stacks the lower triangular portion of its matrix argument as a vector, $\overline{\boldsymbol{C}}$ is the sample mean of RC matrices, regressors $\overline{\boldsymbol{C}}_{t-1:t-h}$ are the averages of past RC matrices between $t-1$ and $t-h$ $(h = 5, 22)$, and α's are scalar parameters. Finally, $\boldsymbol{\varepsilon}_t$ is a vector of disturbances.

The use of the so-called targeting, i.e., $(1-\alpha_D-\alpha_W-\alpha_M)\overline{\boldsymbol{C}}$, as a constant, guarantees that the estimated matrix $\boldsymbol{S}_t$ is positive-definite (PD), while at the same time reduces significantly the number of parameters to be estimated.

We consider three alternative specifications for the distribution of $\boldsymbol{\varepsilon}_t$, ensuring the PD-ness of the RC series:

- Multivariate Normal, which implies the log transformation of RC [9], so that all variables in (1) are expressed in logarithms. We call this model "log vech-HAR";
- Wishart distribution (e.g., [13, 16, 30]), with $\boldsymbol{C}_t|\mathcal{F}_{t-1} \sim W_m(v, \boldsymbol{S}_t/v)$, where $W_m(v, \boldsymbol{S}_t/v)$ denotes the m-dimensional central Wishart distribution with $v \geq m$ degrees of freedom and $m \times m$ PD scale matrix $\boldsymbol{S}_t/v$, i.e., $\mathbf{E}(\boldsymbol{C}_t|\mathcal{F}_{t-1})) = \boldsymbol{S}_t$. We call this model "vech-HAR Wishart";
- Matrix F distribution [45, 50] due to the sudden large movements in RC that might potentially impair estimated dynamics of the series if a "thin-tailed" distribution, such as Wishart, is adopted (e.g., [24, 34]). In this regard, we suppose $\boldsymbol{C}_t|\mathcal{F}_{t-1} \sim F_m(v_1, v_2, ((v_2-m-1)/v_1)\boldsymbol{S}_t)$, where $v_1, v_2 \geq m+1$ are degrees of freedom. When v_2 is finite, the Matrix F distribution exhibits a leptokurtic behavior, typical

of realized measures of variance and covariance. As $v_2 \to \infty$, the convergence to the Wishart distribution prevails. We call this model "vech-HAR Matrix F".

2.2 Nonlinear HE–Vech-HAR Model

The adoption of the scalar vech-HAR model requires imposing a very strong assumption, i.e., all the variances and covariances obey the same dynamics. Bauwens and Otranto [11] suggest a way of adding great flexibility in model dynamics while preserving the parsimonious parameterization, i.e., via a single additional parameter, using the HE function of the matrix.

In practical terms, let us consider again the simple scalar vech-HAR model (1), where each daily RC matrix is specified as a linear function of lagged daily, weekly, and monthly covariances. Considering the HE operator's algebraic properties, [11] propose replacing the scalar impact coefficient α_D of the lagged daily RC on the current RC matrix by a PD matrix of time-varying coefficients determined via a unique new parameter. The new HE–vech-HAR model is given by:

$$\begin{aligned} \mathrm{vech}(\boldsymbol{C}_t) = &(1 - \overline{\alpha}_{D,t} - \alpha_W - \alpha_M)\overline{\boldsymbol{C}} + \alpha_D \exp\odot(\phi\, \mathrm{vech}(\boldsymbol{P}_{t-1} - \mathbf{J}_m)) \odot \mathrm{vech}(\boldsymbol{C}_{t-1}) \\ &+\alpha_W\, \mathrm{vech}(\overline{\boldsymbol{C}}_{t-1:t-5}) + \alpha_M\, \mathrm{vech}(\overline{\boldsymbol{C}}_{t-1:t-22}) + \boldsymbol{\varepsilon}_t, \end{aligned} \tag{2}$$

where $\odot$ denotes the Hadamard product and $\exp\odot$ the HE operator, i.e., element-by-element exponential; ϕ is the new parameter and $\overline{\alpha}_{D,t}$ is the average of $\alpha_D \exp(\phi\, \mathrm{vech}(\boldsymbol{P}_{t-1} - \mathbf{J}_m))$, where $\boldsymbol{P}_{t-1}$ is the lagged realized correlation matrix and $\mathbf{J}_m$ is a $m \times m$ matrix of ones. Thus, if ϕ differs from zero, the impact coefficient becomes both asset pair-specific and time-varying. Indeed, the HE operator applied to a PD matrix guarantees the PD-ness of the resulting matrix, whereas the Hadamard product of two PD matrices is again PD.

The adoption of lagged correlations in the impact parameter can be justified by the strong evidence that volatilities and correlations move together (e.g., [1]). As such, we expect the transfer of the volatility clustering phenomenon to correlations as well. During turbulent periods, the correlations increase, but the level of persistence of such an increase can be distinct across asset pairs. Hence, by adding a dependence of the impact coefficient on lagged correlations, we are able to account for the volatility clustering effect on RC that differs across asset pairs and changes through time.

The vech-HAR model with the HE extension could be estimated by NLS. In this regard, we investigate the asymptotic properties of HE–vech-HAR NLS estimates generated via the Gauss-Newton algorithm. Again, we consider RC in log form, thus following the Normal distribution, conditional Wishart, as well as conditional Matrix F distributed ones. The resulting models are denoted by "log HE–vech-HAR", "HE–vech-HAR Wishart", and "HE–vech-HAR Matrix F", respectively.

2.3 Nonlinear HE–HAR-DRD Model

Inspired by the seminal DCC model of [27], [44] introduce the widely utilized HAR-DRD. We begin by decomposing the RC matrix as:

$$\boldsymbol{C}_t = \boldsymbol{V}_t^{1/2} \boldsymbol{P}_t \boldsymbol{V}_t^{1/2}, \tag{3}$$

where $\boldsymbol{V}_t$ is the diagonal matrix of realized variances (RV) and $\boldsymbol{P}_t$ is the realized correlation matrix, which allows for estimating the full model in two steps (see [27]).

In particular, the first step consists of modelling m univariate volatilities, i.e., $V_{ii,t}$ $(i = 1, \ldots, m)$, or their logs to ensure the positiveness, via the benchmark HAR model [23]. Hence, the OLS estimator applies to:

$$V_{ii,t} = c_i + \alpha_{i,D} V_{ii,t-1} + \alpha_{i,W} \overline{V}_{ii,t-1:t-5} + \alpha_{i,M} \overline{V}_{ii,t-1:t-22} + \varepsilon_{i,t}. \tag{4}$$

Subsequently the benchmark HAR-type structure is adopted to model realized correlations, i.e.,

$$\begin{aligned} \text{vech}(\boldsymbol{P}_t) = (1 - \alpha_D - \alpha_W - \alpha_M)\overline{\boldsymbol{P}} + \alpha_D \, \text{vech}(\boldsymbol{P}_{t-1}) + \alpha_W \, \text{vech}(\overline{\boldsymbol{P}}_{t-1:t-5}) \\ + \alpha_M \, \text{vech}(\overline{\boldsymbol{P}}_{t-1:t-22}) + \boldsymbol{\varepsilon}_t, \end{aligned} \tag{5}$$

where, similarly to (1), regressors are the averages of past realized correlation matrices, $\overline{\boldsymbol{P}}$ is the sample mean of the realized correlation series, and α_i $(i = D, W, M)$ are scalar parameters to be estimated by OLS.

To allow for distinct dynamics of realized correlations, [11] introduce the HE parameterization to (5), i.e.,

$$\begin{aligned} \text{vech}(\boldsymbol{P}_t) = (1 - \overline{\alpha}_{D,t} - \alpha_W - \alpha_M)\overline{\boldsymbol{P}} + \alpha_D \exp\left(\phi \, \text{vech}(\boldsymbol{P}_{t-1} - \mathbf{J}_m)\right) \odot \text{vech}(\boldsymbol{P}_{t-1}) \\ + \alpha_W \, \text{vech}(\overline{\boldsymbol{P}}_{t-1:t-5}) + \alpha_M \, \text{vech}(\overline{\boldsymbol{P}}_{t-1:t-22}) + \boldsymbol{\varepsilon}_t. \end{aligned} \tag{6}$$

Similarly to HE–vech-HAR, HAR-DRD model with the HE extension could be estimated by NLS. As follows, we examine the asymptotic properties of HE–HAR-DRD NLS estimates generated via the Gauss-Newton algorithm for "log HE–HAR-DRD", "HE–HAR-DRD Wishart", and "HE–HAR-DRD Matrix F" specifications.

3 Monte Carlo Experiments

For the subsequent empirical analyses, we employ 3 HAR models under 3 distributional assumptions, i.e., a total of 9 specifications. Specifically, 3 linear vech-HAR (1), 3 nonlinear HE–vech-HAR (2), and 3 nonlinear HE–HAR-DRD specifications (3), (4), (6), distinguished by the distribution: Normal for logs, Wishart, and Matrix F.
To evaluate the asymptotic properties of OLS/NLS, we generate data from each specification referring to a 3–variate time series ($m = 3$) of increasing length $T = 1000, 3000, 5000, 10000$, with the corresponding parameters fixed in the range of estimated values on real data for similar models.[1] In particular, for vech-HAR, we consider the range of values presented in Tables 1, 2, and 3 in [18]; for HE–vech-HAR, the values are taken from Table 14 of [11], but fixing $\phi = 0.15$ to ensure PD matrices[2]; for HE–HAR-DRD, we consider Table 2 of [15] for the HAR part and keep $\phi = 0.15$. The sample covariance matrix for each model is randomly generated from the underlying distribution. For Wishart, we set $\upsilon = 4$ degrees of freedom, coupled with $\upsilon_1 = 22$ and $\upsilon_2 = 35$ for Matrix F. We replicate each experiment $N = 1000$ times.

Concerning the RC in log form, we initially generate a disturbance term $\boldsymbol{\varepsilon}_t$ from a standardized Normal distribution, with the series of $\boldsymbol{C}_t$ obtained iteratively from the model considered, i.e., vech-HAR (1), HE–vech-HAR (2), HE–HAR-DRD (6).

Instead, disturbances do not enter models under Wishart and Matrix F distributions, where RC matrices are generated by the conditional distribution of $\boldsymbol{C}_t$. More in detail, for both cases, we first generate the unconditional $\overline{\boldsymbol{C}}$ discussed above, with the expectation equal to the identity matrix. Then, for the Wishart case, we simulate $\boldsymbol{S}_t$ from one of the three models, such that the RC matrix $\boldsymbol{C}_t$ at time t is generated from $W_3(4, \boldsymbol{S}_t/4)$.

Analogously, for the Matrix F case, we simulate $\boldsymbol{S}_t$ from one of the three models. Then, we generate $\boldsymbol{C}_t$ from $F_3(\upsilon_1, \upsilon_2, ((\upsilon_2 - 3 - 1)/\upsilon_1)\boldsymbol{S}_t)$, with $\upsilon_1 = 22$ and $\upsilon_2 = 35$. Specifically, we obtain RC matrices using the following decomposition property [39]:

$$\boldsymbol{C}_t = ((\upsilon_2 - m - 1)/\upsilon_1)\boldsymbol{S}_t^{1/2}\boldsymbol{M}_t^{1/2}\boldsymbol{L}_t^{-1}\boldsymbol{M}_t^{1/2}\boldsymbol{S}_t^{1/2}, \tag{7}$$

where $\boldsymbol{M}_t$ and $\boldsymbol{L}_t$ are independent Wishart matrices, i.e., $\boldsymbol{M}_t \sim W_m(\upsilon_1, \mathrm{I}_m)$, $\boldsymbol{L}_t \sim W_m(\upsilon_2, \mathrm{I}_m)$. At the same time, we guarantee both symmetry and PD-ness of the generated RC series.

For HE–HAR–DRD specifications, we do not consider the part relative to the RV, so that we generate directly realized correlations $\boldsymbol{P}_t$ from the corresponding conditional distribution, putting $\boldsymbol{S}_t = \boldsymbol{R}_t$, and implementing the same steps previously described.

In the following tables, reporting results of our experiments, the first column shows the *true* value of each parameter. To evaluate the unbiasedness and efficiency, we

[1] The code to generate data and estimate models is written using the R software.

[2] In the paper of [11] this coefficient is negative.

report the average and standard deviation of N estimates for each parameter and T, as well as the degree of biasedness. Finally, we show the p-value of the Jarque-Bera test [38] on the set of N estimates for each coefficient to evaluate the normality.

In what follows, we comment results for each DGP.

3.1 Vech-HAR Model

We run experiments on discussed linear vech-HAR variants to confirm the theoretical asymptotic properties under the Normal DGP and verify their validity under the other two distributional assumptions.

Table 1 shows the corresponding results. For the log and Wishart cases, we confirm the unbiasedness of α_D and α_W already at small sample sizes, whereas α_M converges to the true value at $T = 3000$ for the Wishart case and $T = 5000$ for the log case. The decrease in standard deviations is consistent with the efficiency of the OLS estimator, and the normality of estimates is also satisfied in small samples. A different behavior characterizes the Matrix F case with some puzzling results. The convergence seems to be achieved only asymptotically, and the normality of estimators is ambiguous, in particular for the parameter α_M. In practice, the strong asymmetric and leptokurtic form of the Matrix F distribution may affect the goodness of the OLS asymptotic performance.

3.2 HE–Vech-HAR Model

The NLS estimates for HE–vech-HAR models are obtained by utilizing the Gauss-Newton method, which is based on the Taylor series approximation to obtain the linearized model and solve it via OLS [32]. Such a choice might be supported not only because this algorithm avoids computing the second-order derivatives but also due to the absence of potential issues arising from initial starting values provided for estimations given the specified simulation parameters.

As indicated in Table 2, all NLS estimates of regular HAR parameters within HE–vech-HAR models reach the full convergence for the log and Wishart cases (a small bias persists for α_M in the Wishart-based model).[3] The convergence is faster for the log case, i.e., at $T = 5000$, coupled with the continuous reduction of respective standard deviations. The estimate of the HE term coefficient ϕ is featured by a relatively small but persistent negative bias that does not exceed 2%.

Concerning the Matrix F case, the partial convergence of conventional HAR parameters is attained at $T = 5000$, while the monthly coefficient α_M subsequently fully converges as well. At the same time, the bias does not exceed 1.3%, being less than 0.5% in the majority of cases. Similar comments with respect to the two

[3] We verify with further experiments at $T = 50000$, not presented here, that the monthly coefficient hits the true value.

Table 1 vech-HAR Model: $N = 1000$ MC results in correspondence to three different distributions

Parameter/Sample Size	$T = 1000$	$T = 3000$	$T = 5000$	$T = 10000$
Log				
$\alpha_D = \mathbf{0.45}$				
Sample Mean	**0.45**	**0.45**	**0.45**	**0.45**
SD	(0.00114)	(0.00065)	(0.0005)	(0.00035)
Bias	–0.239%	–0.142%	–0.110%	–0.112%
Jarque-Bera test (p-value)	**0.8037**	**0.5434**	**0.6151**	**0.3466**
$\alpha_W = \mathbf{0.25}$				
Sample Mean	0.24	**0.25**	**0.25**	**0.25**
SD	(0.00173)	(0.00101)	(0.00079)	(0.00055)
Bias	–0.560%	–0.207%	–0.079%	–0.020%
Jarque-Bera test (p-value)	**0.1051**	**0.9129**	**0.2670**	**0.9632**
$\alpha_M = \mathbf{0.15}$				
Sample Mean	0.14	0.14	**0.15**	**0.15**
SD	(0.00174)	(0.00103)	(0.0008)	(0.00055)
Bias	–1.452%	–0.519%	–0.347	–0.177%
Jarque-Bera test (p-value)	**0.9688**	**0.8918**	**0.8063**	**0.4520**
Wishart				
$\alpha_D = \mathbf{0.45}$				
Sample Mean	**0.45**	**0.45**	**0.45**	**0.45**
SD	(0.00115)	(0.00065)	(0.0005)	(0.00034)
Bias	–0.198%	–0.159%	–0.140%	–0.068%
Jarque-Bera test (p-value)	**0.4210**	**0.8894**	**0.2636**	**0.8559**
$\alpha_W = \mathbf{0.25}$				
Sample Mean	**0.25**	**0.25**	**0.25**	**0.25**
SD	(0.00176)	(0.00102)	(0.0008)	(0.00055)
Bias	–0.126%	–0.181%	–0.100%	–0.122%
Jarque-Bera test (p-value)	**0.9815**	**0.6958**	**0.7642**	**0.3186**
$\alpha_M = \mathbf{0.15}$				
Sample Mean	0.13	**0.15**	**0.15**	**0.15**
SD	(0.0018)	(0.00105)	(0.0008)	(0.00055)
Bias	–1.547%	–0.272%	–0.147%	–0.054%
Jarque-Bera test (p-value)	**0.1738**	**0.5051**	**0.8959**	0.0413

(continued)

Table 1 (continued)

Parameter/Sample Size	T = 1000	T = 3000	T = 5000	T = 10000
Matrix F				
α_D = **0.45**				
Sample Mean	0.49	0.47	**0.45**	0.46
SD	(0.00069)	(0.00041)	(0.00037)	(0.00026)
Bias	4.357%	1.616%	0.296%	0.622%
Jarque-Bera test (p-value)	0.0008	**0.1099**	**0.9438**	0.0209
α_W = **0.25**				
Sample Mean	0.23	0.24	**0.25**	0.24
SD	(0.00055)	(0.00042)	(0.00031)	(0.00021)
Bias	–2.474%	–0.686%	–0.254%	–1.409%
Jarque-Bera test (p-value)	0.0222	**0.5801**	0.0322	0.0011
α_M = **0.15**				
Sample Mean	0.13	0.13	0.14	**0.15**
SD	(0.00056)	(0.00030)	(0.00020)	(0.00019)
Bias	–2.097%	–2.183%	–0.986%	–0.406%
Jarque-Bera test (p-value)	**0.3034**	0.0000	0.0084	0.0004

Notes The table reports the sample mean of 1000 OLS estimates of each vech-HAR parameter with the corresponding standard deviation in parentheses (at sample sizes $T = 1000, 3000, 5000, 10000$). The bias is calculated as the difference between the sample mean and the true parameter value, multiplied by 100. We also show the p-value of the Jarque-Bera test for each coefficient and respective sample size, where the entries in boldface indicate the failure to reject the null hypothesis of normality at the 5% significance level

previous cases hold for regular HAR coefficients, while the bias of ϕ has an opposite sign but higher maximum magnitude, i.e., around 6%.

Due to the absence of the overall consistency evidence for NLS estimates of HE–vech-HAR variants and in line with related literature, the underlying distribution of estimated coefficients often diverges from Normal. On the other hand, we confirm the normality hypothesis for all parameter estimates, even at $T = 5000$, in log model.

3.3 *HE–HAR-DRD Model*

The MC results for HE–HAR-DRD variants reported in Table 3 suggest a relatively inferior NLS performance compared to the estimation of HE–vech-HAR models.

The regular HAR coefficients only occasionally hit the *true* value, also under the Normal DGP. In particular, the coefficient α_D seems to exhibit the larger bias for high T. The "HE" term estimator is characterized by not only protracted but also notable bias, exceeding 24% at $T = 10000$ for the Matrix-F case, and 7% for the log case.

Table 2 HE–vech-HAR Model: N = 1000 MC results in correspondence to three different distributions

Parameter/Sample Size	T = 1000	T = 3000	T = 5000	T = 10000
Log				
$\alpha_D = \mathbf{0.21}$				
Sample Mean	**0.21**	**0.21**	**0.21**	**0.21**
SD	(0.00060)	(0.00034)	(0.00026)	(0.00019)
Bias	–0.274%	–0.335%	–0.301%	–0.275%
Jarque-Bera test (p-value)	**0.0579**	**0.4740**	**0.1157**	0.0050
$\alpha_W = \mathbf{0.33}$				
Sample Mean	**0.33**	**0.33**	**0.33**	**0.33**
SD	(0.00099)	(0.00057)	(0.00043)	(0.00031)
Bias	–0.041%	0.047%	0.015%	–0.005%
Jarque-Bera test (p-value)	**0.8834**	**0.9225**	**0.7298**	**0.3233**
$\alpha_M = \mathbf{0.22}$				
Sample Mean	0.21	**0.22**	**0.22**	**0.22**
SD	(0.00115)	(0.00062)	(0.00049)	(0.00035)
Bias	–1.327%	–0.431%	–0.249%	–0.089%
Jarque-Bera test (p-value)	**0.8069**	**0.1961**	**0.3642**	**0.8022**
$\phi = \mathbf{0.15}$				
Sample Mean	0.14	0.13	0.13	0.13
SD	(0.00327)	(0.00185)	(0.00143)	(0.00100)
Bias	–1.005%	–1.621%	–1.757%	–1.836%
Jarque-Bera test (p-value)	0.0000	0.0379	**0.5345**	**0.7134**
Wishart				
$\alpha_D = \mathbf{0.21}$				
Sample Mean	**0.21**	0.18	0.20	**0.21**
SD	(0.00068)	(0.00056)	(0.00043)	(0.00041)
Bias	0.493%	–2.737%	–1.099%	0.202%
Jarque-Bera test (p-value)	0.0000	0.0113	0.0001	0.0000
$\alpha_W = \mathbf{0.33}$				
Sample Mean	0.31	**0.33**	0.34	**0.33**
SD	(0.00096)	(0.00081)	(0.00049)	(0.00059)
Bias	–2.208%	0.068%	0.891%	0.360%
Jarque-Bera test (p-value)	0.0008	0.0000	0.0114	0.0055
$\alpha_M = \mathbf{0.22}$				
Sample Mean	0.19	0.19	0.19	0.21
SD	(0.00173)	(0.00117)	(0.00088)	(0.00073)
Bias	–3.121%	–2.985%	–3.143%	–1.482%
Jarque-Bera test (p-value)	0.0000	0.0000	0.0000	0.0000
$\phi = \mathbf{0.15}$				
Sample Mean	0.11	0.14	0.12	0.13
SD	(0.00399)	(0.00342)	(0.00218)	(0.00145)
Bias	–3.634%	–1.405%	–2.519%	–2.396%
Jarque-Bera test (p-value)	**0.3721**	0.0000	0.0000	0.0000

(continued)

Table 2 (continued)

Parameter/Sample Size	$T = 1000$	$T = 3000$	$T = 5000$	$T = 10000$
Matrix F				
$\alpha_D = \mathbf{0.21}$				
Sample Mean	0.24	0.22	**0.21**	0.22
SD	(0.00031)	(0.00034)	(0.00032)	(0.00020)
Bias	3.269%	0.936%	0.436%	1.250%
Jarque-Bera test (p-value)	0.0491	0.0000	**0.1033**	0.0000
$\alpha_W = \mathbf{0.33}$				
Sample Mean	**0.33**	0.34	0.34	0.32
SD	(0.00056)	(0.00041)	(0.00029)	(0.00021)
Bias	0.325%	0.985%	0.505%	–0.601%
Jarque-Bera test (p-value)	0.0000	0.0182	0.0153	**0.2053**
$\alpha_M = \mathbf{0.22}$				
Sample Mean	0.20	0.20	**0.22**	**0.22**
SD	(0.00073)	(0.00052)	(0.00032)	(0.00027)
Bias	–2.456%	–1.755%	–0.043%	0.052%
Jarque-Bera test (p-value)	0.0000	0.0000	**0.1033**	0.0000
$\phi = \mathbf{0.15}$				
Sample Mean	0.21	0.16	0.17	0.21
SD	(0.00209)	(0.00173)	(0.00128)	(0.00080)
Bias	5.799%	1.183%	2.042%	6.319%
Jarque-Bera test (p-value)	**0.1970**	0.0000	0.0000	0.0000

Notes The table reports the sample mean of 1000 NLS estimates of each HE–vech-HAR parameter with the corresponding standard deviation in parentheses (at sample sizes $T = 1000, 3000, 5000, 10000$). The bias is calculated as the difference between the sample mean and the true parameter value, multiplied by 100. We also show the p-value of the Jarque-Bera test for each coefficient and respective sample size, where the entries in boldface indicate the failure to reject the null hypothesis of normality at the 5% significance level

The estimates do not follow Normal distribution, also asymptotically, excluding α_W and α_M in "log HE–HAR-DRD".

4 Final Remarks

To accurately capture the cross-correlation dynamics among assets and volatility spillovers in financial markets via the HF data, the multivariate HAR-type models have been widely adopted. This class of models appears particularly attractive, being featured by a solid, empirically proved forecasting performance (e.g., [25, 44, 49, 50]), coupled with a rather simple LS estimation method.

As such, this paper presents the empirical evidence on the asymptotic properties of the LS estimator, i.e., consistency and normality, with respect to both linear and nonlinear HAR-type framework. To the best of our knowledge, this is the first study

Table 3 HE–HAR-DRD Model: $N = 1000$ MC results in correspondence to three different distributions

Parameter/Sample Size	$T = 1000$	$T = 3000$	$T = 5000$	$T = 10000$
Log				
α_D = **0.05**				
Sample Mean	0.09	0.07	0.07	0.06
SD	(0.01248)	(0.00265)	(0.00185)	(0.00086)
Bias	4.130%	2.370%	2.028%	1.002%
Jarque-Bera test (p-value)	0.0000	0.0000	0.0000	0.0000
α_W = **0.16**				
Sample Mean	**0.16**	**0.16**	**0.16**	**0.16**
SD	(0.00185)	(0.00088)	(0.00068)	(0.00047)
Bias	0.273%	0.106%	0.149%	–0.029%
Jarque-Bera test (p-value)	**0.5544**	**0.9242**	**0.3741**	**0.9167**
α_M = **0.56**				
Sample Mean	0.55	0.55	0.55	**0.56**
SD	(0.00199)	(0.00101)	(0.00080)	(0.00057)
Bias	–1.056%	–0.927%	–0.604%	–0.303%
Jarque-Bera test (p-value)	**0.5414**	**0.9225**	**0.7957**	**0.2093**
ϕ = **0.15**				
Sample Mean	0.06	0.02	0.12	0.08
SD	(0.04148)	(0.02834)	(0.02192)	(0.01367)
Bias	–8.599%	–12.743%	–3.256%	–7.105%
Jarque-Bera test (p-value)	0.0000	0.0000	0.0000	**0.1189**
Wishart				
α_D = **0.05**				
Sample Mean	0.15	0.03	**0.05**	0.04
SD	(0.00645)	(0.00076)	(0.00066)	(0.00033)
Bias	10.211%	–2.272%	0.039%	–1.166%
Jarque-Bera test (p-value)	0.0000	0.0000	0.0000	0.0000
α_W = **0.16**				
Sample Mean	0.23	0.17	**0.16**	0.17
SD	(0.00131)	(0.00129)	(0.00085)	(0.00086)
Bias	7.146%	1.422%	0.049%	0.971%
Jarque-Bera test (p-value)	0.0000	0.0000	0.0000	0.0000
α_M = **0.56**				
Sample Mean	0.45	0.52	0.54	0.55
SD	(0.00061)	(0.00011)	(0.00006)	(0.00046)
Bias	–11.009%	–4.354%	–2.055%	–1.070%
Jarque-Bera test (p-value)	0.0000	0.0000	0.0000	0.0000
ϕ = **0.15**				
Sample Mean	1.27	–0.47	0.04	0.14
SD	(0.05209)	(0.01711)	(0.00742)	(0.00286)
Bias	111.601%	–62.171%	–11.331%	–0.750%
Jarque-Bera test (p-value)	0.0000	0.0000	0.0000	0.0000

(continued)

Table 3 (continued)

Parameter/Sample Size	$T = 1000$	$T = 3000$	$T = 5000$	$T = 10000$
Matrix F				
$\alpha_D = \mathbf{0.05}$				
Sample Mean	0.08	0.06	0.07	0.07
SD	(0.00030)	(0.00093)	(0.00028)	(0.00036)
Bias	3.285%	1.128%	1.945%	1.829%
Jarque-Bera test (p-value)	0.0000	0.0000	0.0000	0.0000
$\alpha_W = \mathbf{0.16}$				
Sample Mean	0.17	0.15	0.15	**0.16**
SD	(0.00005)	(0.00002)	(0.00008)	(0.00014)
Bias	1.497%	–0.578%	–1.488%	0.471%
Jarque-Bera test (p-value)	0.0000	0.0000	0.0000	0.0000
$\alpha_M = \mathbf{0.56}$				
Sample Mean	0.51	0.52	**0.56**	0.55
SD	(0.00193)	(0.00089)	(0.00081)	(0.00068)
Bias	–5.058%	–3.662%	0.055%	–0.634%
Jarque-Bera test (p-value)	0.0000	0.0000	0.0000	0.0000
$\phi = \mathbf{0.15}$				
Sample Mean	0.51	–0.05	0.18	0.39
SD	(0.00273)	(0.01592)	(0.00925)	(0.00036)
Bias	36.419%	–19.616%	3.304%	24.420%
Jarque-Bera test (p-value)	0.0000	0.0000	0.0000	0.0000

Notes The table reports the sample mean of 1000 NLS estimates of each HE–HAR-DRD parameter with the corresponding standard deviation in parentheses (at sample sizes $T = 1000, 3000, 5000, 10000$). The bias is calculated as the difference between the sample mean and the true parameter value, multiplied by 100. We also show the p-value of the Jarque-Bera test for each coefficient and respective sample size, where the entries in boldface indicate the failure to reject the null hypothesis of normality at the 5% significance level

that not only verifies empirically consistent and asymptotically Normal OLS estimates for benchmark HAR models but also investigates the properties of the NLS for flexible yet parsimonious nonlinear HAR extensions.

In particular, we consider adding the HE extension [10, 11] to HAR benchmarks, which implies the asset pair-specific and time-varying impact parameters of the lagged RC, or correlations, adding great flexibility to scalar HAR dynamics via a unique additional parameter. Thus, it enables to adequately counteract the well-known "curse of dimensionality" issue of the RC modelling framework. Not to mention great computational advantages in terms of the simplicity, stability, and cost on behalf of HE–HAR models with respect to all ML-based RC models.

Based on $N = 1000$ MC replication outcomes, we aim at verifying consistent, efficient, and asymptotically Normal OLS/NLS estimates for benchmark multivariate HAR specifications (with covariance targeting), i.e., vech-HAR models, as well as their HE version, under the three different distributions. Moreover, we consider the HE–HAR-DRD specification under analogous distributional assumptions, as it provides the possibility of a practically convenient 2-step estimation, separating the estimation of the RV from the estimation of the correlation part. All nonlinear HE extensions are estimated by NLS, utilizing the Gauss-Newton algorithm.

Our experiments confirm both consistency and asymptotic normality of the OLS for vech-HAR variants, except for the Matrix F DGP. In fact, using a stronger leptokurtic and asymmetric distribution than Wishart to generate data, we notice a (small) loss of the satisfaction of asymptotic properties. Similar results are obtained for regular HAR parameters in HE–vech-HAR models. The nonlinear coefficient of the HE parameterization shows a degree of biasedness, which increases with the degree of nonlinearity of the distribution generating data. This problem further amplifies for HE–HAR-DRD specifications, where even some conventional HAR estimates exhibit biasedness and non-normality.

Given the breadth of the HAR modelling framework and their widespread and straightforward adoption in the context of the accurate RC matrix forecasting, these results provide a valuable extension to existing ones in the literature and immediate practical implications. In general, we believe that it is important to highlight advantages and problems in the performance of HAR estimators and to draw the researchers' attention to potential pitfalls in using them in empirical studies. Thus, our findings suggest that researchers and practitioners can confidently adopt linear HAR models estimated by OLS to obtain the RC estimates. In addition, they might implement the NLS for the HE–HAR class with caution.

However, an avenue we have not explored in this study is building up alternative flexible but still parsimonious multivariate HAR extensions subject to the NLS estimation. E.g., incorporating the smooth transition structure within benchmark HAR specifications as a way to explicitly account for the asymmetric behavior in RV and RC (see [42, 47]). Hence, it remains completely an open empirical question whether the consistency and asymptotic normality of the NLS could be confirmed for upcoming nonlinear HAR models for the RC matrix, which arguably reflect interesting topics for further research.

Acknowledgements Otranto acknowledges financial support from the Italian PRIN 2022 grant "Methodological and computational issues in large scale time series models for economics and finance" (20223725WE).

References

1. Andersen, T.G., Bollerslev, T., Diebold, F.X., Ebens, H.: The distribution of realized stock return volatility. J. Financ. Econ. **61**(1), 43–76 (2001)
2. Andersen, T.G., Bollerslev, T., Diebold, F.X., Labys, P.: Modeling and forecasting realized volatility. Econometrica **71**(2), 579–625 (2003)
3. Andreou, E.: On the use of high frequency measures of volatility in MIDAS regressions. J. Econ. **193**(2), 367–389 (2016)
4. Andreou, E., Ghysels, E., Kourtellos, A.: Regression models with mixed sampling frequencies. J. Econ. **158**(2), 246–261 (2010)

5. Asai, M., McAleer, M., Yu, J.: Multivariate stochastic volatility: a review. Econ. Rev. **25**(2–3), 145–175 (2006)
6. Barndorff-Nielsen, O.E., Shephard, N.: Econometric analysis of realized covariation: high frequency based covariance, regression, and correlation in financial economics. Econometrica **72**(3), 885–925 (2004)
7. Barndorff-Nielsen, O.E., Hansen, P.R., Lunde, A., Shephard, N.: Designing realized kernels to measure the ex post variation of equity prices in the presence of noise. Econometrica **76**(6), 1481–1536 (2008)
8. Barndorff-Nielsen, O.E., Hansen, P.R., Lunde, A., Shephard, N.: Multivariate realised kernels: consistent positive semi-definite estimators of the covariation of equity prices with noise and non-synchronous trading. J. Econ. **162**(2), 149–169 (2011)
9. Bauer, G.H., Vorkink, K.: Forecasting multivariate realized stock market volatility. J. Econ. **160**(1), 93–101 (2011)
10. Bauwens, L., Otranto, E.: Nonlinearities and regimes in conditional correlations with different dynamics. J. Econ. **217**(2), 496–522 (2020)
11. Bauwens, L., Otranto, E.: Modeling realized covariance matrices: a class of Hadamard exponential models. J. Financ. Econ. **21**(4), 1376–1401 (2023)
12. Bauwens, L., Laurent, S., Rombouts, J.V.: Multivariate GARCH models: a survey. J. Financ. Econ. **21**(1), 79–109 (2006)
13. Bauwens, L., Storti, G., Violante, F.: Dynamic conditional correlation models for realized covariance matrices. CORE DP **60**, 104–108 (2012)
14. Bollerslev, T., Engle, R.F., Wooldridge, J.M.: A capital asset pricing model with time-varying covariances. J. Polit. Econ. **96**(1), 116–131 (1988)
15. Bollerslev, T., Patton, A.J., Quaedvlieg, R.: Modeling and forecasting (un)reliable realized covariances for more reliable financial decisions. J. Econ. **207**(1), 71–91 (2018)
16. Bonato, M., Caporin, M., Ranaldo, A.: A forecast-based comparison of restricted Wishart autoregressive models for realized covariance matrices. Eur. J. Financ. **18**(9), 761–774 (2012)
17. Bose, A., Mukherjee, K.: Estimating the ARCH parameters by solving linear equations. J. Time Ser. Anal. **24**(2), 127–136 (2003)
18. Buccheri, G., Corsi, F.: HARK the SHARK: realized volatility modeling with measurement errors and nonlinear dependencies. J. Financ. Econ. **19**(4), 614–649 (2021)
19. Chang, Y., Park, J.Y.: Index models with integrated time series. J. Econ. **114**(1), 73–106 (2003)
20. Chen, Z., Liu, W., Wang, C.D., Wu, W.-Q., Wu, Y.-H.: Nonlinear least squares estimation of Log-ACD models. Acta Math. Appl. Sin. Engl. Ser. **34**(3), 516–533 (2018)
21. Chiriac, R., Voev, V.: Modelling and forecasting multivariate realized volatility. J. Appl. Econ. **26**(6), 922–947 (2011)
22. Ciuperca, G.: Asymptotic behaviour of the LS estimator in a nonlinear model with long memory. J. Korean Stat. Soc. **40**(2), 193–203 (2011)
23. Corsi, F.: A simple approximate long-memory model of realized volatility. J. Financ. Econ. **7**(2), 174–196 (2009)
24. Creal, D., Koopman, S.J., Lucas, A.: A dynamic multivariate heavy-tailed model for time-varying volatilities and correlations. J. Bus. & Econ. Stat. **29**(4), 552–563 (2011)
25. Cubadda, G., Guardabascio, B., Hecq, A.: A vector heterogeneous autoregressive index model for realized volatility measures. Int. J. Forecast. **33**(2), 337–344 (2017)
26. Dias, G.F.: The Nonlinear Iterative Least Squares (NL-ILS) Estimator: An Application to Volatility Models. School of Economics and Finance, Queen Mary University, DP (2013)
27. Engle, R.F.: Dynamic conditional correlation: a simple class of multivariate generalized autoregressive conditional heteroskedasticity models. J. Bus. & Econ. Stat. **20**(3), 339–350 (2002)
28. Ghysels, E., Santa-Clara, P., Valkanov, R.: There is a risk-return trade-off after all. J. Financ. Econ. **76**(3), 509–548 (2005)
29. Ghysels, E., Santa-Clara, P., Valkanov, R.: Predicting volatility: getting the most out of return data sampled at different frequencies. J. Econ. **131**(1–2), 59–95 (2006)
30. Golosnoy, V., Gribisch, B., Liesenfeld, R.: The conditional autoregressive Wishart model for multivariate stock market volatility. J. Econ. **167**(1), 211–223 (2012)

31. Granger, C.W., Joyeux, R.: An introduction to long-memory time series models and fractional differencing. J. Time Ser. Anal. **1**(1), 15–29 (1980)
32. Greene, W.H.: Econometric Analysis. Pearson Education Limited, London (2018)
33. Hamadeh, T., Zakoïan, J.-M.: Asymptotic properties of LS and QML estimators for a class of nonlinear GARCH processes. J. Stat. Plan. Inference **141**(1), 488–507 (2011)
34. Harvey, A.C.: Dynamic Models for Volatility and Heavy Tails: with Applications to Financial and Economic Time Series. Cambridge University Press, Cambridge (2013)
35. Hosking, J.: Fractional differencing. Biometrika **68**(1), 165–176 (1981)
36. Hwang, E., Shin, D.W.: Infinite-order, long-memory heterogeneous autoregressive models. Comput. Stat. & Data Anal. **76**, 339–358 (2014)
37. Hwang, S., Kim, T.Y.: Power transformation and threshold modeling for ARCH innovations with applications to tests for ARCH structure. Stoch. Process. Appl. **110**(2), 295–314 (2004)
38. Jarque, C.M., Bera, A.K.: Efficient tests for normality, homoscedasticity and serial independence of regression residuals. Econ. Lett. **6**(3), 255–259 (1980)
39. Konno, Y.: A note on estimating eigenvalues of scale matrix of the multivariate F-distribution. Ann. Inst. Stat. Math. **43**(1), 157–165 (1991)
40. Lai, T.L.: Asymptotic properties of nonlinear least squares estimates in stochastic regression models. Ann. Stat. **22**(4), 1917–1930 (1994)
41. Lanne, M., Saikkonen, P.: Non-linear GARCH models for highly persistent volatility. Econ. J. **8**(2), 251–276 (2005)
42. McAleer, M., Medeiros, M.C.: A multiple regime smooth transition heterogeneous autoregressive model for long memory and asymmetries. J. Econ. **147**(1), 104–119 (2008)
43. Müller, U.A., Dacorogna, M.M., Davé, R.D., Pictet, O.V., Olsen, R.B., Ward, J.R.: Fractals and intrinsic time: a challenge to econometricians [Unpublished manuscript]. Olsen & Associates, Zürich (1993)
44. Oh, D.H., Patton, A.J.: High-dimensional copula-based distributions with mixed frequency data. J. Econ. **193**(2), 349–366 (2016)
45. Opschoor, A., Janus, P., Lucas, A., Van Dijk, D.: New HEAVY models for fat-tailed realized covariances and returns. J. Bus. & Econ. Stat. **36**(4), 643–657 (2018)
46. Pollard, D., Radchenko, P.: Nonlinear least-squares estimation. J. Multivar. Anal. **97**(2), 548–562 (2006)
47. Qu, H., Chen, W., Niu, M., Li, X.: Forecasting realized volatility in electricity markets using logistic smooth transition heterogeneous autoregressive models. Energy Econ. **54**, 68–76 (2016)
48. Rohan, N., Ramanathan, T.: Nonparametric estimation of a time-varying GARCH model. J. Nonparametric Stat. **25**(1), 33–52 (2013)
49. Taylor, N.: Realized volatility forecasting in an international context. Appl. Econ. Lett. **22**(6), 503–509 (2015)
50. Vassallo, D., Buccheri, G., Corsi, F.: A DCC-type approach for realized covariance modeling with score-driven dynamics. Int. J. Forecast. **37**(2), 569–586 (2021)
51. Wang, Q.: Least squares estimation for nonlinear regression models with heteroscedasticity. Econ. Theory **37**(6), 1267–1289 (2021)
52. Wooldridge, J.M.: Estimation and inference for dependent processes. In: Handbook of Econometrics. Elsevier (1986)
53. Wu, C.-F.: Asymptotic theory of nonlinear least squares estimation. Ann. Stat. **9**(3), 501–513 (1981)

Spatial Explorative Analysis of Thyroid Cancer in Sicilian Volcanic Areas

Francesca Bitonti and Angelo Mazza

Abstract In the most advanced countries the epidemiological transition has led non-communicable diseases to dominate the causes of death (ruling out the recent Covid-19 pandemic). Therefore, the spatial methodologies traditionally implemented to study infectious diseases have been also exploited to analyze those non-communicable diseases whose causes include environmental risk factors. In this work, we aim to explore the spatial distribution of thyroid cancer (TC) near Mt Etna (Sicily), a possible environmental risk factor for TC. Although numerous studies have already detected an increase in TC incidence in proximity of active volcanoes, none of them has included a geographic approach to investigate the phenomenon, so far. For this reason, we aim to implement a spatial analysis to explore the phenomenon. In particular, by georeferencing data from the Cancer Registry of Eastern Sicily, we provide a visualization of the phenomenon through geographical maps, highlighting areas with significantly higher incidence than expected. Subsequently, we exploit the local Moran's I index to assess the presence of clusters of higher-risk areas, considering both the proximity to the volcano and the non-uniform distribution of the population across the territory. Our results seem to confirm the established belief that the presence of a volcano can lead to an increase in TC incidence.

Keywords Thyroid cancer · Spatial analysis · Disease incidence · Volcanoes · Sicily

---

F. Bitonti (✉) · A. Mazza
Department of Economics and Business, University of Catania, Catania, Italy
e-mail: francesca.bitonti@unict.it

A. Mazza
e-mail: a.mazza@unict.it

M. Bini et al. (eds.), *Advanced Methods in Statistics, Data Science and Related Applications*, Springer Proceedings in Mathematics & Statistics 467,
https://doi.org/10.1007/978-3-031-65699-6_13

1 Introduction

In the most advanced countries experiencing the third phase of the epidemiologic transition, infectious disease pandemics (ruling out the recent Covid-19 pandemic) are replaced by degenerative diseases as major causes of death [23]. Specific patterns of human behaviour (e.g., violence, accidents, or substance abuse) overtake infectious agents as the major contributor to morbidity and mortality. In these contexts, the use of the spatial analysis methodologies, typical approaches to the analysis of infectious diseases, has also spread within the study of non-communicable diseases, in which environmental risk factors are often included among the possible contributing causes [1, 7, 12, 13, 22, 24].

Thyroid cancer (TC), the object of this study, is the most widespread endocrine neoplasm, whose incidence has steadily increased worldwide in recent decades [9, 11, 19]. Epidemiological studies suggest that the aetiology of TC includes iodine consumption, familiarity, and exposure to ionising radiation [16, 27]. Kung et al. [18] identified the elements present in volcanic gases as possible further TC etiological agents. Other studies have confirmed a higher TC incidence than national averages in the Hawaiian Islands [14, 17], Iceland [3, 15], the Philippines [8, 10], New Caledonia [28] and French Polynesia [6, 9], all regions whose common denominator is the presence of active volcanoes. Goodman et al. [14] compared the standardised TC incidence rates of different ethnic groups residing in Hawaii with those of the same groups residing in other geographic areas. Also in Sicily, a region located in South Italy, higher TC incidence rates than the regional and national averages were detected in the vicinity of an active volcanic areas, [20, 21, 29, 30]. In particular, a double incidence compared to that of the region was observed near Mt Etna [26].

All the mentioned studies reinforce the hypothesis of a TC-volcano relationship but lack a geographic approach for analyzing the phenomenon of interest. We believe that epidemiological data from volcanic areas require a geographical investigation capable of offering a new perspective on TC risk. When concepts of proximity and spatial variations are discussed, geographic tools and approaches from spatial statistics cannot be ignored. Our work represents an attempt to address this gap in the literature by introducing a geographic perspective into the study of TC distribution over space. Specifically, after georeferencing data from the Cancer Registry of Eastern Sicily using the Google Maps Geocoding API interface, we created maps describing TC risk at the census tract level in the provinces of Messina, Catania, Enna, and Siracusa for the period 2003–2016. The chosen risk indicator is the Standardized Incidence Ratio (SIR), calculated through indirect standardization. Using additional maps, we have shown which sections record an increase in incidence compared to the expected, which is statistically significant. Subsequently, to assess the presence of high-risk cluster areas, we applied the local Moran's I indicator. The local Moran's I statistic can detect the presence of spatial autocorrelation at the sub-area level, which might not emerge at the global level. Although TC case maps and cluster analysis cannot prove the underlying cause-effect mechanisms of the investigated phenomenon, we rely on these methodologies to provide additional evidence regarding the TC-volcano relationship and to support decision-making processes in the healthcare sector. Our results show the presence of higher-risk areas suggesting a possible effect of proximity to both Mt Etna and Mt Vulcano, although the latter exhibits lower activity compared to the former. Nevertheless, given the exploratory nature of our work, a more in-depth study would be necessary to gain a deeper understanding of the phenomenon.

This work is organized as follows: the next section describes the applied methodology and the exploited data sources; the third section discusses the main findings and the last one provides the conclusions of the analysis.

2 Methodology and Analysed Data

The spatial distribution of TC cases can be described through the inhomogeneous Poisson point process. In this model, the expected number of events $N(u)$ in a generic sub-area U of the region under study R follows a Poisson distribution whose intensity λ varies over space. For a generic point u we will have $\lambda = \lambda(u)$ and the expected number of events $E[N(u)]$ is

$$E[N(U)] = \int_U \lambda(u)du$$

The Standardised Incidence Ratio (SIR) is computed at census tract level and is exploited to produce maps of TC risk. The SIRs were calculated by indirect standardization [31]. The SIR is the ratio between observed O_i and expected TC cases E_i in each census tract i

$$SIR = \frac{O_i}{E_i}$$

The expected number of cases is computed as the product of the population at risk P_i (and therefore the entire resident population) in the given census tract i and the general incidence rate r_+ for the entire investigated area

$$E_i = P_i r_+$$

Hence it follows that the SIR of a single census tract is thus computed as

$$SIR = \frac{O_i}{P_i \frac{O_+}{P_+}}$$

where O_+ corresponds to the total number of observed TC cases and P_+ is the total resident population of eastern Sicily. When the characteristics of the population determine a subdivision into strata with different risk levels (for instance, by sex and age group), a proper weight is associated with each stratum based on the risk of the specific stratum. In this case, the general incidence rate r_+ is replaced by a different rate computed for each stratum j as $r_j = \frac{\sum_i O_{ij}}{\sum_i P_{ij}}$. Any deviation of the observed SIR distribution from the theoretical one, obtained if the allocation of TC cases over the territory were random, would depend on other variables than the demographic structure of the resident population. Comparing the two distributions enables to verify, for each sub-area, whether potential environmental risk variables determine significant departures in TC risk from the expected one. The inferential test is based on the Poisson distribution and admits that under the null hypothesis, the SIR is equal to 1 and that an observed value of SIR significantly greater than 1 implies an increase in risk [4].

The local Moran's I was proposed by Anselin [2] and it is defined by the following formula

$$I_i = \frac{(y_i - \overline{y})}{S_i^2} \sum_{j=1, j\neq i}^{n} \left(w_{ij}(y - \overline{y}) \right)$$

where n is the number of geographical units, y_i is the value of the variable y in region i, $\overline{y}$ is the sample mean of the variable, y_j is the value of the variable y in unit j, S_i^2 is the sample variance of the variable y, and w_{ij} is a weight that can be defined as the inverse of the distance between the various geographical units. There are several alternatives to define w_{ij}, some contemplate the choice of a limit distance to define the neighbourhood of a given geographical unit: the units falling within the limit distance take on a positive weight, while the external ones take on a weight equal to zero. Positive values of local Moran's I for a given region indicate that the region is surrounded by regions that have mostly similar values of the variable under study. The emerging spatial groups are defined as "high-high" (region with a high value surrounded by regions with high values) or "low-low" (region with low value surrounded by regions with low values). Spatial outliers result in the composition of two cluster types: "high-low", in the presence of a high value surrounded by neighbours with low values, or "low–high", vice versa. To test the statistical significance of the local Moran's statistic we performed conditional permutation proposed by Anselin in 1995 [2].

The data available come from the Cancer Registry of Eastern Sicily and refer to TC cases diagnosed in the period 2003–2016 to residents in the four provinces of interest with age between 5 and 95 years. Overall, 7,085 cases are included. The data concerning the resident population in the census tracts of the four provinces of interest come from the 15th General Population Census carried out by Istat in 2011 (https://www.istat.it/it/archivio/104317). Google Maps API was exploited to geocode the residential addresses of TC cases (https://cloud.google.com/maps-platform/).

3 Results and Discussion

The TC cases distribution is influenced by the spatial arrangement and the age and sex structure of the resident population. It's well-known that adult females are the most affected by TC [25]; furthermore, the population ageing in inner areas, where depopulation dynamics are in place, tends to accentuate the imbalance between sexes, given the higher life expectancy of women compared to men [32]. In an attempt to distinguish the effects of the demographical and environmental components on the TC cases spatial distribution, we propose SIR maps at the census tract level and together with the geographical representation of the statistical significance. Figure 1a, b respectively report the SIR and the results of the one-tailed test with significance level $\alpha = 0.05$ highlighting the census tracts with SIR significantly greater than 1. From the mere SIR calculation (Fig. 1a) different areas of risk emerge. These areas are located in the area nearby Mt Etna, but also in the non-volcanic provinces, especially in those of Enna and Messina. The area south-east of Mt Etna and different sections located mainly in the Messina province prove to be statistically significant from the one-tailed test (Fig. 1b). Finally, whether in the non-volcanic provinces the census sections with SIR greater than 1 are randomly arranged, in the province of Catania the high-risk sections mainly concentrate in an area close to Mt Etna, leaving the rest of the province almost empty. To detect the location of the TC cases clusters over the area and to visually assess their arrangement in relation to the proximity of Mt Etna, the local Moran's I on the SIR was computed. Figure 2a, b show the local statistic for all TC cases with a bandwidth of 10 km and the pseudo p-values obtained from the conditioned permutation procedure, respectively. Two high-risk areas appear: the first in the south-eastern side of Mt Etna and the other to the north, between the Aeolian Islands and near Barcellona Pozzo di Gotto. The location of the high-risk areas along the NW–SE axis could suggest that the presence of persistent winds toward the SE direction, documented by Boffetta et al. [5], can carry the toxic volcanic substances and therefore pollute the atmosphere of the territories positioned along the axis itself. A similar consideration could be applied to the Aeolian Islands and the sections near Barcellona Pozzo di Gotto which are in proximity of Mt Vulcano, an active volcano presenting limited activity compared to that of Mt Etna.

4 Conclusive Remarks

In this research, we applied spatial analysis methodologies typically used for infectious diseases to the study of TC, a prevalent endocrine neoplasm. The findings presented significant insights into the spatial distribution of TC cases in eastern Sicily. Through the SIR mapping and local Moran's I analysis, it became evident that specific areas, particularly those near Mt Etna, exhibited significantly higher TC risk. The results also highlighted the influence of demographic factors, emphasizing the higher susceptibility of adult females and the impact of population aging on TC

(a)

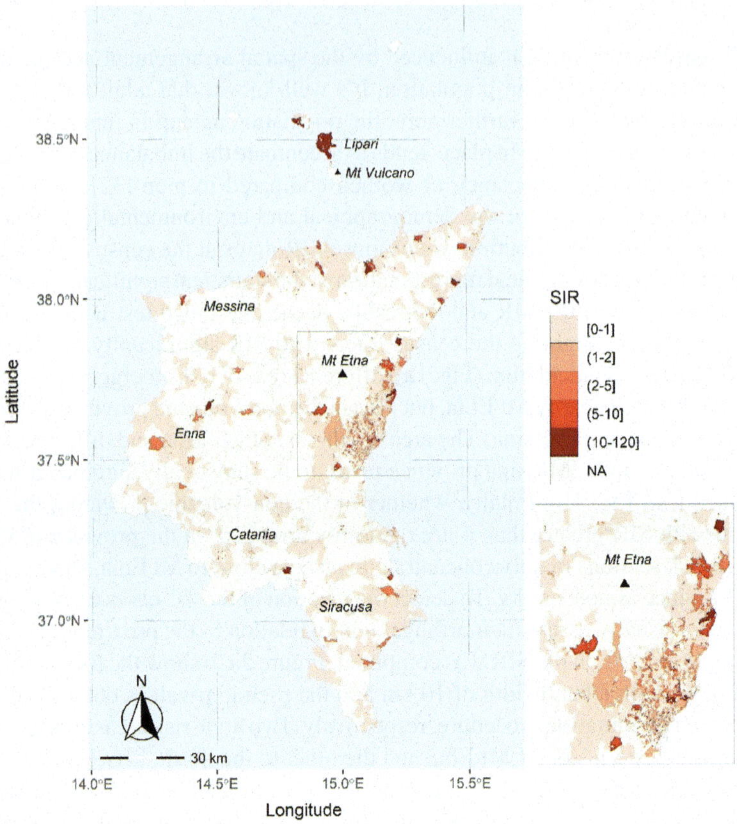

Fig. 1 SIR distribution by census tracts (**a**) and representation of their statistical significance (**b**). *Source* authors' original elaboration

distribution. Notably, the study demonstrated the existence of distinct high-risk clusters, indicating a potential correlation between TC incidence and volcanic activities needing further investigation.

The identification of geographically concentrated high-risk areas for TC carries substantial implications for public health interventions and resource allocation. This research underscores the importance of considering environmental risk factors, especially volcanic activity, in understanding disease patterns. These findings advocate for targeted health education and screening programs in the identified high-risk regions. Furthermore, this study prompts policymakers and urban planners to incorporate health considerations, including geological factors, into regional development strategies.

This study lays the foundation for future research endeavours in several directions. Firstly, the consideration of some multiple testing corrections would allow

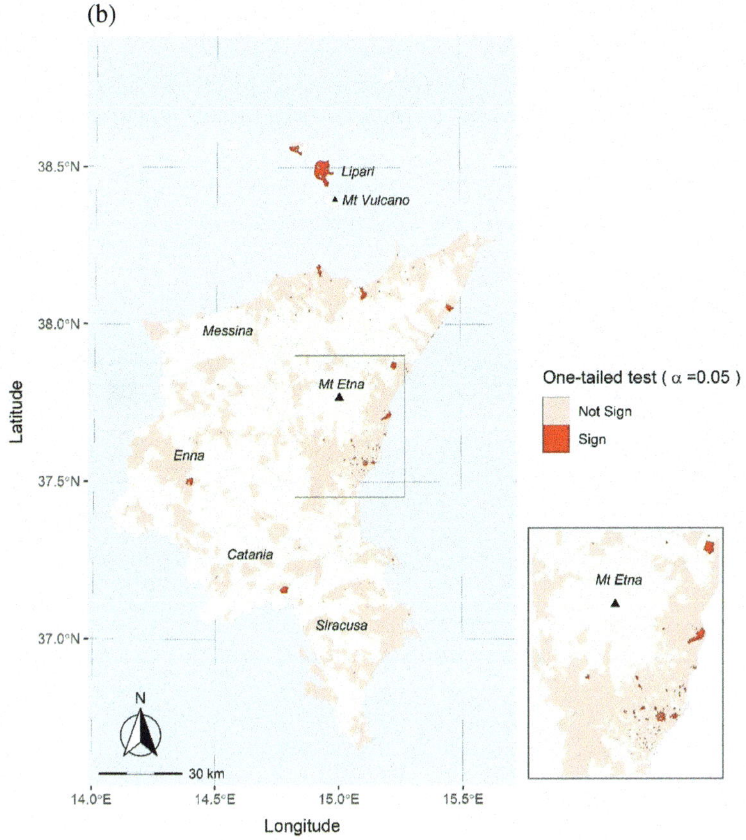

Fig. 1 (continued)

to control for the risk of type I error when assessing the statistical significance of the SIR. Secondly, the consideration of a more flexible distribution than the Poisson could shape more properly the TC cases occurrence. A viable alternative could be the negative binomial which relaxes the assumption of equality between mean and variance, proper of the Poisson distribution. Finally, the implementation of an ad-hoc procedure to explicitly include the direction of the winds as a predictor for the increased TC incidence could allow to assess the causality between volcanic activity and TC.

In conclusion, this research not only enhances our understanding of the spatial distribution of thyroid cancer but also exemplifies the significance of integrating environmental factors into disease mapping and analysis. As we move forward, addressing the complex interplay between geological, demographic, and genetic elements will be crucial in developing effective strategies for TC prevention and management, not only in eastern Sicily but potentially in other volcanic regions worldwide.

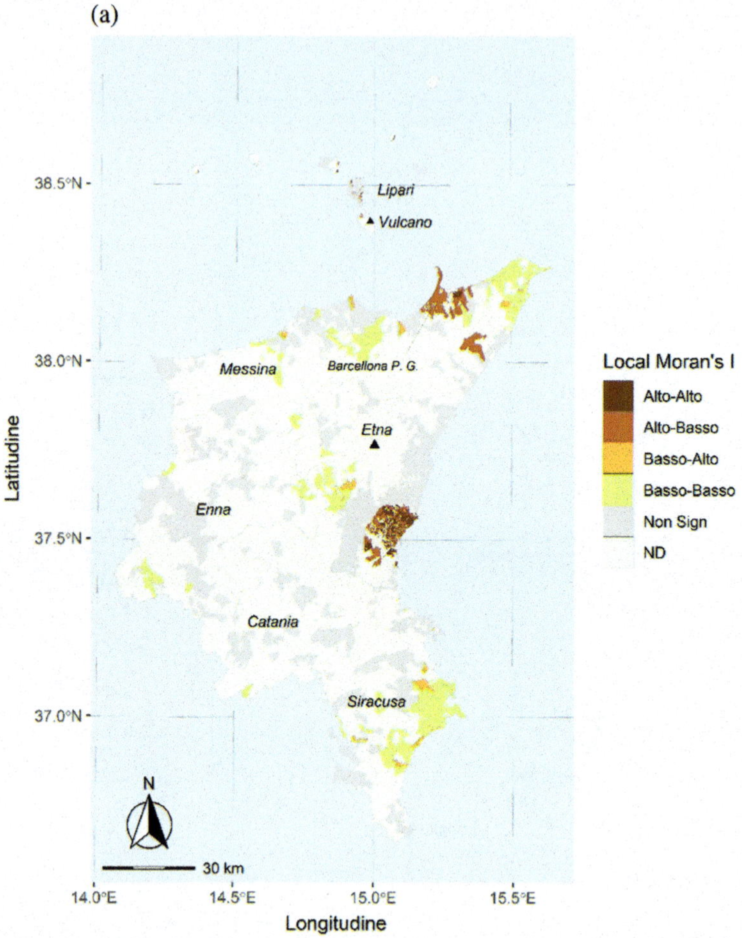

Fig. 2 Cluster map based on the Local Moran's I (**a**) and related pseudo p-values (**b**) computed by conditional permutation method. *Source* authors' original elaboration

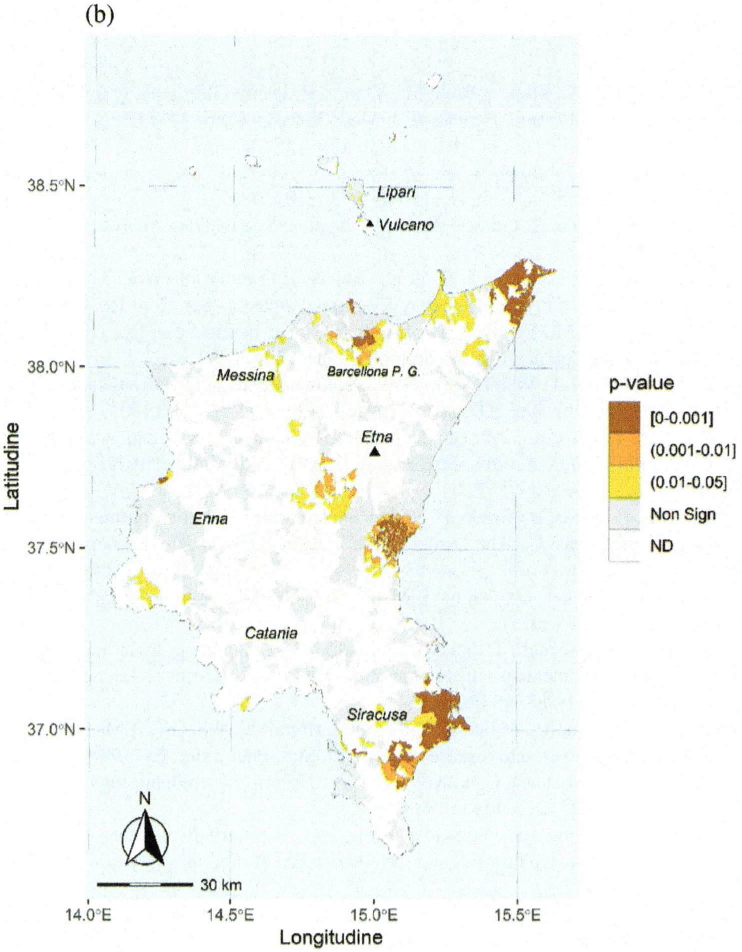

Fig. 2 (continued)

References

1. Anand, S., Staniec, A., Montez-Rath, M., Vlahos, P.: Using GIS mapping to track hot spots of kidney disease in California. New Engl. J. Med. **382**(23), 2265–2267 (2020)
2. Anselin, L.: Local indicators of spatial association-LISA. Geogr. Anal. **27**, 93–115 (1995)
3. Arnbjörnsson, E., Arnbiörnsson, A., Ólafsson, A.: Thyroid cancer incidence in relation to volcanic activity. Arch. Environ. Health **41**(1), 36–40 (1986)
4. Bivand, R.S., Pebesma, E., Gómez-Rubio, V.: Applied Spatial Data Analysis with R. Springer, New York (2008)
5. Boffetta, P., Memeo, L., Sciacca, S., et al.: Exposure to emissions from Mount Etna (Sicily, Italy) and incidence of thyroid cancer: a geographic analysis. Sci. Rep. **10**, 21298 (2020)
6. Bray, F., Colombet, M., Ferlay, J., et al. (eds.): Cancer incidence in five continents, vol. XI. Lyon: International Agency for Research on Cancer (2017)
7. Bristow, R.E., Chang, J., Vieira, V.M., et al.: Spatial analysis of advanced-stage ovarian cancer mortality in California. Am. J. Obstet. Gynecol. **213**(1), 43.e1-43.e8 (2017)
8. Caguioa, P.B., Bebero, K.G.M., Bendebel, M.T.B., Saldana, J.S.: Incidence of thyroid carcinoma in the Philippines: a retrospective study from a tertiary university hospital. Ann. Oncol. **30** (2019)
9. Curado, M.P., Edwards, B., Boyle, P., et al.: Cancer incidence in five continents, vol. IX (2019)
10. Duntas, L.H., Doumas, C.: The "rings of fire" and thyroid cancer. Hormones **8**(4), 249–253 (2009)
11. Fitzmaurice, C., Dicker, D., Naghavi, M., et al.: The global burden of cancer 2013. JAMA Oncol. **1**(4), 505–527 (2015)
12. Geraghty, E.M., Balsbaugh, T., Nuovo, J., Tandon, S.: Using geographic information systems (GIS) to assess outcome disparities in patients with type 2 diabetes and hyperlipidemia. J Am Board Fam Med **23**(1), 88–96 (2010)
13. Ghosh, M., Natarajan, K., Waller, L.A., Kim, D.: Hierarchical bayes GLMs for the analysis of spatial data: an application to disease mapping. J. Stat. Plan. Infer. **75** (1999)
14. Goodman, M.T., Yoshizawa, C.N., Kolonel, L.N.: Descriptive epidemiology of thyroid cancer in Hawaii. Cancer **61**, 1272–1281 (1988)
15. Hrafnkelsson, J., Tulinius, H., Sigvaldason, H., et al.: Papillary thyroid carcinoma in Iceland: a study of the occurrence in families and the coexistence of other primary tumours. Acta Oncol. **28**(6), 785–788 (1989)
16. Kitahara, C.M., Sosa, J.A.: The changing incidence of thyroid cancer. Nat. Rev. Endocrinol. **12**(11), 646–653 (2016)
17. Kolonel, L.N., Hankin, J.H., Hinds, M.W., et al.: An epidemiologic study of thyroid cancer in Hawaii. Cancer Cause. Control **1**, 223–234 (1990)
18. Kung, T.M., Ng, W.L., Gibson, J.B.: Volcanoes and carcinoma of the thyroid: a possible association. Arch. Environ. Health **36**(5), 265–267 (1981)
19. Liu, Y., Su, L., Xiao, H.: Review of factors related to the thyroid cancer epidemic. Int. J. of Endocrinol. (2017)
20. Malandrino, P., Russo, M., Vigneri, R., et al.: Increased thyroid cancer incidence in a basaltic volcanic area is associated with non-anthropogenic pollution and biocontamination. Endocrine **53**, 471–479 (2016)
21. Malandrino, P., Scollo, C., Mian, C., et al.: Descriptive epidemiology of human thyroid cancer: experience from a regional registry and the "volcanic factor." Front. Endocrinol. **4**(65), 1–7 (2013)
22. Musa, G.J., Chiang, P.H., Hoven, C.W., et al.: Use of GIS mapping as a public health tool–from cholera to cancer. Health Serv. Ins. **6** (2013)
23. Omran, A.R.: The epidemiologic transition: a theory of the epidemiology of population change. Milbank. Meml. Fund. Q. **49**(4), 509–538 (1971)
24. Palagiano, C., Pesaresi, C.: La salute nel mondo: Geografia medica e qualità della vita. Carocci, Roma (2011)

25. Parkin, D.M., Bray, F., Ferlay, J., Pisani, P.: Global cancer statistics, 2002. CA-Cancer J. Clin. **55**(2), 74–108 (2005)
26. Pellegriti, G., De Vathaire, F., Vigneri, R., et al.: Papillary thyroid cancer incidence in the volcanic area of Sicily. J. Natl. Cancer I. **101**, 1575–1583 (2009)
27. Seib, C.D., Sosa, J.A.: Evolving understanding of the epidemiology of thyroid cancer. Endocrin. Metab. Clin. N. Am. **48**(1), 23–35 (2019)
28. Truong, T., Rougier, Y., Guénel, P., et al.: Time trends and geographic variations for thyroid cancer in New Caledonia, a very high incidence area (1985–1999). Eur. J. Cancer Prev. **16**(1), 62–70 (1985)
29. Vigneri, R., Malandrino, P., Vigneri, P., et al.: Heavy metals in the volcanic environment and thyroid cancer. Mol. Cell. Endocrinol. **457**, 73–80 (2017)
30. Vigneri, R., Malandrino, P., Vigneri, P.: The changing epidemiology of thyroid cancer: why is incidence increasing? Curr. Opin. Oncol. **27**, 1–7 (2015)
31. Waller, L.A., Gotway, C.A.: Applied Spatial Statistics for Public Health Data. Wiley, Hoboken (2004)
32. World Population Ageing (2017). Retrieved from https://www.un.org/en/development/desa/population/publications/pdf/ageing/WPA2017_Highlights.pdf

Adjusted Calibration Estimators for Sparse Spatial Data

Ivan Sciascia

Abstract We developed an adjusted calibration estimator for spatial data analysis where the units of a population N have low spatial density. The calibrated estimator is based on linear equations applied on auxiliary variables $\mathbf{x_i}$ and constrained on known population total $\mathbf{X}$. The linear model explains the relation between the study variable y and the auxiliary variables $\mathbf{x_i}$. In a context of spatial data analysis with low density we propose to adjust the calibrated estimator by a distance function. The Traveling Salesman Problem (TSP) is a basic routing problem stated for visit to P cities with the shortest closed tour. Different algorithms have been developed to resolve heuristically the TSP. This article developed a correction to the calibrated estimator based on a TSP. The performances of the adjusted estimator are evaluated using an application study on a forest landscape.

Keywords Regression estimators · Survey sampling · Forest survey · Traveling Salesman Problem

1 Introduction

The Geographical Information System (GIS) improves the availability of remote sensed ancillary variables $\mathbf{x_i}$ useful for parameter estimation in the spatial data domain. Variables measurements can be used to improve the precision of the estimation procedures based on filed measurements in GIS based ecological studies.

The linear regression estimation model can be used when we know the parameters (total or mean) of auxiliary variables in population. This method is called calibrated estimation and it applies a linear model ξ to explain the underlying relationship between a study variable y and auxiliary variables $\mathbf{x_i}$.

The original work [1] introduces the method focusing on linear equations with constraints on the known parameters of the auxiliary variables [2] to obtain the calibrated weights starting from the weights of the Horvitz-Thompson estimator. Sarndal

I. Sciascia (✉)
Department of Life Sciences and Systems Biology, Viale Mattioli, 25, Torino, Italy
e-mail: ivan.sciascia@unito.it

M. Bini et al. (eds.), *Advanced Methods in Statistics, Data Science and Related Applications*, Springer Proceedings in Mathematics & Statistics 467,
https://doi.org/10.1007/978-3-031-65699-6_14

[3] describes, after 15 years from his co-contributed work, the several developments of the proposed estimator while Fuller [4] considers a more general overview focused on the general regression estimators (GREG).

Between recent contributions to methodological development originated by calibration in survey sampling, the frameworks of [5, 6] involve semiparametric and nonparametric tool trough different propositions of model-assisted calibration estimators while [7] considers the optimal calibration estimators and [8] consider the adaptive calibration.

The calibrated estimators have been applied in different contexts where geospatial data analysis links field and remotely sensed measurements. Frequently this survey samplying studies are applied on forestry resources. References [9] consider the application of semiparametric methods for natural resource estimation involving nonlinear models more able to capture complex relationship between variables. [10] apply a multiphase estimation model to forestry resources describing trough conceptual tools like spatial grids and difference estimators how to better use the field measurements coupled with remotely sensed measurements and the relationship between field and remotely sensed measurements is explained by General Additive Models (GAM). On the forestry resources studies [11] apply different well known parametric and nonparametric estimators comparing their performances.

Our study is applied on forestry resources survey sampling. We combined the calibrated estimation with a method stated to resolve a routing issue. The Traveling Salesman Problem (TSP) is used for apply an adjustement in the model inserting a correction term according to a distance calculation in the spatial domain. The route in a TSP is the short way for the salesman to travel when he must visit P cities and come back home to close the tour. In survey sampling for forestry studies it can be an useful information. If we consider a net structure applicable to spatial data where the cities are the sample units and the distances between units compose the route, we can transpose the routing problem paradigm to forest landscape where nodes are the trees coordinates.

In the methodological section the spatial structures are designed, the low density constraint is defined [12], the sampling method is chosen and the distance function is built on the shortest route of a TSP. We evaluated the estimator performances with an application example on a forest survey data [13].

In the discussion paragraph the comparisons with other calibration perspectives are considered and the conclusive paragraph resumed the peculiarities of the estimator.

2 The Method

The design features involve the sampling grid, the constraint on the space sampling units, the method for building a route, the correction to the calibrated estimator. The formalizations of the classical calibrated estimator are proposed in paragraph 2.5 and they establish the preliminary remarks.

In the first paragraph of this section we described the design assumptions that involve the constraint of spatial low-density of the population N and the procedures of geospatial sampling due to the structuring of a spatial grid composed by square *areas* with square *sectors* inside and we built a sampling spatial structure to extract sample with known inclusion probability.

After designed the spatial sampling grid we extracted the first (and lonely extracted) unit of the sample s for the sparse population N. Then another unit is chosen with routing information.

In logistics studies the terms network routing are referred to services scheduling problems where transport resources must serve customers positioned on the territory. The customers and the relative distances are modeled by a network which is a graph with elements, nodes and arcs, associated additional resources like length.

A Traveling Salesman Problem (TSP) is the simplest version of a network routing calculating the shortest closed tour for a traveling salesman [14].

One of the nodes of the graph can represent the point of departure of the traveling salesman and the other nodes are the clients. The arcs represents the shifts and have associated an information on distance. For large numbers of nodes heuristic algorithms were developed to resolve in approximated mode the TSP, obtaining a good solution, not necessary optimal. We utilize one of the heuristic algorithms called nearest neighbor.

The nearest neighbor rule (NNR) is a simple heuristic to solve the TSP constructing a related tour with relative computed distance. The NNR starts with a node x_i arbitrarily chosen; between other nodes not yet visiting the NNR choses the closest one and puts it at the end of the partial sequence. The NNR works in the same way with the other nodes until the sequence is complete with all nodes visited then the NNR returns from the last node x_n to the first node x_1. The vehicle routing problem involves combinatorial optimization issues and a TSP rule must chose in some way the shortest tour.

In survey sampling we used the information from distances applying the TSP between sampled units. We assumed that the length of a tour for a sample s of the population N can be indicative of the population density.

If we consider to apply the TSP with NNR rule to spatial data analysis we could use supplementary information coupling field measurements, remote sensing and nodes routing solving.

2.1 The Construction of the Spatial Grid with Spatial Constraint

We imagined to superimpose to forest territory a simple spatial grid shown in Fig. 1 with following assumptions:

sector: minimal structural square unit of size in plan $\frac{g^2}{4}$

area: intermediate structural square unit of size in plan g^2

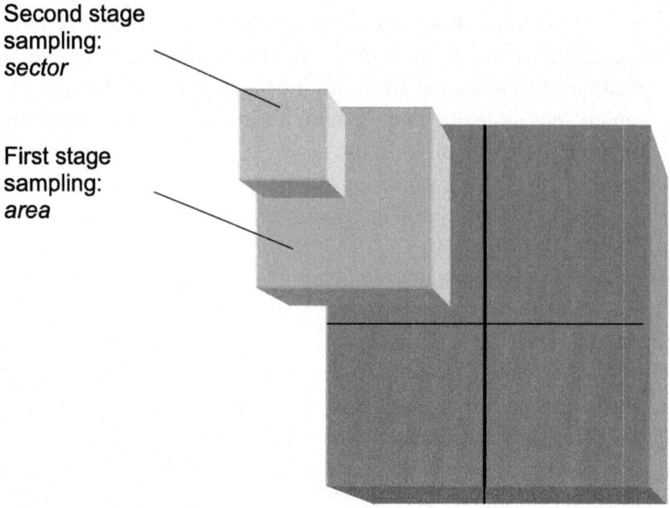

Fig. 1 Two-stage sampling

grid: overall square structure of size in plan $4g^2$
where g is the *area*'s side.

This *grid* is composed by da 4 *areas* and 16 *sectors*, 4 *sectors* in each *area* and overall dimensions in plan $4g^2$.

It is defined for the *grid* the following space constraint:

$$\sharp(z_i) \leq 2$$

where $\sharp(z_i)$ is the number of units included in the minimum space unit z_i: the *sectors*.

2.2 Two-Stage Sampling

If a population N is distributed on the territory and is possible to superimpose the previous described spatial *grid* covering all the population units and respecting for each *sector* the previous described space constraint, by a survey sampling it is possible to estimate a parameter (mean or total) of a study variable y_i with the support of auxiliary variables $\mathbf{x_i}$. It is chosen to build a small sample $n = 2$ where only the first unit is sampled.

According to the *grid* structure it is designed a two-stage sampling: in the first stage an *area* is extracted with simple random sampling from the *grid*; in the second stage a *sector* is extracted with simple random sampling from the extracted *sector*. One unit, two units or no unit can belong to the *sector* according to the space constraint.

If in the extracted *sector* there are two units, one unit is arbitrarily chosen; if no units belongs to the *sector* another *sector* is sampled again at the second stage. The first unit of the sample is the one belonging to the extracted *sector*. This sampling procedure is rigid and loose information because it considers the sector as sample unit and then the relative inclusion probability neglecting that one, two or no unit can belong to the *sector*. We will consider the particular case of no unit in the sector samples and we will discuss the peculiarities for the building of the adjusted calibrated estimator.

2.3 The Choice of the Second Sample Unit

If we give relative importance to distances between population units covered by the *grid* we can start to use linear approximations. The second unit of the sample is chosen by considering the minimum distance from the first sampled unit. This choice is done within the *area* of the *sector* extracted. The inclusion probability of the units in the sample are the same and are assumed to be that of the *sector*. The distance between sample units is utilized to start to build a distance function useful to correct the calibrated estimator.

Note that if distance is linearly approximated, then it can be computed by using the coordinates of the units. For (x_1, y_1), (x_2, y_2), the two units in the sample, the distance is calculated as:

$$d_i = \sqrt{(x_2 - x_1)^2 + (y_2 - y_1)^2} * 2$$

d_i is the length necessary to close a tour for the two units.

2.4 The Nearest Neighbor Rule

After constructing the small sample in the grid space the information in the relative *area* is searched using a tour from nearest neighbor rule. For the grid case proposed the NNR is as follows:

Initialization of the rule

$$M = \{1, 2, 3, 4, ..., j\} \text{ set of the units in the extracted } \textit{area}$$

$$K = \{1, i\} \text{ set of the units in the extracted } \textit{sector}$$

$$i^0 \in K \text{ sampled unit}$$

$$T = M \setminus i^0 \text{ set of the residual units in the extracted } \textit{area}$$

$$S = \{i^0, ..., j^x\} \text{ partial sequence}$$

Choosing the next city

$$i^l \text{ last unit in the partial sequence in S}$$

We find the unit j^* closest to the last from those in T resolving:

$$\min_{j \in T} c_{i^l, j}$$

where c = distance

In the case of alternative optimal we arbitrarily chose a unit among the best units.

Expansion of the sequence.

We put on j^* at the end of the partial sequence $S \longleftarrow (S, j^*)$ and we eliminate it from the units to visit below $T \longleftarrow T \setminus j^*$.

Conclusion of the algorithm If $T = \emptyset$ that is there are no residuals units in the *area*, we put on the initial unit at the end of the sequence $S \longleftarrow (S, i^0)$ to close the tour.

2.5 The Construction of the Adjusted Calibration Estimator

Considering a spatial population of N units covered by space *grid* according to here proposed sampling procedure a sample $n = 2$ is extracted. Associated with the ith unit of the population is a study variable y_i and an auxiliary variable x_i (for reason of description simplicity here we consider one auxiliary variable). The values of x_i are known by remote sensing for the entire population while the values of the variable y_i is known by field measurement so in estimation procedure the variable y_i is known for the two unit of the sample n. According with the sampling procedure here described we consider the *sector* inclusion probabilities for the sample units. We estimated the population total $Y = \sum_{i=1}^{N} y_i$ with calibration technique.

The formulations resume the step in classical calibration technique to obtain the calibrated estimator starting from Horvitz-Thompson estimator.

The Horvitz-Thompson estimator:

$$\hat{Y}_{HT} = \sum_{i=1}^{n} w_i y_i$$

where w_i is the weight of the ith observation.

The Deville and Sarndal estimator:

$$\hat{Y}_c = \sum_{i=1}^{n} c_i y_i$$

where c_i is the calibrated weight of ith observation according to constrained optimization.

The constrained optimization for the calibrated estimator is:

$$c_i : min \sum_{i=1}^{n} G_i(c_i, w_i)$$

$$subject\ to \sum_{i=1}^{n} c_i x_i = X$$

$G_i(c_i, w_i)$ is the generic distance function among c_i and w_i; considering the Euclidean distance among c_i and w_i:

$$G_i(c_i, w_i) = (c_i - w_i)^2 / w_i$$

resolving the optimization problem by analytical the general regression estimator (GREG) is obtained:

$$\hat{Y}_c = \sum_{i=1}^{n} c_i y_i = \hat{Y} + \hat{\beta}_c (X - \hat{X})$$

where β_c is:

$$\hat{\beta}_c = \sum_{i=1}^{n} w_i x_i y_i / \sum_{i=1}^{n} w_i x_i^2$$

The calibrated estimator is corrected considering the tour distance in TSP resolved in the *area* where the *sector* extracted belongs. The tour distance is interpreted for the units in the *area* considering it like an indicator of the general distribution of the total units in the *grid*. In some way the *area* TSP tour is an indicator of the density in the *grid*.

The area TSP tour is the "real situation" estimable for the grid. We can imagine the "ideal situation" of minimal *grid* density and to compare them to derive a *distance function* to correct the calibrated estimator taking in accounts TSP tour information. Different assumptions on *distance function* will be considered in the rest of the framework. The correction come from partial tour and total tour considerations on *area* TSP; the following notations are utilized:

d_i: distance between ith sampled unit and its ith closest neighbor in the *area*.

$$d_i = \sqrt{(x_2 - x_1)^2 + (y_2 - y_1)^2} * 2$$

d_t: TSP tour in the *area*, multiplied for two plus *grid* diagonal. This calculation is done to estimate a TSP tour in the *grid* assuming to multiply for two the TSP tour in the area and come back along the diagonal to return to the ith sampled unit:

$$d_t = 2S + \sqrt{(2g)^2 + (2g)^2} = 2S + 2\sqrt{2}g \tag{1}$$

where g is the base of the *area* and S is the tour length.

D_t: length of the diagonal of the grid multiplied for two. It is assumed to compare a specific territorial distribution of units d_t with an ideal territorial disposition D_t where the ideal TSP tour assume the double of the value of the diagonal of the grid like if territorial disposition of the units is alongside the diagonal of the grid:

$$D_t = 4\sqrt{2}g \tag{2}$$

From these premises it is proposed the formulation of the following estimator $\hat{Y}_{SD}$, sparse data estimator:

$$\hat{Y}_{sd} = \hat{Y}_c + \alpha(d_t - D_t) \tag{3}$$

where

$$\alpha = \frac{\sum_{i=1}^{n} w_i d_i Y_i}{\sum_{i=1}^{n} w_i d_i^2} \tag{4}$$

or in simplified form:

$$\alpha = \frac{\sum_{i=1}^{n} Y_i}{2d_i} \tag{5}$$

The distance function $f(d) = \alpha(D_t - d_t)$ that characterize the estimator for sparse data is the addend to the calibration estimator. This proposed function resumes three information deduced by a routing problem: by the coefficient α it is considered the distance between the two units in the sample n, by the distance D_t it is considered the "real" territorial distribution of the population units N where the *area* TSP tour indicated with S is an "estimator" of the *grid* TSP tour, by the distance d_t it is considered the "ideal" territorial distribution of the population units N, where all the units are distributed alongside the diagonal of the *grid*. When a *sector* with no units is sampled, a new *sector* is sampled again and it is added to distance D_t the length of the diagonal of the *sector*.

3 Application on a Forest Landscape

The application have been done on a territory containing a distribution of trees field measured. The units are choosen to respect the constraint for sparse data here considered. In this application study we built all 16 samples n_i where i are the 16 *sectors* of the *grid*. We considered the variable Y and the variable topographic variable X computing a linear correlation coefficient ϱ of 0.87. We measured y_i on field and x_i are remote sensed. We estimate for the population N the total $Y = \sum_{i=1}^{N} y_i$ for each sample n_i by three estimators: $\hat{Y}_{HT}$ (Horvitz-Thompson estimator),

Table 1 Relative efficiencies with w_i=16

Relative efficiency	Area1	Area2	Area3	Area4	Relative efficiency on 16 samples
$\frac{MSE_{Y_{HT}}}{MSE_{Y_C}}$	8,344	118,651	186,158	2,813	8,838
$\frac{MSE_{Y_{SD}}}{MSE_{Y_C}}$	0,124	24,079	11,562	0,831	1,376

$\hat{Y}_C$(calibrated estimator), $\hat{Y}_{SD}$ (estimator for sparse spatial data) then evaluating the relative efficiencies respect to $\hat{Y}_C$. The observation weights are considered for all units of the samples $w_i = 16$.

The spatial convention adopted in this example is the following:
sectors 1,2,5,6 belongs to area 1, north-west of the grid
sectors 3,4,7,8 belongs to area 2, north-est of the grid
sectors 9,10,13,14 belongs to area 3, south-west of the grid
sectors 11,12,15,16 belongs to area 4, south-est of the grid.

The complete results considered on 16 tests describe a relative efficiency of the estimator $\hat{Y}_{SD}$ that is worse of the efficiency of the estimator $\hat{Y}_C$. Considering the calculation of the relative efficiency for distinct areas $\hat{Y}_{SD}$ results better than $\hat{Y}_C$ in two areas while it is worse in the other two. Although the relative efficiency is 24,079 in the area 2 and 11,562 in the area 3, in the area 4 the two estimators are similar in performances with a relative efficiency of 0,831. In area 1 there is the best performance of $\hat{Y}_{SD}$ with a relative efficiency of 0,124.

The application results are described in Table 1. Considered the example further study should be done on the performances of $\hat{Y}_{SD}$ regarding the key elements:

- the role of the coefficient α and its dependence from the values of y_i on the two sample observations and from the distance d_i between the unit in the sample. For example an improvement of the performances could derive from the substitution of the distance d_i with another distance deduced by TSP resolution;
- the role of the difference $(D_t - d_t)$ here built assuming a "real" distribution that duplicates the TSP tour adding the diagonal of the grid and an "ideal" distribution that duplicates the diagonal of the grid, but the same difference could derive from other assumptions on real and ideal distributions;
- the role of the sampled sectors that doesn't contain units: the proposed method provide to sample again when an empty sector is extracted and to add the diagonal of the *sector* to the distance D_t.

4 Discussion

We used calibrated estimation [1] to build an estimator applicable to spatial data. Geoforestry data are constructed from ground measurements and remotely sensed data with modern GIS technologies such as aerial photos or laser Lidar systems. Field measurements and remote sensing measurements allow the estimation of ecological characteristics of forest areas with an increase in the precision of the estimate.

The calibrated method is a regression estimator with calibration constraints on the totals of remotely sensed variables that are known. The estimator we proposed combines the calibrated estimator with a measurement of a routing problem calculating the distance between nodes of a spatial distribution of points. This routing problem is solved by an algorithm that calculates a distance by applying rules to minimize the total path between a series of nodes and return to the starting node.

This calculated distance can be an indicator of the density of points in a selected area. We considered a sparse distribution by imposing the constraint of a maximum of two units in a measurement space sector. The proposed estimator was applied to a forest area studied in the past [13] and the relative performances were analysed. Overall in the grid the calibrated estimator prevails in efficiency both with respect to the Horvitz-Thompson estimator and with respect to the estimator proposed but the partial results in two of four sectors demonstrate how the correction made leads to a better efficiency of the estimate.

5 Conclusion

The analysis of the characteristics of spatial distributions are useful for estimates in the geo-referenced sciences. For example, the study of forest areas uses measurements taken on the ground and remote sensing measurements. Field measurements are sample-based and provide sampling patterns and can be combined with remote sensing measurements which are less time and costs consuming and use modern GIS technologies.

From aerial photographs and Lidar data it is possible to extract information to increase the precision of the estimate both in forest inventories and in the estimation of limited forest areas. In our case study we were interested in applying a method called calibration which uses linear regression and the knowledge of totals over a study area of remotely sensed data. Using data collected in the field in an Alpine landscape, we wanted to test the relative efficiency on the estimate of the tree diameter as study variable using remotely sensed auxiliary variables, imagining the comparison with the classical Horvitz Thompson estimator.

We used a solution to a routing problem in a network of spatially distributed points to adjust the estimator on the population of trees of a limited area of woods. The results demonstrate that the original calibrated estimator is overall more efficient, but considering sectors of the spatial area the corrected estimator demonstrates better efficiency.

This study described that remotely sensed information can be used to improve estimation efficiency. However, the study has a limited scope as it considered a limited area of forest in the Italian Alps. Further studies could be conducted using larger forest areas and using both topographic variables and variables of anthropogenic influence as auxiliary variables.

References

1. Deville, J.C., Sarndal, C.E.: Calibration estimators in survey sampling. J. Am. Stat. Assoc. **87**(418), 376–382 (1992)
2. Sarndal, C.E., Swensson, B., Wretman, J.: Model-Assisted Survey Sampling. Springer, New York (1992)
3. Sarndal, C.E.: The calibration approach in survey theory and practice. Surv. Methodol. **33**(2), 99–119 (2007)
4. Fuller, W.A.: Regression estimation for survey samples. Surv. Methodol. **28**(1), 5–23 (2002)
5. Wu, C., Sitter, R.R.: A model-calibration approach to using complete auxiliary information from survey data. J. Am. Stat. Assoc. **96**(453), 185–193 (2001)
6. Montanari, G.E., Ranalli, M.G.: Nonparametric model-calibration estimation in survey sampling. J. Am. Stat. Assoc. **100**(472), 1429–1442 (2005)
7. Wu, C.: Optimal calibration estimators in survey sampling. Biometrika **90**(4), 937–951 (2003)
8. Clark, R.G., Chambers, R.L.: Adaptive calibration for prediction of finite population totals. Surv. Methodol. **34**(2), 163–172 (2008)
9. Breidt, F.J., Opsomer, J.D., Johnson, A.A., Ranalli, M.G.: Semiparametric model-assisted estimation for natural resource surveys. Surv. Methodol. **33**(1), 35–44 (2007)
10. Opsomer, J.D., Breidt, F.J., Moisen, G.G., Kauermann, G.: Model-assisted estimation of forest resources with generalized additive models. J. Am. Stat. Assoc. **102**(478), 400–409 (2007)
11. Bertini, R., Chirici, G., Corona, P., Travaglini, D.: Confronto di metodi parametrici e non-parametrici per la spazializzazione della provvigione legnosa tramite integrazione di misure a terra, dati telerilevati e informazioni ancillari. Forest@ **4(1)**, 110–117 (2007)
12. Sciascia, I.: Calibration with spatial data constraints. In: Classification and Data Mining, Springer Series Studies in Classification, Data Analysis and Knowledge Organization, 89–94, Springer (2013)
13. Sciascia, I, Garbarino, M., Vacchiano, G., Motta, R.: Comparing model-assisted estimators of structural variables in forest surveys. In: SIS Scientific Meeting Proceedings, Roma (2012)
14. Brandimarte, P., Zotteri, G.: Logistica di distribuzione. CLUT, Torino (2004)
15. Frosini, B.V., Montinaro, M., Nicolini, G.: Il campionamento da popolazioni finite. UTET Libreria, Torino (1999)

An Empirical Tool to Classify Industries by Regional Concentration and Spatial Polarization

Diego Giuliani, Maria Michela Dickson, Flavio Santi, and Giuseppe Espa

Abstract Traditional measures of geographical concentration of industries based on regional data (such as Gini, Herfindhal, and Ellison–Glaeser indices) do not consider the information about the spatial positions of regions. This implies their insensitivity to regions' spatial order and inability to account for neighboring effects. As an attempt to cope with this limitation, a recent stream of literature has focused on developing measures that quantify the degree of concentration of an industry while adjusting for spatial connections among regions. Following the idea that a single measure cannot fully describe the characteristics of an industry in terms of both concentration and spatial interactions, this paper proposes an alternative approach that measures the two dimensions jointly and allows for the classification of economic sectors into meaningful types of geographical configurations.

Keywords Geographical concentration · Spatial polarization · Spatial autocorrelation · Bootstrap confidence regions

1 Introduction

The phenomenon of the spatial concentration of economic activity is of paramount importance in the fields of regional sciences, urban studies, economic geography and spatial economics. Indeed, the problem of developing proper statistical measures of spatial economic concentration has a long history that can be traced back, at least,

D. Giuliani (✉) · M. M. Dickson · F. Santi · G. Espa
Department of Economics and Management, University of Trento, via Inama 5, Trento, Italy
e-mail: diego.giuliani@unitn.it

M. M. Dickson
e-mail: mariamichela.dickson@unitn.it

F. Santi
e-mail: flavio.santi@unitn.it

G. Espa
e-mail: giuseppe.espa@unitn.it

M. Bini et al. (eds.), *Advanced Methods in Statistics, Data Science and Related Applications*, Springer Proceedings in Mathematics & Statistics 467,
https://doi.org/10.1007/978-3-031-65699-6_23

to [11] and is still open nowadays. The main contributions to this problem can be grouped into three generations of indices [2].

The first generation indices are based on the use of regional (areal) data, that is, data aggregated according to a predefined discretization of the geographical space, and consists of adaptations of standard concentration and entropy measures (such as Gini, Herfindhal and Theil indices). These measures proved to be subject to some critical downsides: (i) they do not control for the overall agglomeration of manufacturing and level of industry concentration, thus hindering the comparability among industries with different market structures; (ii) the use of areal data makes them sensitive to the Modifiable Areal Unit Problem (MAUP) [16], and hence possibly biased; (iii) they are invariant to the spatial configuration of the areal units and, therefore, do not account for the spatial dependence among neighbouring regions.

The second generation indices, such as those proposed by [3, 5, 14], have been developed specifically to deal with the first of the three pitfalls that characterize the previous generation. They follow the so-called "dartboard" approach [5] that consists on comparing the degree of spatial concentration of a given industry with the one that would occur if all firms (plants) of that industry were randomly located across the observed areal units. These measures represent an advancement because they allow for assessing genuine spatial concentration while controlling for overall agglomeration, ensuring comparability among different sectors.

Finally, the third generation indices [4, 7, 13, among others] also overcome the limits related to using areal data. Indeed, since they treat the geographical space as continuous, without imposing any arbitrary spatial partitioning, and rely on the distances between individual economic activities, they are robust to MAUP and adequately account for spatial dependence. Implementing these measures requires processing data about the exact geographical position of every single business unit. Unfortunately, however, the availability of micro-geographic firm-level data is still quite limited, and often researchers can only rely on areal data.

In light of the limited applicability of distance-based measures of spatial economic concentration, a recent stream of literature [6, 10, 17, among others] has focused on modifying the areal data-based indices to account for spatial connections among areal units. While these new tools are still sensitive to MAUP, they do not suffer from the third limitation.

This paper follows the same principle of adjusting areal data-based indices but in a different direction. According to the idea that a single measure cannot fully describe the characteristics of an industry in terms of both spatial concentration and spatial dependence, this paper proposes an alternative approach that measures the two dimensions jointly and allows for the classification of industries into meaningful types of geographical configurations.

The structure of the paper is as follows: in the second section, we illustrate our proposed approach to classify industries into valid geographical configurations according to proper measures of geographical concentration and spatial dependence. The subsequent section shows, through an illustrative application based on synthetic paradigmatic data, how the proposed approach works and highlights its usefulness. Finally, the last section offers concluding remarks and indicates directions for future developments in this field.

2 Classifying Industries Considering Regional Concentration and Spatial Polarization of Economic Activities

To develop an approach to classify industries into valid geographical configurations, we follow the idea formalized by [1] and reprised by [9, 18, 19] of grouping industries according to the combination of the values of a geographical concentration index and a spatial autocorrelation index. More specifically, let $\mathbf{s} = [s_1, s_2, \ldots, s_n]$ represent the vector containing the regional shares of employment in a given industry, where n is the total number of regions in the economy of interest. Then, let $C(\mathbf{s})$ denote a proper index of geographical concentration of $\mathbf{s}$ characterized by a benchmark value, B_c, such that $C(\mathbf{s}) > B_c$ if the industry is *geographically concentrated* while $C(\mathbf{s}) < B_c$ if the industry is *geographically dispersed*. Moreover, let $A(\mathbf{s})$ indicate a proper index of spatial autocorrelation of $\mathbf{s}$ characterized by a benchmark value, B_a, such that $A(\mathbf{s}) > B_a$ if the industry is *positively spatially autocorrelated* while $A(\mathbf{s}) < B_a$ if the industry is *negatively spatially autocorrelated*.

By considering both indices together, seven different notable geographical configurations can be readily identified:

1. When the employment of an industry is both geographically concentrated and positively spatially autocorrelated, that industry can be classified as a *strongly polarized concentrated* industry (*sPC*) as it is characterized by concentration within regions but also by strong polarization of regions in which the industry is concentrated.
2. An industry that tends to agglomerate mainly by concentrating within regions without sprawling over a relatively large number of neighboring regions can be described as a *mildly polarized concentrated* industry (*mPC*). This geographical configuration implies that industry employment is geographically concentrated but not spatially autocorrelated.
3. An industry that is overrepresented in a few isolated regions, as evidenced by geographically concentrated and negatively spatially autocorrelated employment, can be appropriately denoted as a *weakly polarized concentrated* industry (*wPC*).
4. An industry that is geographically dispersed and positively spatially autocorrelated may be categorized as a *strongly polarized dispersed* industry (*sPD*) as it is depicted by employment dispersion in neighboring regions.
5. An industry that is geographically dispersed but not spatially autocorrelated may be labeled as a *mildly polarized dispersed* industry (*mPD*) as it shows employment dispersion within regions that are not spatially polarized.

6. An industry that is geographically dispersed and negatively spatially autocorrelated could be treated as a *weakly polarized dispersed* industry (*wPD*) as it shows employment dispersion in a few isolated regions.
7. Finally, a *randomly located* industry (*Rand*) is an industry in which employment does not follow any notable geographical pattern as it is randomly spatially distributed.

Table 1 and Fig. 1 summarize the relationship between the geographical configurations and the values of $C(\mathbf{s})$ and $A(\mathbf{s})$.

Table 1 Geographical configurations and corresponding values of $C(\mathbf{s})$ and $A(\mathbf{s})$

Geographical configuration	$C(\mathbf{s})$	$A(\mathbf{s})$
sPC	$>B_c$	$>B_a$
mPC	$>B_c$	$\approx B_a$
wPC	$>B_c$	$<B_a$
sPD	$<B_c$	$>B_a$
mPD	$<B_c$	$\approx B_a$
wPD	$<B_c$	$<B_a$
Rand	$\approx B_c$	$\approx B_a$

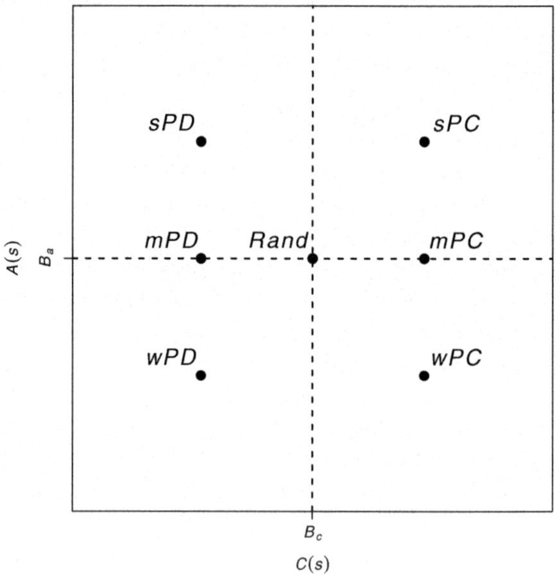

Fig. 1 Graphical representation of the relationship between the geographical configurations and corresponding values of $C(\mathbf{s})$ and $A(\mathbf{s})$

2.1 Index of Geographical Concentration

To classify industries according to the proposed framework, a good candidate as an index of geographical concentration, $C(\mathbf{s})$, is the Ellison and Glaeser (EG) index [5]. In the class of second-generation areal data-based spatial concentration indices, EG is the one with properties closest to those of an "ideal" index; indeed, it controls for overall agglomeration of manufacturing, it is robust to industrial concentration from small numbers of firms and allows comparability across industries and levels of spatial aggregation of data [4]. For a given industry, it can be expressed as

$$\gamma_{EG} = \frac{G - H(1 - \mathbf{x}'\mathbf{x})}{(1 - H)(1 - \mathbf{x}'\mathbf{x})},$$

where H is a Herfindahl index measuring concentration of the industry firm's employment distribution,[1] $G = (\mathbf{s} - \mathbf{x})'(\mathbf{s} - \mathbf{x})$ is the Gini statistic measuring the raw industry geographical concentration and $\mathbf{x}' = [x_1, x_2, \ldots, x_n]$ is a vector containing the regional shares of total employment.

Reference [5] show that $E[\gamma_{EG}] = 0$ when the industry's activity is randomly distributed among regions, that is when the distribution of economic activities in the industry (summarized by G) corresponds to the overall distribution of economic activities (summarized by $H(1 - \mathbf{x}'\mathbf{x})$). Therefore, the benchmark value of γ_{EG} representing spatial randomness is 0. Positive values of γ_{EG} measure industry employment concentration beyond that due to randomness while negative values measure industry employment dispersion beyond that due to randomness.

2.2 Index of Spatial Autocorrelation

The natural choice for an index $A(\mathbf{s})$ of spatial autocorrelation of regional employment is the popular Moran's I index [15]. For a given industry, it can be expressed as

$$I = \frac{n}{S_0} \frac{\mathbf{q}'\mathbf{W}\mathbf{q}}{\mathbf{q}'\mathbf{q}},$$

where $\mathbf{q} = \mathbf{s} - \bar{\mathbf{s}}$, $\mathbf{W} = (w_{ij})$ is a nonnegative matrix and $S_0 = \sum_{i=1}^{n} \sum_{j=1}^{n} w_{ij}$ is the sum of all the elements of $\mathbf{W}$.

$\mathbf{W}$ represents the so-called spatial weight matrix and is such that w_{ij} indicates how close is region j to region i; in particular, a large value of w_{ij} means a strong spatial connectivity between i and j. The benchmark value of I representing the absence of spatial autocorrelation is $-1/(n-1)$. Values higher than $-1/(n-1)$

[1] Therefore, $H = \mathbf{z}'\mathbf{z}$, where $\mathbf{z} = [z_1, z_2, \ldots, z_K]$ is a vector containing the K firms' shares of industry's total employment.

measure positive spatial autocorrelation while values lower than $-1/(n-1)$ indicate negative spatial autocorrelation.

2.3 *Bivariate Confidence Regions*

To assign an industry to one of the seven geographical configurations, one must look at how far the EG and Moran's I indices are from their respective benchmarks. Verifying if they are *far enough* requires taking into account the statistical significance of the estimated γ_{EG}- and I-values. However, since the indices can be correlated, separate hypothesis tests may not be appropriate. To take into account that γ_{EG} and I are related, we suggest a procedure to obtain their bivariate confidence region. However, since the joint sampling distribution of the two indices is unknown, and their covariance is analytically intractable, the confidence region can be derived by the bootstrap method. In particular, the joint bootstrap distribution can be obtained through block-wise resampling, defining the industry-region combinations as blocks. For each replication, a sample of firms is drawn randomly and independently with replacement from each block. Following [20], the bootstrap confidence regions are then obtained by deleting the bootstrap cloud of points according to Tukey's depth until the chosen confidence level is reached.

3 An Illustrative Application

To verify that the proposed approach provides the desirable results, we apply it to a simulated dataset which is meant to emulate the geographical configurations postulated in Sect. 2. The details of how we generated the data are as follows.

We specified $n = 49$ regions, as cells of a 7-by-7 lattice, and 7 industries, each with a total employment of 5000 and a total number of firms of 200 units. For each industry, the values of the vector $\mathbf{z}$ of the 200 firms' employment shares in the industry's total employment have been generated from a Beta distribution, $\mathrm{B}(\alpha_1, \beta_1)$, and then scaled by their sum to add to one. Therefore, $5000\mathbf{z}$ provides the number of employees for each of the 200 firms in the industry. Secondly, for each industry, the 200 firms are randomly assigned to the 49 regions (say the cells) with probabilities generated from a Beta distribution, $\mathrm{B}(\alpha_2, \beta_2)$, and then scaled by their sum to add to one. The Beta random variable is particularly useful in this setting as it allows to control explicitly the shape of the distribution. The values of γ_{EG} depend, indeed, on the difference in skewness between the distribution of the industry's regional shares of employment and the distribution of the industry's firms' shares of employment. Therefore, by setting specific values for α_1, β_1, α_2 and β_2, it is possible to obtain regionally distributed firm-level employment data characterized by levels of geographical concentration that are consistent with the seven geographical configurations.

Thirdly, for some industries, to make **s** characterized by a prescribed level of spatial autocorrelation measured by a given value of the Moran's I index, I^*, the cells have been randomly swapped according to the cell-swapping algorithm by [8]. More precisely, for the geographical configurations characterized by positive spatial autocorrelation, that is when $I^* > 0$, we applied the following steps,

1. Compute the Moran's I index
2. Swap two regions chosen at random.
3. Recompute the Moran's I index.
4. If the value of the Moran's I index has increased but is not greater than or equal to I^* then maintain the swap otherwise undo it.
5. Repeat steps 2–4 until the Moran's I index is greater than or equal to I^*.

On the other hand, for the geographical configurations characterized by negative spatial autocorrelation, that is when $I^* < 0$, we applied the following steps,

1. Compute the Moran's I index
2. Swap two regions chosen at random.
3. Recompute the Moran's I index.
4. If the value of the Moran's I index has decreased but is not lower than or equal to I^* then maintain the swap otherwise undo it.
5. Repeat steps 2–4 until the Moran's I index is lower than or equal to I^*.

By setting properly the parameters, the data for the 7 industries have been generated to recreate the seven archetypal geographical configurations. In particular,

- *Industry 1* (randomly located):
  $\alpha_1 = 20$, $\beta_1 = 20$, $\alpha_2 = 2$, $\beta_2 = 2$.
- *Industry 2* (strongly polarized dispersed):
  $\alpha_1 = 20$, $\beta_1 = 20$, $\alpha_2 = 1$, $\beta_2 = 1$, $I^* \geq 0.4$.
- *Industry 3* (strongly polarized concentrated):
  $\alpha_1 = 20$, $\beta_1 = 20$, $\alpha_2 = 2$, $\beta_2 = 6$, $I^* \geq 0.4$.
- *Industry 4* (mildly polarized dispersed):
  $\alpha_1 = 20$, $\beta_1 = 20$, $\alpha_2 = 1$, $\beta_2 = 1$.
- *Industry 5* (mildly polarized concentrated):
  $\alpha_1 = 20$, $\beta_1 = 20$, $\alpha_2 = 1$, $\beta_2 = 6$.
- *Industry 6* (weakly polarized dispersed):
  $\alpha_1 = 20$, $\beta_1 = 20$, $\alpha_2 = 1$, $\beta_2 = 1$, $I^* \leq -0.4$.
- *Industry 7* (weakly polarized concentrated):
  $\alpha_1 = 30$, $\beta_1 = 30$, $\alpha_2 = 2$, $\beta_2 = 6$, $I^* \leq -0.4$.

In all configurations, the Moran's I index has been computed using a row-standardized spatial weight matrix based on the rook contiguity criterion, that is $\mathbf{W}$ has been specified according to the following criterion,

$$w^*_{ij} = \begin{cases} 1 & \text{if cell } j \text{ shares a common side with cell } i \\ 0 & \text{otherwise} \end{cases}$$

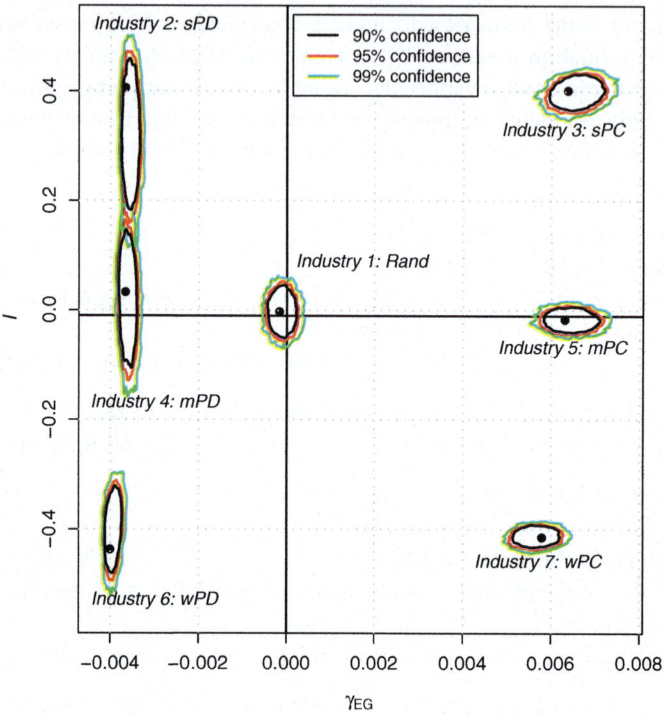

Fig. 2 Bivariate confidence regions to classify industries into geographical configurations

and

$$w_{ij} = \frac{w_{ij}^*}{\sum_{j=1}^{n} w_{ij}^*}.$$

Figure 2 depicts the 90, 95 and 99% bootstrap confidence regions of the estimated γ_{EG}- and I-values for the seven industries based on 9999 replications. The graph shows that the procedure assigns the industries to the correct geographical configurations.

4 Conclusion

In this paper, we propose a novel method based on the use of regionally aggregated data to assess the phenomenon of the spatial concentration of economic activity. The novelty lies in departing from the traditional approach of summarizing the spatial distribution pattern of an industry using a unique index. To fully characterize the geographical configuration of one industry is indeed necessary to take into consideration at least two different aspects: the regional concentration of the industry's activity and the spatial dependence among regions. Since these two dimensions can interact in different non-obvious ways, accounting for them with a unique measure can inevitably lead to a loss of relevant information about the geographical configuration of the industry. In an attempt to address this problem, this study proposes a framework to classify industries into meaningful spatial distribution patterns based on the joint analysis of geographical concentration, on the one hand, and spatial interactions, on the other hand. We argue that the empirical identification of relevant types of geographical configurations can be more informative, for industrial policies, than a single numerical value.

The procedure could be potentially extended in two promising directions. The first pertains to the possibility of using this approach even in those circumstances where firm-level employment data are not available but data about the number of firms per industry and region are available. Indeed, [14] developed an unbiased estimator of the Ellison and Glaeser index that circumvents the problem of the lack of firm-level data as it can be computed using the regional shares in terms of the number of firms by industry. However, computing the confidence regions would not be straightforward in this case since simple bootstrap methods do not work for regionally aggregated data, as they cannot represent an independent sample because of spatial autocorrelation.

The second possible extension concerns the inclusion of other interesting types of geographical configurations that consider the coagglomeration and spatial association between pairs or groups of industries, which may be assessed through the EG index of coagglomeration [14] and the bivariate Moran's I index [12]. This would lead to an over-complicated framework characterized by four dimensions; hence, further studies would be necessary to identify proper approaches to summarize the results.

References

1. Arbia, G.: The role of spatial effects in the empirical analysis of regional concentration. J. Geogra. Syst. **3**(3), 271–281 (2001)
2. Combes, P.P., Overman, H.G.: The spatial distribution of economic activities in the European union. In: J.V. Henderson, J.F. Thisse (eds.) Handbook of Regional and Urban Economics, vol. 4, pp. 2845–2909. Elsevier (2004)
3. Devereux, M.P., Griffith, R., Simpson, H.: The geographic distribution of production activity in the UK. Reg. Sci. Urban Econ. **34**(5), 533–564 (2004)
4. Duranton, G., Overman, H.G.: Testing for localization using micro-geographic data. Rev. Econ. Stud. **72**(4), 1077–1106 (2005)
5. Ellison, G., Glaeser, E.L.: Geographic concentration in U.S. manufacturing industries: a dartboard approach. J. Political Econ. **105**(5), 889–927 (1997)
6. Ferrante, M., Magno, G.L.L., Cantis, S.D., Hewings, G.J.: Measuring spatial concentration: a transportation problem approach. Pap. Reg. Sci. **99**(3), 663–682 (2020)
7. Giuliani, D., Arbia, G., Espa, G.: Weighting Ripley's k-function to account for the firm dimension in the analysis of spatial concentration. Int. Reg. Sci. Rev. **37**(3), 251–272 (2014)
8. Goodchild, M.F.: Algorithm 9: simulation of autocorrelation for aggregate data. Environ. Plan. A: Econ. and Space **12**(9), 1073–1081 (1980)

9. Guillain, R., Gallo, J.L.: Agglomeration and dispersion of economic activities in and around Paris: an exploratory spatial data analysis. Environ. Plan. B: Plan. Des. **37**(6), 961–981 (2010)
10. Guimarães, P., Figueiredo, O., Woodward, D.: Accounting for neighboring effects in measures of spatial concentration. J. Reg. Sci. **51**(4), 678–693 (2011)
11. Krugman, P.: Geography and Trade. MIT Press, Cambridge, MA (1992)
12. Lee, S.I.: Developing a bivariate spatial association measure: an integration of Pearson's r and Moran's i. J. Geogra. Syst. **3**, 369–385 (2001)
13. Marcon, E., Puech, F.: Measures of the geographic concentration of industries: improving distance-based methods. J. Econ. Geogr. **10**(5), 745–762 (2009)
14. Maurel, F., Sédillot, B.: A measure of the geographic concentration in French manufacturing industries. Reg. Sci. Urban Econ. **29**(5), 575–604 (1999)
15. Moran, P.: Notes on continuous stochastic phenomena. Biometrika **37**(1/2), 17–23 (1950)
16. Openshaw, S., Taylor, P.: A million or so correlation coefficients: three experiments on the modifiable areal unit problem. Stat. Appl. Spat. Sci. **21**, 127–144 (1979)
17. Panzera, D., Cartone, A., Postiglione, P.: New evidence on measuring the geographical concentration of economic activities. Pap. Reg. Sci. **101**(1), 59–79 (2022)
18. Sohn, J.: Information technology in the 1990s: more footloose or more location-bound? Pap. Reg. Sci. **83**(2), 467–485 (2004)
19. Sohn, J.: Industry classification considering spatial distribution of manufacturing activities. Area **46**(1), 101–110 (2014)
20. Yeh, A.B., Singh, K.: Balanced confidence regions based on Tukey's depth and the bootstrap. J. R. Stat. Soc. Series B Stat. Methodol. **59**(3), 639–652 (1997)

Spatial Statistics Analysis Using Microdata: An Application Agricultural Sector

Daniela Fusco, Maria Antonietta Liguori, Valerio Moretti, and Francesco Giovanni Truglia

Abstract The increasing availability of geo-referenced microdata has led to important changes in the production of official statistics. These changes were driven also by the need of policy makers to have information related to territorial characteristics. Agriculture is the sector that more than any other is linked to the territory and an analysis of the sector cannot be separated from spatial analysis. The aim of the research is to analyse, through spatial statistical analysis, the territorial connections of farms, identified at the microdata level. The study focuses on a cohort of farms identified at a microdata level, using statistical and administrative sources, which resist the economic changes of the last twenty years. Given the territorial peculiarities that characterize Italy, the analysis was carried out at the regional level in order to identify the territorial links of these companies within their geographical areas and to understand any behavioural differences between the Italian regions.

Keywords Spatial analysis · Microdata · Agriculture · Autocorrelation

1 Introduction

The attention to specifying, estimating, and testing for the presence of spatial interaction in the mainstream of applied and theoretical econometrics can be attributed to two major factors [1]: the growing interest in the interaction of an economic agent with other heterogeneous agents in the system and the need to handle spatial data. This has been stimulated by the explosive diffusion of Geographic Information Systems (GIS) and the associated availability of geocoded data [6].

As the volume of geospatial data has been increasing exponentially, the European Statistical System (ESS) has undergone a comprehensive process of reform covering most aspects of its statistical production. This new statistical approach involved each study sector, particularly agriculture. Agriculture is essentially a spatial phenomenon,

D. Fusco (✉) · M. A. Liguori · V. Moretti · F. G. Truglia
Istat, Rome, Italy
e-mail: dafusco@istat.it

M. Bini et al. (eds.), *Advanced Methods in Statistics, Data Science and Related Applications*, Springer Proceedings in Mathematics & Statistics 467,
https://doi.org/10.1007/978-3-031-65699-6_19

which is not independent from location: the effectiveness of practices is correlated with natural resource assets used on the farm, the spatial pattern of practice use will be determined by the distribution of those physical characteristics [12]. The integration between statistical and geospatial data is fundamental to improve the monitoring of development outcomes, e.g. intensification of agricultural management or other changes in the farming system: land use and land cover changes are the result of the interplay between socio-economic, institutional and environmental factors [9]. Driving forces, which guide these changes, are generally subdivided into two groups: proximate causes and underlying causes. Proximate causes are the activities and actions that directly affect land use, e.g. wood extraction or road building. Underlying causes are the 'fundamental forces' that underpin the proximate causes, including demographic, economic, technological, institutional and cultural factors [8].

In Italy, the result of socio-economic changes is an increase in the average farm size that went from 5.5 hectares in 2000 to 8.4 in 2017. The livestock sector too is affected by a rationalization of resources and a conspicuous increase in intensive farming, leading to a progressive growth in the average size of the animals per farm [7]. In this period, some farms, unable to resist changes, died and others were born. The linkage at micro data level identifies a panel of 531,536 farms surveyed for the first time in 2000 and listed in 2017. The aim of the analysis is to explore and describe the territorial dimension of "resilient" farms, which have resisted socio-economic changes, through spatial statistical analysis. We want to test if the territorial proximity influence the resilience of farm.

The basic hypothesis is that the presence of territorial clusters formed by municipalities with a high density of farms favoured this feature.

2 Data Used

The reference panel was obtained through the combination of the Statistical Register of Italian farms (Farm Register 2017) with the results of the V and VI Agricultural Census (2000 and 2010).

The statistical unit, both for censuses and for the farm register, is the farm unit. This makes the data completely comparable even if the methods are different: the censuses are total surveys, while the register is the result of the integration between administrative and statistical sources. Through the integration of administrative sources (provided spontaneously by the farm manager) and statistical sources (sector surveys, sample surveys, etc.) the information of the Farm Register are obtained limiting the response burden.

For the identification of panel data, only the farms that have not undergone changes of ownership during the reference period were considered, in accordance with the hypothesis that this would be equivalent to a substantial continuity of the company's mission [5]. In the first instance, the farms survived between the two censuses were identified. The V and VI Agricultural Census did not use the same farm identification system. Therefore a statistical matching across the two data sources was necessary.

The linkage of the statistical units was based on three variables, which identify the farms:

1. Unique Code Farm.
2. Address of the headquarter.
3. Name of the farm.

The first step of the matching model selected for the linkage was to link the Unique Code applying a deterministic model of equality. Then the address and the name were linked by applying a function of the distance of the strings via an indicator normalized between 0 and 1. It measures how information contained in a cell (in this case, the address) is similar to the content of another cell. The value of the index is positively correlated with the degree of similarity in information.

Finally, the result was deterministically linked to the 2017 Farm Register through the Unique Code Farm only.

The dataset generated during the current study is not publicly available as it contains proprietary information covered by statistical confidentiality. Information on how to obtain it, in aggregated form, and reproduce the analysis is available from the corresponding author on request.

3 Methodology: Global and Local Autocorrelation Index

The concept of spatial autocorrelation is one of the most important issues of spatial statistics and it derives directly from the first law of geography by Tobler: "All things are related, but nearby things are more related than distant things" [13].

For what concerns the variables of interest, the spatial analysis is conducted at the municipality level through the Moran spatial autocorrelation index [11] in its global and local version (LISA).

By adopting a matrix notation, the global Moran index (I) for a generic x variable can be formalized as follows:

$$I = \frac{N}{S_0} \frac{z'Wz}{z'z}$$

where:

z is the vector of deviations of variable x from its mean M in each territorial unit $i : z_i = x_i - M$, i ranges from 1 to N;

W is the contiguity matrix (of dimension $N{*}N$) whose elements, w_{ij}, are equal to 1 if the ith and jth territorial units are contiguous, or equal to 0 if they are not[1];

S_0 is a scalar corresponding to the sum of all the elements of matrix W, expressing the number of contiguous territorial units.

[1] The criteria to determine the contiguity of two units may be different (e.g. sharing a border line, distance between the gravity centre, travel times, etc.); in the present analysis two municipalities are contiguous if they share a boundary segment.

Index I generally assumes values between –1 (negative auto-correlation) and 1 (positive auto-correlation). In this latter case, it indicates a spatial process of aggregation (contiguous municipalities present similar values for the x variable). On the contrary, in the first case there is a repulsive process (contiguous municipalities do not present similar values). In case of absence of autocorrelation, the index is equal to:

$$-\frac{1}{N-1}$$

The Moran index can be split into as many parts as the units of analysis are, thus obtaining the local index (LISA). It measures the contribution of the ith unit to the global autocorrelation and can be so formalized:

$$\mathbf{L_i} = \frac{\mathbf{z_i}}{\sum_i \mathbf{Z_i^2}} \sum_j \mathbf{w_{ij} z_j};$$

Spatial autocorrelation is a property of spatial data that exists whenever there is a systematic pattern in the value recorded at different locations in an area [10]. By the Moran index, the municipalities can be clustered as follows (in addition to the cluster of municipalities not giving a significant contribution to global autocorrelation and to the cluster of municipalities not confining with other municipalities—minor island):

1. High-High (HH)—contiguous units with values of the variable x (wage level) higher than the mean;
2. Low-Low (LL)—contiguous units with values of the variable x lower than the mean;
3. High-Low (HL)—units with values of the variable x higher than their contiguous units;
4. Low–High (LH)—units with values of the variable x lower than their contiguous units.

The first two clusters are composed of homogeneous units (referring to the variable of interest), while the last two clusters concern abnormal cases or enclaves [2].

4 Results

Figure 1 shows the density of farms survived during the last 17 years, defined "resilient". The density is expressed in quintiles and darker colours highlight a higher concentration: South-Tyrrhenian and Adriatic coastlines, part of Veneto, Friuli-Venezia Giulia, Lazio, Liguria and Piemonte.

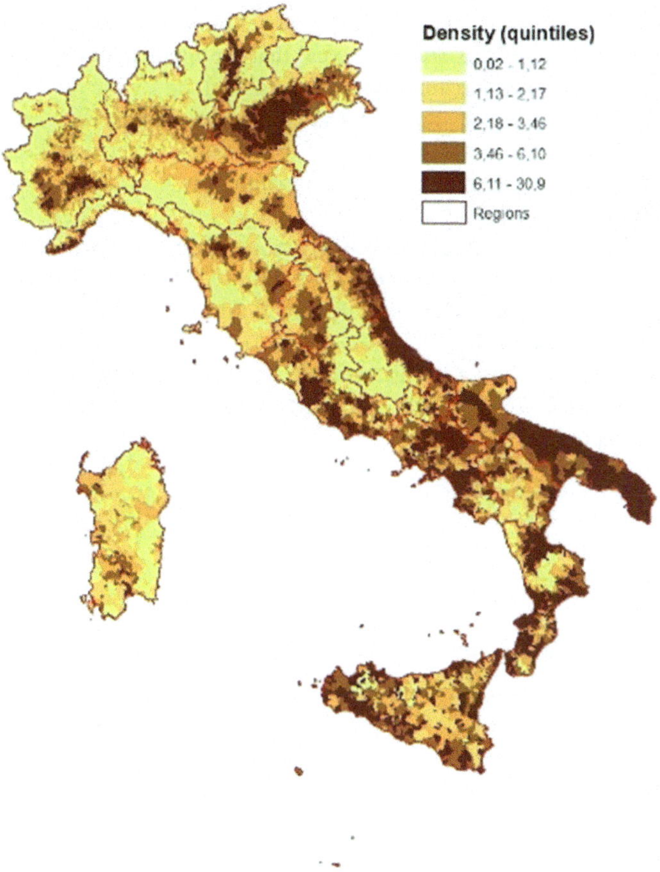

Fig. 1 Density distribution of "resilient" farms for municipalities

5 The Clusters at National Level

Spatial autocorrelation has been visualized adopting Moran scatter plot [3] and considering the density of resilient farms as abscissa and the spatial lag of density values in all j units contiguous to unit i as ordinate. In the graph, the Moran Index corresponds to the slope of linear regression. Positive autocorrelation corresponds to spatial clusters in upper right and lower left quadrants. Lower right and upper left quadrants are classified as spatial outliers.

The Moran's index (Fig. 2) is 0.585. It's very significant ($p < 0{,}001$): the positive sign and the intensity of this statistic reveal that the municipalities, with a similar density value, are located territorially close to each other.

The scatter plot in Fig. 2 shows that most of the municipalities are in the first quadrant: municipalities with high density are close to other municipalities with similar density. In any case, the other municipalities are near the intersection of the Cartesian axes.

Figure 3 identifies several clusters of HH municipalities, located in almost all of Puglia; on the coast of Abruzzo and part of Molise; in Veneto and Trentino Alto Adige; in Campania, Calabria and Sicily. Small clusters are in Liguria, Piemonte, Emilia Romagna and Lazio. These are the territories where there is the highest intensity of resilient farms. LL clusters are located mainly along Alps, Apennines and Sardegna. These areas can be considered as places where this type of structure has not whidespread.

Fig. 2 Moran scatter plot

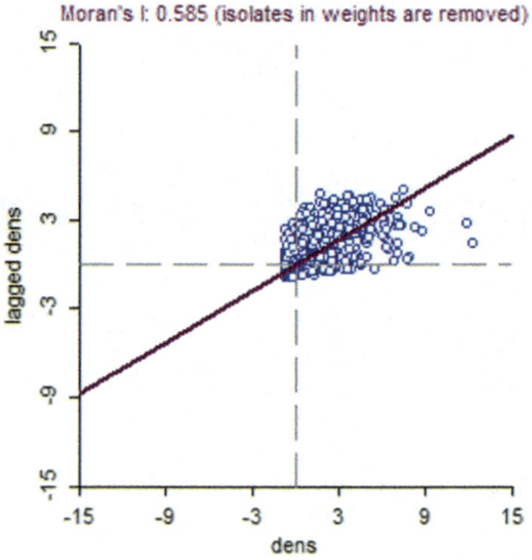

Fig. 3 Local Indicator of Spatial Association—LISA

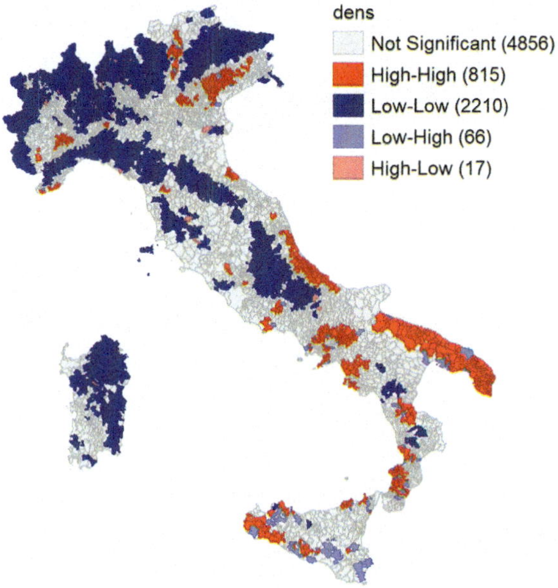

6 Regional Focuses: Puglia, Abruzzo E Molise

Based on the results obtained for the Italian territory, the analysis focused on the areas with the most significant HH clusters: the Moran index was calculated for Puglia, Abruzzo and Molise, individually and for the whole area, in order identify the presence of territorial proximity of the phenomenon.

Figure 4 shows the density of resilient farms in the area considered. As with Fig. 4, denser areas of farms are highlighted in more intense colors.

The results show a strong territorial correlation, as the Moran scatter plot well explained (Fig. 5). The Moran index is equal to 0.615 and most of the municipalities are located in the first quadrant, the others remain close to the Cartesian axes. Also in this case a positive spatial correlation is reported and the global Moran's index for this area is higher than the index calculated for the entire Italian territory, demonstrating that the influence of neighboring territories is stronger in these places: even more so in these three regions, contiguous municipalities have similar density values.

The Fig. 6 shows that HH and LL clusters of municipalities are more widespread respect to the other clusters, suggesting that homogeneous geo-statistical configurations have developed in these territories. The HH clusters, formed by agglomerations of municipalities with a high density of farms, are mainly located in Puglia and in particular on the far end of Lecce, between Brindisi and Taranto and in the upper part of the province of Bari. The only HH cluster not located in Puglia region it is near Pescara. They represent the territories in which there is a greater number of farms that have been resisting for at least seventeen years and in these areas, the territory plays an important role of involvement. The LL clusters, formed by municipalities

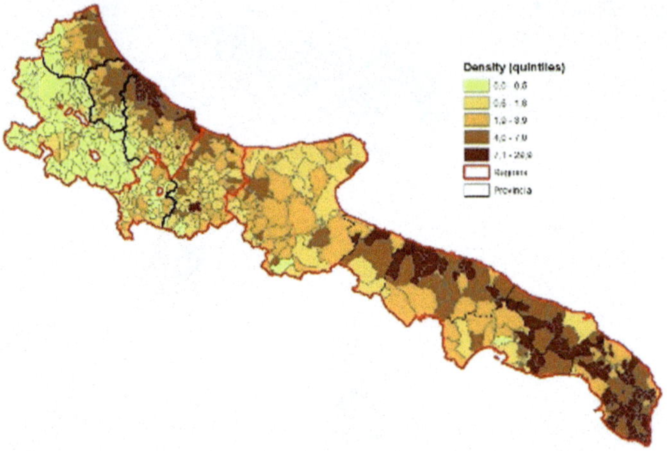

Fig. 4 Density distribution of "resilient" farms for municipalities in the areas of Puglia, Molise, Abruzzo

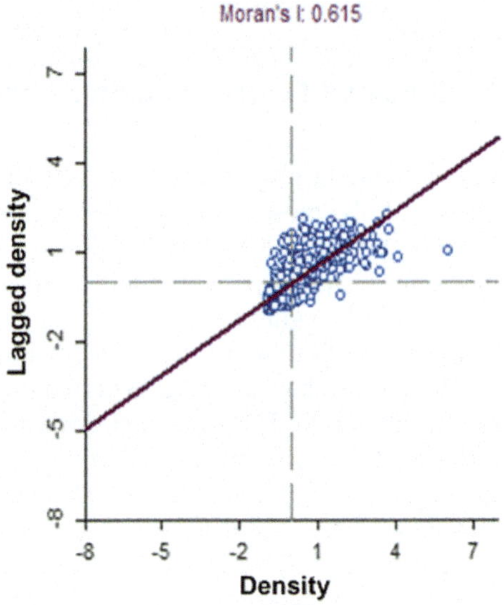

Fig. 5 Moran scatter plot—Abruzzo, Molise, Puglia areas

with low density, flanked by municipalities with very similar densities, are mainly located along the Abruzzo and Molise Apennines and in the province of Foggia.

The analysis carried out on the three regions separately, at the scope to analyze the phenomenon at regional level. It shows positive territorial correlations with significant differences. Particularly, the Moran index calculated for Abruzzo is equal to 0.758 and, as is illustrated in the scatter plot (Fig. 7), almost all points are very close to the bisector. In the dispersion graphs of Puglia and Molise there is a greater

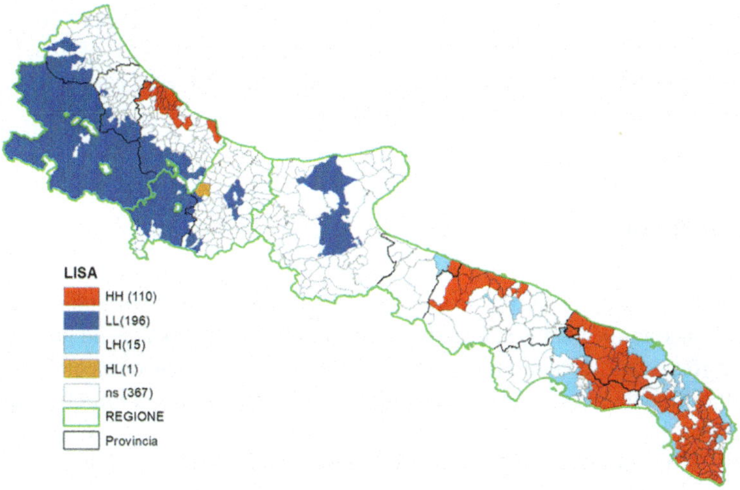

Fig. 6 Local Indicator of Spatial Association—LISA. Puglia, Molise and Abruzzo areas

spreading of the municipalities; in fact, the indicators are equal to 0.414 for Puglia and 0.324 for Molise.

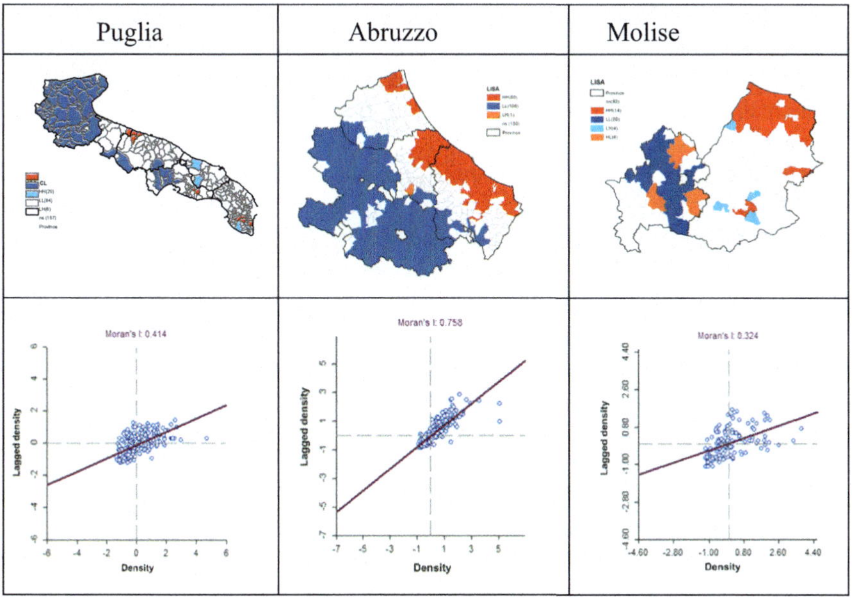

Fig. 7 Local Indicator of Spatial Association—LISA and Moran scatter plot—Puglia, Abruzzo e Molise

The analysis focused on the single regions separately, show clusters of municipalities not appeared in the previous analysis conducted on the aggregated area. Besides, Fig. 7 shows HH-type clusters along the Abruzzo and Molise coasts and LL-type clusters from the Gargano area to the Apennines.

7 Conclusion

The micro data availability allows to explore the mechanisms that influence the farms resistance economic changes. Changes in the number of farms depend on the way in which farms adapt to economic changes within the socio-environmental context in which they find themselves operating. The in–out process is a very important factor in agriculture: this mechanism is fundamental for the global competitiveness and profitability of the agricultural sector [4]. A cohort of farms survived at the last 17 years was identified. The results show the importance of territory as a physical support where socio-economic information is spread out. The results highlight the potential of spatial autocorrelation to prove the surviving spreads via some mechanisms of 'contamination' or spatial spill-overs.

There are some differences in the territorial distribution of this phenomenon, not only between north and south, but also between the Tyrrhenian and Adriatic coast lines where the number of farms survived at the last twenty years is grown.

We choose the Moran index, global and local variant, as a statistical measure of this event [14]. It registers units' autocorrelation with respect to the intensity of a statistical variable (density of farms surviving) in contiguous territorial units (municipalities). The statistical results demonstrate that the farms surviving is strictly tied to the territorial proximity.

The value of the Moran index is very significant. This situation indicates the presence of a process of contagion that leads to the formation of territorial aggregations with higher than average levels of farms surviving. The LISA index highlights four main types of territorial aggregations, the two most common of which are displayed, in Fig. 3, in blue (Low-Low aggregation) and red (High-High aggregation) where the creation of municipalities' cluster (territorial spill-over) is strongest. In particular, we note the presence of a great spill-over effect in south of Italy, principally in Puglia region and Abruzzo coast, where the agricultural sector is linked to deep socio-cultural roots. The smaller presence of the spill-over effects in northern Italy probably depends on a greater farm liveliness in these areas, in fact, we can observe little clusters of HH just in North-Est.

Based on the results obtained at the national level, it was decided to focus on the regions with the greatest presence of spill-overs, i.e. Puglia, Abruzzo and Molise, at first considering the area at an aggregate level, then on each region. The Moran index in these areas is higher than the index calculated for the entire Italian territory, demonstrating that the influence of neighboring territories is stronger in these places, although there are differences between the three regions, with a particular spill-over effect in Abruzzo.

References

1. Anselin, L.: GeoDa 0.9 "User's Guide, Spatial Analysis Laboratory." Department of Agricultural and Consumer Economics and CSISS, p. 125. University of Illinois (2003)
2. Anselin, L.: Spatial econometrics. In: Baltagi B.H. (ed.) A Companion to Theoretical Econometrics, pp. 310–330. Blackwell Publishing (2003)
3. Anselin, L.: The Moran scatterplot as an ESDA tool to assess local instability in spatial association. In: Fischer, M., Scholten, H.J., Unwin, D. (eds.) Spatial Analytical Perspectives on GIS, pp. 111–126. Routledge, London (2019)
4. Breustedt, G., Glauben, T.: Driving forces behind exiting from farming in Western Europe. J. Agric. Econ. **58**(1), 115–127 (2007)
5. Cardillo, C., Fusco D., Moretti, M., Russo, C.: "Farms' structural adjustment to the increasing competitive pressure: specialization versus de-specialization in Italian agriculture". Rivista di Economia Agraria, Anno LXXI **1**(Supplement) (2016)
6. Eurostat "Merging statistics and geospatial information. Experiences and observations from National Statistical authorities". Publications Office of the European Union, Luxembourg (2019)
7. Fusco, D., Liguori, M.A., Moretti, M., Truglia, F.G.: Spatial statistics analysis using microdata: an application at agricultural sector. In: Balzanella, A., Bini, M., Cavicchia, C., Verde, R (eds.) 51th Scientific Meeting of the Italian Statistical Society. Book of the Short Papers. Published by Pearson (2022)
8. Geist, H.J., Lambin, E.F.: Proximate causes and underlying driving forces of tropical deforestation. Bioscience **52**(2), 143–150 (2002)
9. Lesschen, J.P., Verburg, P.H., Staal, S.J.: "c" LUCC Report Series 7. Edited by: The International Livestock Research Institute, Nairobi, Kenya & LUCC Focus 3 Office, Wageningen University, the Netherlands (2005)
10. Mazziotta, C., Mazziotta, M., De Castris, M.: Spatial Autocorrelation of Alternative Infrastructural Indicator—Atti della XLIII riunione scientifica (2006)
11. Moran, P.A.P.: The interpretation of statistical map. J. R. Stat. Soc. B 243–251 (1948)
12. OECD.: Adoption of technologies for sustainable farming systems Wageningen workshop proceedings. Wageningen workshop proceedings (2001)
13. Tobler, W.R.: A computer movie simulating urban growth in the Detroit region. Econ. Geogr. **46**(2), 234–240 (1970)
14. Truglia, F.G, Zeli, A.: Spatial analysis of economic and social determinants of vote: the case of the European Parliament and constitutional referendum votes in Italy. Italian Political Science Review/Rivista Italiana di Scienza Politica, page 1 of 18 (2019) https://doi.org/10.1017/ipo.2019.29

References

Experimental Design, Robust Methods, Bayesian Optimization, and Statistical Matching

Intertemporal Statistical Matching for Causal Inference in the Context of Multivariate Time-Series Data

Daniele Bondonio and Paolo Chirico

Abstract This paper develops an intertemporal statistical matching (SM) approach for causal inference in the context of multivariate time-series data. While SM models have been developed and implemented exclusively for the matching of cross-sectional units (or observations from different cross-sectional time series, holding constant a same temporal unit), this paper shows that, under specific impact identification conditions, SM can be also applied for estimating causal inference by means of matching across different temporal units, holding constant a same cross-sectional time series. We apply our inter-temporal SM model, with propensity score (PS) and Mahalanobis-distance (MAHD) specifications, to the data collected from a large number of air-pollution-measurement stations in Northern Italy in a period before and during the stringent national covid-19 lockdown of March–May 2020. The estimated causal effects from the PS and MAHD matching are then compared to the results from a random forest regression model, with a discussion on the advantages of our intertemporal SM approach. The findings from our empirical application are significant for informing on the air-quality improvements that can be expected from future measures aimed at drastically reducing fossil-fuel emission in areas with high concentration of air pollutants.

Keywords Statistical matching · Air pollution · Coronavirus lockdown

1 Introduction

Statistical matching (SM) is a widely used empirical strategy for estimating causal inference in the context of quasi-experimental designs in which cross-sectional units are sorted into treated and untreated units based on the exposure of the program

D. Bondonio · P. Chirico (✉)
DIGSPES, Universty of Piemonte Orientale, Alessandria, Italy
e-mail: paolo.chirico@uniupo.it

D. Bondonio
e-mail: daniele.bondonio@uniupo.it

M. Bini et al. (eds.), *Advanced Methods in Statistics, Data Science and Related Applications*, Springer Proceedings in Mathematics & Statistics 467,
https://doi.org/10.1007/978-3-031-65699-6_3

intervention of interest [1, 13, 14, 17, 22, 31]. For all known SM estimations, the temporal periods in which the outcome variables are measured have to remain equal across all treated and untreated units. This is a strict requirement for all quasi-experimental designs, in which the existence of the comparison group of untreated units is aimed at controlling for the secular trends that may affect the outcome variable (Y) independently from the treatment. For this reason, intertemporal SM matching (i.e. matching across units with Y measured at different times) is not usually considered a viable impact identification strategy.

The stringent covid-19 lockdowns of the first half of 2020 provide an exceptional source of abrupt exogenous decrease on fossil-fuel human activities (transportations and productions) that offers a unique opportunity to quantify the effect of human activities on air pollution in urbanized areas. What is unique about the impact identification conditions related to the covid-19 lockdowns, however, is also the fact that the available data are in terms of a number of multivariate time-series, one for each air-pollution measurement station, with units of observations represented by the days in which air pollution and weather characteristic are measured. These features of the data represent a scenario in which the abrupt change in the secular trend (i.e. the lockdown-induced sharp decrease in traffic and other fossil-fuel activities) is the treatment of interest, while all the major confounding factors to be controlled for (i.e. the weather characteristics) are observable in the data and they can be safely assumed of not being subject to unobserved secular trend (within the short pre-post-treatment periods of times considered in the analysis) besides the changes that are considered the treatment effect to be estimated.

This paper aims at showing that in such a data scenario, an intertemporal SM approach in which, separately for each air-pollution measurement station, the treated days (corresponding to the lockdown periods) are matched to previous non-lockdown days (untreated units) that share the same control variables is a viable empirical option which offers a number of advantages with the respect to the empirical strategies adopted in the recently-emerging literature on the air-quality impact of covid-19 lockdowns.

We apply our intertemporal SM approach, defined in terms of Propensity Score (PS) and Mahalanobis-distance (MAHD) matching [9, 21, 23, 26], to the data collected from a large number of air-pollution-measurement stations in five regions of Northern Italy in a period before and during the stringent national covid-19 lockdown of March–May 2020. The robustness of the estimated causal effects from the intertemporal PS and MAHD models is then tested and confirmed with an extensive sensitivity analysis which also includes a re-estimation by means of a random forest regression model [5], in addition to the adoption of variety of alternative matching parameters.

Because of the high concentration of air-pollutants in Northern Italy and the severe stringency of the Italian covid-19 lockdown, the results from this empirical application are of great external validity in terms of the local air-quality improvements that can be expected from future fossil-fuel reduction policies.

The reminder of the paper is organized as follows. Section 2 discusses the existing literature on the causal impact of covid-19 lockdowns on air pollution. Section 3 illustrates the impact identification properties of intertemporal SM. Section 4 describes the empirical application on covid-19 lockdown and air pollution in Northern Italy. Section 5 discusses the results and offers concluding remarks.

2 Existing Literature

Since spring 2020, an increasingly large number of studies have been using covid-19 lockdowns data on a variety of locations around the world in order to estimate the reduction of air pollution caused by an abrupt decrease of traffic and human activities. A number of these papers fail to properly address the challenges of a reliable causal inference that requires, above all, a suitable controlling of the possible differences in the distribution of the weather characteristics of the pre-lockdown days and the lockdown days (treated units) [2, 3, 7, 28]. These papers identify the causal effect of the lockdowns only under the very stringent assumption of a perfectly equal distribution of the weather characteristics between the lockdown and the pre-lockdown periods.

Another stream of the literature makes use of panel-data regression approaches that have to face the issue of error autocorrelation, and that require strong functional form assumptions with respect to the way in which the observable confounding factors affect air-pollution [8, 12, 29, 32].

Finally, a random forest regression model has also been used with the aim of identifying the causal impact of a covid-19 lockdown separately from the effects of meteorological confounding factor in an empirical application on data from an Austrian municipality [20].

3 Impact Identification Properties of Intertemporal Statistical Matching

The intertemporal statistical matching approach developed in this paper can be described as follows. Let $\{Y_t\}$ be a stochastic process on which an intervention lasting for a time interval L_1 has a potential to produce a causal effect on Y_t for each time $t \in L_1$. The causal effect of the intervention should be determined as the difference:

$$\mathrm{E}\left[Y_t^{(1)} \mid t \in L_1\right] - \mathrm{E}\left[Y_t^{(0)} \mid t \in L_1\right] \tag{1}$$

where $Y_t^{(1)}$ is the observable outcome at time t if time t were affected by the intervention, and $Y_t^{(0)}$ if not affected. In the case of the covid-19 lockdown analyses, or, more broadly, for causal inference estimation of policies aimed at reducing fossil-fuel emissions, the temporal units t are represented by the days in which air-pollution and weather characteristics are measured. The difference (1) can be considered as an intertemporal version of the standard *Average Treatment effect on the Treated* (ATT) [17].

As $Y_t^{(0)}$ is not observable in the intervention period L_1, the estimation of ATT is not immediate. Nevertheless if the existing multivariate process $\{X_t\}$ of observable covariates (i.e. the weather controls) makes $\{Y_t\}$, given X_t, mean-stationary in a suitable period including the intervention period L_1 and a previous period L_0, named control period, i.e.:

$$\mathrm{E}\left[Y_t^{(0)} \mid X_t = x, t \in L_1\right] = \mathrm{E}\left[Y_t^{(0)} \mid X_t = x, t \in L_0\right] \quad \forall x \tag{2}$$

then the intertemporal ATT (1) can be identified as:

$$\mathrm{E}_{1t}\left[\mathrm{E}\left[Y_t \mid X_t = x, t \in L_1\right] - \mathrm{E}\left[Y_t \mid X_t = x, t \in L_0\right]\right] \tag{3}$$

where E_{1t} denotes expectation with respect to the distribution of X_t in the intervention period L_1. For the case of propensity score (PS) matching, the intertemporal local ATTs are identifiable in terms of conditioning on propensity score $[\varphi_t = P(t \in L_1 \mid X_t)]$, and the difference (3) can be expressed as:

$$\mathrm{E}_{1t}\left[\mathrm{E}\left[Y_t \mid \varphi_t, t \in L_1\right] - \mathrm{E}\left[Y_t \mid \varphi, t \in L_0\right]\right] \tag{4}$$

Such an intertemporal ATT can be estimated by the following matching procedure:

1. each day of L_1 is matched, if possible, with one or more days of L_0 having similar X_t[1];
2. for each matched day in L_1, an estimate of the "counterfactual" value $y_t^{(0)}$ is determined as the Y-value of the corresponding matched day from L_0[2];
3. the intertemporal ATT is estimated as the difference of the sample means:

$$A\hat{T}T = \bar{y}_{S_1} - \bar{y}_{S_1}^{(0)} \tag{5}$$

   where S_1 is the set of the matched days in L_1;
4. the estimated ATT is validated if an adequate balancing is achieved and tested for the covariates X_t of the days in the intervention period L_1 and those of the matched control days in L_0;
5. when the analysis focuses on multiple multivariate time-series, the procedure 1–4 is repeated separately for each time-series, yielding a series of local ATTs (one for each multivariate time series) that are then aggregated into a global ATT estimate, obtained as a weighted average of the local ATTs with weights proportional to the number of treated days in L_1 that are on the common support in each time-series.

---

[1] Some math methods require a minimum level of similarity, without which the matching does not occur [31].

[2] In case of multiple matched days from L_0, $y_t^{(0)}$ is the average of the Y-values of the matched days from L_0.

Compared to the panel-data regression approaches adopted elsewhere in the literature of covid-19 lockdowns [8, 12, 29, 32], such an intertemporal PS matching approach shares the same well-known advantages [9, 15] of the standard cross-sectional PS matching in strongly diminishing the sensibility of the estimated impacts to the choice of functional form that links the effects of the controls on the outcome variable.

With respect to other recent literature on matching methods for causal inference with time-series cross-sectional data [16], our intertemporal SM approach is suited for the conditions posed specifically by national policy/regulatory interventions and when the focus of the analysis is on air-pollution outcomes. For these interventions, often, the available data are in terms of multiple multivariate time-series (one for each cross-sectional unit of observation represented by the locations of the air-pollution measuring station) and the treatment happen on the same calendar dates for all the time series. This is the typical scenario of national covid-19 lockdowns and, more in general, of policies and regulations aimed at reducing air pollution with an identical starting and end date across all different geographic locations. In such cases, the multivariate time-series data deriving from the different cross-sectional locations in which air pollution is measured contains treated days (i.e. those after the starting date of the policy) and pre-treatment control days (i.e. those before the starting date of the policy). With our empirical strategy, the matching is performed intertemporally between different time units (calendar days) holding constant a same multivariate time-series, yielding estimates of local ATTs (one for each cross-sectional time-series) that are then aggregated into a global ATT estimate. This is very different than in [16] in which the treatment status in a same time period is different across the various cross-sectional time series, and hence the matching is performed across observations (with different treatment status) of different cross-sectional time series holding constant the same time period.

4 Empirical Application: The Causal Effect of National Covid-19 Lockdown on Air Pollution in Northern Italy

Our intertemporal SM model described in the previous section is applied to estimate the causal effect of the covid-19 national lockdown in March–May 2020 on the urban air pollution measurements from the regions of Piedmont, Lombardy, Emilia-Romagna, Veneto, Friuli Venezia Giulia, located along the Po-river valley in Northern Italy. These regions represent the most industrialized and densely-populated area of Italy (with an average density of 252 in-hab. per km2). Because of these characteristics, and the specific orographic features of the Po Valley (that determine a systemic lack of sustained ventilation), some of the highest levels of air-pollution in Europe are consistently recorded in this area [6].

Due to the large death toll experienced in Italy in the very early stages (beginning of March 2020) of the pandemic, the ensuing national covid-19 lock-down was one of longest and with the most stringent constrictions of all Europe and most of the western world, with a mandatory at-home-stay order continuously enforced during the entire period March 10th–May 17th, 2020 and with a complete lockdown of schools, universities and all non-essential services and production activities.

For these reasons, a reliable estimation of the causal effects of the Italian covid-19 lockdown on the air pollution of the Norther-Italian regions bears a significant relevance and external validity. This is in terms of empirical evidence on the air-quality gains that could be achieved by means of future polices aimed at drastically reducing fossil-fuel vehicular traffic and other human activities in an area with a systemically-high concentration of air pollutants.

4.1 Data

The air pollution data used in the empirical application are collected from 83 different measurement stations of the Regional Environmental Agencies (ARPA). The focus of the analysis is on the five regions located along the Po-river Valley (Piedmont, Lombardy, Emilia-Romagna, Veneto, Friuli Venezia Giulia) and on the average daily levels of nitrogen dioxide (NO2) and particulate matter (PM10). Within these five regions, the ARPA measurement stations included in the analysis are those located in the cities that are the head of the provincial districts. The provincial areas not affected by the peculiar orography of the Po valley (which is prone to systematically-high air pollution) were not included in the analysis.[3]

The observation period covers the entire length of the Italian nationwide covid-19 lockdown, 69 days from March 10th to May 17th, 2020 (i.e. the intervention period L_1), together with the 366 days that preceded the start of the lockdown (the control period L_0). This latter period (that includes the pool of non-lockdown days from which the matching procedure selects the actual control group) is such that it includes the March–May 2019 calendar days corresponding to the same dates of the 2020 lockdown period, along with a pool of other calendar days (e.g. those in the fall 2019) from which it is also potentially possible to find common support conditions in terms of meteorological characteristics. A longer L_0 period was instead ruled out due to the possibility that it could entail the existence of time trends that invalidate the stationarity of Y_t given X_t.

These specific features posed by the Italian national covid-19 lockdown and the available ARPA-station time series data are conducive to straightforward stable unit treatment value assumptions (SUTVA) [24, 25] conditions for estimating causal effects on air pollution outcomes. This is because the implementation of the lockdown started and ended at the same dates for all the cross-sectional units of the analysis (i.e. the ARPA measuring stations). Thus for any treated calendar days, there is no potential of treatment-effect spillovers from the treated units to the non-treated

[3] These are the areas at high elevation, within the Alps mountain range, located in the provinces of Aosta, Sondrio, Trento e Bolzano.

units: being the lockdowns status identical for all the measuring stations at exactly the same days, there is no possibility that the effect of the lockdown reverberates also on the non-treated stations and vice-versa. Moreover, in our empirical design, the non-treated calendar days are drown solely from the pre-lockdown period. While some temporal spillover from the treated days into the non-treated days can occur if the latter are drown from the post-lockdown period (with some reductions in traffic and pollutant activities that may carry-over from the lockdown period into the subsequent untreated period), the same cannot be true if the non-treated days are drown solely from the pre-lockdown period.[4]

The data used in the analysis also include a set of control variables (X_t) that capture the main meteorological characteristics that have been considered both in the recent literature on covid-19 lockdown and air quality [8, 11, 20, 29, 32] and in the specific meteorological literature applicable to the Po-river basin [30]. These controls X_t are: average daily temperature (TEMP), rainfall (RAIN), maximum daily wind speed (WIND). The resulting database is in terms of 83 different multivariate time series (one for each ARPA measuring station).

Table 1 illustrates the mean and standard deviation of the distribution of NO2 and PM10 (pooling the data from all 83 different stations), during the 2020 lockdown days and the same corresponding dates of 2019. While any difference in fossil-fuel activities (e.g. vehicular-, air-traffic, industrial production) between the two periods is indeed the treatment of interest, not to be controlled for in the analysis, these descriptive statistics do not provide yet any reliable indication of the causal effect of the lockdown on air-pollution. This is because such a simple comparison between the treated and non-treated days does not control for the possible obvious differences in the distribution of the meteorological variables between the two periods. Indeed, the data from Table 2 show, for example, that the lockdown period of 2020 experienced less than half of the rain than the corresponding period of 2019. For this reason, any causal inference drawn from a descriptive-statistics comparison between the two periods would be highly biased, particularly for the PM10 outcomes that has a high potential of being influenced by rain and wind conditions.

4.2 *Estimation of Intertemporal Propensity-Score and Mahalanobis-Distance Matching*

The first step in the implementation of the intertemporal SM estimation is the exclusion (within each multivariate time series from the 83 different ARPA measuring station) of the days in the control period L_0 that are outside of the common support in terms of meteorological characteristics X_t (i.e. WIND, RAIN and TEMP).

---

[4] In this regard, a10 days grace period before the starting date of the lockdown is excluded from the analysis. Because the first officially-reported case of covid-19 in Italy is dated at the end of February 2020, no significant reduction of traffic and other pollutant activities took place before the beginning of March 2023 in anticipation of the lock-down effect.

Table 1 NO2 and PM10 during 2020 lockdown versus same period of 2019 ($\mu g/m^3$)

Pollutant	Period	Obs.	Mean	SE	95-Confid. Int.	
NO2	Lockdown 2020	5,663	15.26	0.15	14.97	15.55
	Same period 2019	5,671	25.45	0.21	25.03	25.87
	Diff		–10.19	0.26	–10.70	–9.68
PM10	Lockdown 2020	5,480	23.90	0.24	23.44	24.36
	Same period 2019	5,580	20.27	0.16	19.95	20.58
	Diff		3.63	0.29	3.07	4.19

Table 2 Meteorological conditions during 2020 Lockdown vs same period of 2019

Factor	Period	Obs.	Mean	SE	95-Confid. Int.	
WIND (m/s)	Lockdown 2020	5,621	2.20	0.02	2.16	2.24
	Same period 2019	5,674	2.28	0.02	2.24	2.32
	Diff		–0.08	0.03	–0.13	–0.02
TEMP (C°)	Lockdown 2020	5,714	13.74	0.05	13.63	13.85
	Same period 2019	5,715	12.55	0.04	12.48	12.62
	Diff		1.19	0.07	1.06	1.32
RAIN (mm)	Lockdown 2020	5,558	1.35	0.07	1.22	1.48
	Same period 2019	5,585	2.85	0.10	2.65	3.05
	Diff		–1.50	0.12	–1.74	–1.26

The causal effect of the covid-19 lockdown on the average level of NO2 and PM10 is then estimated by means of an intertemporal propensity-score (PS, radius) and a MAHD matching specification. For estimating both the PS and the MAHD, the meteorological covariates used in the analysis also include a set of lagged terms capturing the weather characteristics (WIND, RAIN, TEMP) of the previous day ($t - 1$). This is in line with the data operationalization choices adopted elsewhere in the literature on covid-19 lockdown and air pollution [8, 20, 29], and it is a solution that is justified on the base of both the specific meteorological literature (on micro-weather characteristics and air-pollution dispersion, e.g. [30]) and the descriptive statistics on the data showing that the daily air-pollution values of any given location may indeed be affected also by the condition of the previous day (particularly if the latter are in terms of heavy rain or strong wind).

The locational characteristics of the ARPA-measuring station (i.e. distance from major roads, highways, airports, surrounding production activities, agricultural practices and extraction activities near the measuring site etc.), instead, do not need to be included among the controls. This is because such characteristics are already inherently perfectly balanced in the estimation procedure, due to the fact that the lockdown L_1 days from any given ARPA station are matched solely with the similar control L_0 days that belong to the same identical ARPA station (i.e. the matching procedure is replicated separately within each of the different ARPA-station multivariate time series, yielding 83 different local ATTs, one for each of the time-series).

The stringency of the lockdown rules entails that the intensity of fossil-fuel vehicular traffic and other activities was inherently the same between weekends and holidays and working days of the lockdown period L_1. In the period L_0, however, this is not the case, and a different degree of air-pollution reduction could be expected based on whether or not the matched L_0 days are working days versus weekends or holidays. Because, in the pre-lockdown period, the working days are more representative of the typical level of fossil-fuel-emitting vehicular traffic and production actives, the main specifications of the PS radius and MAHD models are estimated by limiting the control period L_0 to the sole working days of the week. This choice bears the greatest external validity of the results, as it enables the analysis to estimate the causal impact of the covid-19 lockdown in terms of the air-pollution difference with respect the typical degree of vehicular traffic and production activities of standard non-lockdown days.

The PS matching estimation procedure is based on a probit radius specification with a tolerance $\delta = 0.1$, while the MAHD estimation is based on a caliper $c = 1$. These values of δ and c are those that maximize the statistical efficiency of the analysis under the constrain of satisfying the balancing of the control variables between the intervention days L_1 and the matched controlled days L_0. This latter testing is implemented by means of the Rubin's R and B indices [19, 26], with thresholds for acceptance in terms of $B < 25$ and $0.5 < R < 2$.

Due to the error-autocorrelation issues common to any time-series data, the local ATTs are estimated with standard errors obtained from a boot-strap procedure. The final (global) ATT estimate is then obtained as the weighted average of the different local ATTs from each of the 83 ARPA-station time series, with weights proportional to the number of lockdown days L_1 on common support within each series.

4.3 Results

Tables 3 and 4 illustrate the balancing of the control variables achieved by the intertemporal PS and MAHD estimations.[5] This is in terms of percentage bias and reduction of percentage bias of the pooled distribution (across all the ARPA-stations time series) of the L_1 days and the matched L_0 days.

[5] The results in this section are obtained using the module "PSMATCH2" of Stata [19].

Table 3 Balancing of control variables, PS radius matching ($\delta = 0.1$)

Variable	Unmatched/ matched	Mean treated	Mean control	%Bias	%Reduct bias
WIND (m/s)	U	2.15	1.9	18.9	
	M	2.09	2.06	2.8	85
RAIN (mm)	U	1.28	2.87	–24.7	
	M	1.27	1.48	–2.7	89.2
TEMP (C°)	U	13.85	11.93	39.2	
	M	13.89	13.88	0.2	99.4
WIND-1 (m/s)	U	2.15	1.9	18.2	
	M	2.09	2.05	2.8	84.8
RAIN-1 (mm)	U	1.24	3.12	–26.9	
	M	1.28	1.43	–2.8	89.7
TEMP-1 (c°)	U	13.68	11.94	34.7	
	M	13.73	13.74	–0.3	99.1

Table 4 Balancing of control variables, MAHD matching ($c = 1$)

Variable	Unmatched/ matched	Mean treated	Mean control	%Bias	%Reduct bias
WIND (m/s)	U	2.23	2.02	13.7	
	M	1.91	1.87	2.7	80.2
RAIN (m/m)	U	1.24	2.76	–23.8	
	M	0.35	0.37	–0.4	98.3
TEMP (c°)	U	13.87	11.92	40.0	
	M	14.25	13.90	7.0	82.4
WIND-1 (m/s)	U	2.23	2.03	13.3	
	M	1.96	1.95	0.7	94.7
RAIN-1 (m/m)	U	1.21	2.95	–25.9	
	M	0.31	0.36	–0.8	97.1
TEMP-1 (c°)	U	13.71	11.93	35.2	
	M	13.70	13.24	9.0	74.5

For all the control variables, the intertemporal PS and MAHD matching drastically reduces the initial %bias between the intervention period L_1 and the control period L_0. Such reduction in the %bias ranges from 85% to 99.1%, and from 74.5% to 98.3% for the PS and MAHD matching, respectively. The after-matching remaining %bias between the treated (lockdown) L_1 days and the matched control (pre-lockdown) L_0 days is (for both the PS and MAHD specifications) below 3% for all the control variables, with the only exception of the TEMP and TEMP-1 variables, in the MAHD specification, which have a remaining % bias of 7% and 9%, respectively, still comfortably below the standard 10% threshold.

Table 5 Rubin's R and B indices for PS radius and MAHD matching

Sample	Ps R2	LR chi2	p>chi2	Mean bias	Med bias	B	R
PS radius:							
Unmatched	0.051	1059.72	0	27.1	25.8	57.2*	0.51
Matched	0.001	8.34	0.214	1.9	2.7	6.2	0.99
MAHD:							
Unmatched	0.048	677.72	0	25.3	24.9	55.5*	0.51
Matched	0.003	17.13	0.009	3.4	1.7	13.5	1.08

* if B>25%, R outside [0.5, 2]

The results of the overall balancing test, in terms of Rubin's R and B indices, for the PS and MAHD intertemporal matching estimations are reported in Table 5. The after-matching values of the B indices are 6.2 and 13.5, while those of the R indices are 0.99 and 1.08, respectively for the PS and MAHD specifications. These values are well within the limits ($B < 25$ and $0.5 < R < 2$) that indicate a fully acceptable balancing of the control variables between the intervention days L_1 and the matched controlled days L_0.

Table 6 reports the estimated causal effect on NO2 and PM10 (in terms of global ATTs) of the covid-19 lockdown produced by the intertemporal PS and MAHD matching procedures with respect to the counterfactual state of same meteorological conditions, identical location of the measuring stations, and the presence of the standard fossil-fuel emitting activities of the pre-lockdown working days of the week.

In terms of NO2 pollution (average daily levels), the covid-19 lockdown is estimated to have caused a reduction of 14.11 and 12.96 μg per cubic meter ($\mu g/m^3$), for the intertemporal PS and MAHD matching estimations, respectively. The statistical precision of the point estimates is quite high, with narrow confidence intervals [(–14.46, –13.76) and (–13.57, –12.35) $\mu g/m^3$, for the PS and MAHD matching respectively] and statistical significance within the 5% level. In terms of percentage change, with respect of the baseline daily average level of air pollution recorded in the same period of 2019 (25.45 $\mu g/m^3$, Table 1), these estimated impacts translate into a reduction of NO2 of 50.9% (PS matching) and 55.4% (MAHD matching).

For what it concerns PM10 pollution, the covid-19 lockdown is estimated to have caused a much smaller reduction of 0.79 and 1.05 $\mu g/m^3$, for the inter-temporal PS and MAHD matching specifications, respectively. The confidence intervals of these estimates are (–1.36,–0.21) $\mu g/m^3$ (PS matching) and (–2.10, 0.01) $\mu g/m^3$ (MAHD matching). With respect to the average PM10 pollution level of the same period of 2019 (20.27 $\mu g/m^3$, Table 1), these estimated impacts of the covid-19 lockdown translate into the modest reduction of 3.9% (PS matching) and 5.2% (MAHD matching).

4.4 Sensitivity Analysis and Comparison with Random-Forest Regression Estimates

The robustness of the intertemporal PS and MAHD matching estimates reported in Table 6 is first tested by replicating the analysis with different values of the tolerance δ and caliper c and with the inclusion of weekends and holidays among the pool of potential control days L_0 that are eligible to be matched with the intervention (lockdown) days L_1.

When adopting different values of the tolerance δ (from $\delta = 0.01$ to $\delta = 0.1$) for the intertemporal PS matching and re-estimating the model specifications with and without the exclusion of the L0 week-ends and holidays, the estimated global ATT range from –12.04 to –14.11 and from 0.06 to –0.79, for NO2 and PM10 pollution, respectively. For the MAHD matching, different values of the caliper c (from "no-caliper" to $c = 1.5$, with and without the exclusion of L_0 weekends and holidays), yield global ATT estimates which range from –13.93 to –12.96 (NO2 pollution) and from –1.79 to –1.05 (PM10 pollution).[6] Taken together, these results show that, for both the NO2 and PM10 outcomes, the findings from the main specification of our intertemporal PS and MAHD approaches are robust to a wide variation of the matching parameters.

The sensitivity analysis on the results of Table 6 is also completed by replicating the estimation of both the NO2 and PM10 impacts with a random-forest regression (RFR) model. This is aimed at assessing how the results from our intertemporal PS and MAHD matching compare with respect to another emerging causal-inference strategy that has been adopted in the literature on covid-19 lockdowns and air-pollution time-series data [10, 18].

Table 7 contains the results of RFR estimation, based on the same control variables of the intertemporal PS and MAHD matching.[7] The RFR model is applied separately to each ARPA-measuring-station multivariate time series data, yielding a different local impact (ATT) estimate for each ARPA-station time series. Similarly to the procedure adopted for our intertemporal SM approach, the local ATTs are then averaged-out into a global ATT estimate. The training period of our RFR model is

Table 6 Causal effect of covid-19 lockdown on NO2 and PM10, global ATTs ($\mu g/m^3$)

Method: pollutant	PS matching:				MAHD matching:			
	$\hat{ATT}$	SE	95-Confid. Int.		$\hat{ATT}$	SE	95-Confid. Int.	
NO2	–14.11	0.18	–14.46	–13.76	–12.96	0.31	–13.57	–12.35
PM10	–0.79	0.29	–1.36	–0.21	–1.05	0.54	–2.1	0.01

[6] For both the PS and MAHD specifications, the reported ranges of global ATTs are those formed from the local ATTs for which the parameters δ and c enabled a satisfactory balancing ($B < 25$ and $0.5 < R < 2$) of the controls to be achieved.

[7] RFR is performed using the module "rforest" of Stata [27].

Table 7 Random forest regression, global ATTs ($\mu g/m^3$)

Pollutant	ATT	SE	inf95CI	sup95CI
NO2	–13.21	0.11	–13.43	–12.99
PM10	–0.49	0.23	–0.94	–0.05

the same control period L_0 (training set) used for the intertemporal PS and MAHD matching, while the predicted NO2 and PM10 values for the days in the period L_1 (intervention) represent the counterfactual estimates for each of the lockdown days. The local ATTs for each of the ARPA station time-series are then estimated as the average difference between the observed and predicted NO2 and PM10 values for the L_1 days.

The results from Table 7, show that the results from replicating the analysis with such a RFR model are of the same magnitude and degree of statistical significance than those obtained by our intertemporal PS and MAHD matching, when the same controls and the same L_0 and L_1 periods are used.

5 Discussion and Concluding Remarks

This paper shows that statistical matching (SM), based on either a propensity score (PS) or Mahalanobis-distance (MAHD), can be applied also in an inter-temporal setting in the context of multivariate time series. This is when the following conditions apply: *i*) the treatment of interest takes place in a specific group of temporal units (e.g. days) that are exposed to the policy intervention of interest, while the rest of the temporal units of the time series are unexposed to the treatment; *ii*) the outcome of interest can be influenced by both human activities and the specific local daily weather conditions at the site where the outcome data are measured; *iii*) the treatment of interest entails an abrupt temporal change in the human activities affecting the outcome of interest (i.e. any change in the secular trend of the time series is part of the treatment effect to be estimates); *iv*) the specific daily local weather conditions affecting the outcome of the analysis are all fully observable, without any systematic difference in unobservable confounding factors between the treated and non-treated temporal units.

All these conditions are in place when estimating the causal effect on air-pollution of covid-19 lockdowns or, more in general, of policy or regulatory interventions entailing a reduction of fossil-fuel emitting activities. Because of the future relevance of these policies, we believe that the intertemporal SM approach presented in this paper has the potential to play an important role in a large array of future causal-inference studies.

In the existing emerging literature on the air-pollution effects of covid-19 lock-down, causal-inference estimation has been attempted by means of before-after descriptive-statistics comparisons, panel-data regression and random forest regression models. Before-after comparisons have to rely on rarely-encountered identical distributions of meteorological confounding factors between the treatment (lockdown) and control period (pre-lockdown). Obviously, these comparisons do not constitute a generalizable approach to causal inference. Compared to panel data regressions, instead, intertemporal SM shares the well-known advantages of the standard cross-sectional SM models in limiting the sensitivity of the results to the specific functional form assumptions on the way in which the observable confounding factors affect the outcome variable (see [4] for a literature review and summative discussion on the issue).

With respect to random forest regressions, finally, the main advantage of intertemporal SM lies on the fact that the existing scientific literature on determinants of air pollution clearly identifies the short list of meteorological conditions to be controlled for. Under these circumstances, the advantages of intertemporal SM are in terms of straightforward testing for the balancing of the meteorological controls and ease of calibrating the matching parameters in order to achieve the largest statistical efficiency conditional to a satisfactory balancing of the confounders.

The intertemporal SM approach developed in this paper has been applied to the estimation of the causal effect on NO2 and PM10 pollution of the March–May-2020 national covid-19 lockdown in five regions along the Po-river valley in Northern Italy. Because of the severe stringency of the Italian lockdown, the large number and geographical coverage of the measuring stations, and the extremely poor air quality of the regions in the Po-river valley, this empirical application bears great external validity in terms of indicating the degree of local air-quality improvements that future fossil-fuel decreasing policies could possibly achieve in areas with high levels of pollutants.

The results of the analysis show that the covid-19 lockdown caused a large reduction in the daily average levels of NO2 (between –50.9% and –55.4%). In terms of daily average levels of PM10, instead, the covid-19 lockdown is responsible for a much smaller reduction in pollution, estimated to be between –3.9% and –5.2%. These findings are robust to a wide range of sensitivity analyses and provide empirical support to what is postulated in the theoretical literature [17] in terms of the different sources of emissions that are separately responsible for NO2 and PM10 pollution. This is because the Italian covid-19 lockdown mainly entailed a drastic decrease in all transportation activities, while it exerted little effect on non-industrial combustion. While for NO2 pollution, vehicular, air, marine and railroad traffic is expected in the literature to contribute for about 61.8% to the total emissions (with 40.3% represented by vehicular traffic alone), traffic sources are instead expected to contribute only for 16.6% of the total PM10 emissions (with 54.6% of emissions expected from non-industrial combustion). For these reasons, the results from our analysis provide a robust empirical confirmation to the prediction that the main air-quality benefits from future policies aimed at reducing fossil-fuel traffic emissions are likely to be in terms of a strong decrease of NO2, but not of PM10 pollution.

Acknowledgements This work was supported by funding from the University of Piemonte Orientale. We thank Giuseppe Gruttad'Auria, Andrea Nigido, Giulia Pasculli and Michela Rosselli for their contribution in the data acquisition process.

Competing Interests The authors have no conflicts of interest to declare that are relevant to the content of this chapter.

References

1. Angrist, J.: Estimating the labor market impact of voluntary military service using Social Security data on military applicants. Econometrica **66**, 249–288 (1998)
2. Anil, I., Alagha, O.: The impact of COVID-19 lockdown on the air quality of Eastern Province. Saudi Arabia. Air Qual. Atmos. Health **14**, 117–128 (2020)
3. Arregoce's, H.A., Rojano, R., Restrepo, G.: Impact of lockdown on particulate matter concentrations in Colombia during the COVID-19 pandemic. Sci. Total Environ. **764**, 142874 (2021)
4. Black, D.: Matching as a regression estimator. IZA World of Labor: 186 (2015). https://doi.org/10.15185/izawol.186
5. Breiman, L.: Random forests. Mach. Learn. **45**(1), 5–32 (2001)
6. Coker, E.S., Cavalli, L., Fabrizi, E., Guastella, G., Lippo, E., Parisi, M.L., Pontarollo, N., Rizzati, M., Varacca, A., Vergalli, S.: The effects of air pollution on COVID-19 related mortality in Northern Italy. Environ. Resour. Econ. **76**, 611–634 (2020)
7. Collivignarelli, M.C., Abba, A., Bertanza, G., Pedrazzani, R., Ricciardi, P., Miino, M.C.: Lockdown for CoViD-2019 in Milan: what are the effects on air quality? Sci. Total Environ. **732**, 139280 (2020)
8. Dang, H.A.H., Trinh, T.A.: Does the covid-19 lockdown improve global air quality? New cross-national evidence on its unintended consequences. J. Environ. Econ. Manag. **105**, 102–126 (2021)
9. Dehejia, R.H., Wahba, S.: Propensity score-matching methods for non experimental causal studies. Rev. Econ. Stat. **84**(1), 151–161 (2002)
10. EEA: Air quality in Europe 2022. Report no. 05/2022 (2022)
11. Grange, S.K., Carslaw, D.C., Lewis, A.C., Boleti, E., Hueglin, C.: Random forest meteorological normalisation models for Swiss PM10 trend analysis. Atmos. Chem. Phys. **18**(9), 6223–6239 (2018)
12. He, G., Pan, Y., Tanaka, T.: The short-term impacts of covid-19 lockdown on urban air pollution in china. Nat. Sustain. **3**(12), 1005–1011 (2020)
13. Heckman, J.J., Ichimura, H., Todd, P.: Matching as an econometric evaluation estimator. Rev. Econ. Stud. **65**(2), 261–294 (1998)
14. Holland, P.W.: Statistics and causal inference. J. Am. Stat. Assoc. **81**(396), 945–960 (1986)
15. Iacus, S.M., King, G., Porro, G.: Multivariate matching methods that are monotonic im-balance bounding. J. Am. Stat. Assoc. **106**(493), 345–361 (2011)
16. Imai, K., Kim, I.S., Wang, E.H.: Matching methods for causal inference with time-series cross-sectional data. Am. J. Polit. Sci. **67**(3), 587–605 (2023)
17. Imbens, G.W., Rubin, D.B.: Causal Inference in Statistics, Social, and Biomedical Sciences. Cambridge University Press, Cambridge (2015)
18. ISPRA: Italian Emission Inventory 1990-2019. Informative Inventory Report 2021. Roma, IT: Istituto Superiore per la Sicurezza e la Ricerca Ambientale (2021)
19. Leuven, E., Sianesi, B.: PSMATCH2: Stata module to perform full Mahalanobis and propensity score matching, common support graphing, and covariate imbalance testing (2003)

20. Lovrić, M., Pavlović, K., Vuković, M., Grange, S.K., Haberl, M., Kern, R.: Understanding the true effects of the COVID-19 lockdown on air pollution by means of machine learning. Environ. Pollut. **274**, 115900 (2021)
21. Rosenbaum, P.R., Rubin, D.B.: The central role of the propensity score in observational studies for causal effects. Biometrika **70**(1), 41–55 (1983)
22. Rubin, D.B.: Estimating causal effects of treatments in randomized and non-randomizedstudies. J. Educ. Psychol. **66**(5), 688 (1974)
23. Rubin, D.B.: Bias reduction using Mahalanobis-metric matching. Biometrics 293–298 (1980)
24. Rubin, D.B.: Randomization analysis of experimental data: The Fisher randomization test comment. J. Am. Stat. Assoc. **75**(371), 591–593 (1980)
25. Rubin, D.B.: Comment: Neyman (1923) and causal inference in experiments and observational studies. Stat. Sci. **5**(4), 472–480 (1990)
26. Rubin, D.B.: Using propensity scores to help design observational studies: application to the tobacco litigation. Health Serv. Outcomes Res. Method. **2**(3), 169–188 (2001)
27. Schonlau, M., Zou, R.Y.: The random forest algorithm for statistical learning. Stand. Genomic Sci. **20**(1), 3–29 (2020)
28. Seo, J.H., Jeon, H.W., Sung, U.J., Sohn, J.R.: Impact of the COVID-19 outbreak on air quality in Korea. Atmosphere **11**(10), 1137 (2020)
29. Song, Y., Li, Z., Liu, J., Yang, T., Zhang, M., Pang, J.: The effect of environmental regulation on air quality in china: a natural experiment during the covid-19 pandemic. Atmos. Pollut. Res. **12**(4), 21–30 (2021)
30. Sozzi R.: La micrometeorologia e la dispersione degli inquinanti in aria. Agenzia per la Protezione dell'Ambiente e per i servizi Tecnici (APAT http:/www.sinanet.apat.it), Roma (2004)
31. Stuart, E.A.: Matching methods for causal inference: a review and a look forward. Stat. Sci.: Rev. J. Inst. Math. Stat. **25**(1)
32. Wang, J., Xu, X., Wang, S., He, S., He, P.: Heterogeneous effects of covid-19 lockdown measures on air quality in northern china. Appl. Energy **282**, 116–179 (2021)

Measuring Strength of Randomized Clinical Trials

Piero Quatto and Enrico Ripamonti

Abstract The outcomes of randomized clinical trials (RCTs) are commonly assessed using the Fragility Index (FI). Althouh the FI offers supplementary information to the p-value, it comes with inherent limitations. This study reexamines the fragility analysis, placing it within a more comprehensive framework to enhance its ability to complement p-value information reliably. Drawing on the fundamental principles of fragility analysis, we introduce a Strength Index (SI) for RCTs (Quatto et al. Beyond the fragility index. Mimeo (2023)). On this basis, we recommend a triplet of indicators for the analysis of strength. In particular, the SI is applicable in both significant and non-significant cases, offering straightforward calculation and addressing some of the issues associated with the FI. Being a normalized index, we propose a threshold for its assessment in practical applications. Importantly, the SI is utilized in the examination of recent clinical trials, providing a valuable alternative perspective.

Keywords Randomized clinical trials · Fragility index · p-value

1 Introduction

Examining the fragility of a randomized clinical trial (RCT) is a highly debated topic. A Fragility Index (FI) has been proposed by Feinstein [1] and has been re-considered by Walsh et al. [2] in case of a significant result. Also, a relative version known as Fragility Quotient (FQ) has been proposed [3] as well as a version of FI to be applied in case of non-significant results, known as Reverse Fragility Index (RFI) [4]. The applications of the FI are numerous and extensively distributed across many fields of medical research [4, 5].

P. Quatto
Department of Economics, Management, and Statistics, University of Milan-Bicocca, Milan, Italy
e-mail: piero.quatto@unimib.it

E. Ripamonti (✉)
Department of Economics and Management, University of Brescia, Brescia, Italy
e-mail: enrico.ripamonti@unibs.it

M. Bini et al. (eds.), *Advanced Methods in Statistics, Data Science and Related Applications*, Springer Proceedings in Mathematics & Statistics 467,
https://doi.org/10.1007/978-3-031-65699-6_8

Even though the FI has been strongly criticized [6], the analysis of fragility seems widely recognized as a valid tool for the assessment of RCTs [7–10]. Such perspective is in line with one of the statements of the American Statistical Association, namely that scientific conclusions and business or policy decisions should not be based only on whether a p-value passes a specific threshold [11]. Nevertheless, fragility should be inserted in a broader framework than the current one. Herein we will introduce the analysis of strength, which will be put forward as a tool to assess the p-value.

The analysis of fragility operates through the identification of the number of events that it is necessary to change (e.g., number of deaths in one arm of the RCT) in order to subvert the p-value from significant to non-significant or vice versa. If such number of events is large enough, then the experimental result can be classified as *strong*. By contrast, if only few events will be enough to subvert the p-value, then the experimental result could be classified as *fragile*. The main shortcoming of the analysis of fragility is that it is difficult to properly quantify how strong or, conversely, how fragile is an experimental result.

The article is organized as follows. In Sect. 2 we present the main characteristics and features of the fragility index. In Sect. 3 we sum up our main proposal to go beyond the fragility index by means of a triplet of indicators. This relies on the work by Quatto et al. [12]. In Sect. 4 three illustrative examples are discussed. Finally, in Sect. 5 we draw our main conclusions.

2 The Fragility Index

Adopting a Fisher's exact test on a 2×2 table, Walsh et al. [2] characterize the Fragility Index (FI) as the smallest number of occurrences that need to shift from non-events to events to convert a significant outcome into a non-significant one. The FI is computed by virtually introducing events to the group with fewer events, until $p > \alpha$ occurs. Most of the literature follows this definition, but there are also slight variations, such as, for example, in Dettori et al. [13] where it is proposed to calculate the change that occurs from event to non-event.

In case of a non-significant result, the adopted procedure to obtain the RFI is to subtract events from the group that contains a smaller number of events until $p \leq \alpha$ is reached with the use of a Fisher's exact test [4].

The FI, in its extreme simplicity, presents some weaknesses that can be summarized as follows: (i) it can only be calculated for dichotomous data; (ii) it does not present with a threshold above which the result can be considered robust, according to a common terminology in the literature; (iii) it grows as the sample size grows, virtually suggesting that experiments with greater sample size may be less fragile than others; (iv) it is negatively correlated with the p-value, therefore large p-values are considered fragile and small p-values are considered robust. In addition, it has been recently stated that (v) FI can also be computationally expensive to run, typically requiring multiple iterations of Fisher's exact test to converge [10]. Many authors have criticized the use of FI (or FQ). Schröder et al. [14] argue that the use of these indexes should be avoided, Condon et al. [15] recommend use of FI with parsimony, while Niforatos et al. [16] suggest caution in the interpretation. This criticism is founded upon some inherent weaknessess of FI, as highlighted by Potter [6] and as it will become evident from the rest of the paper.

Walter et al. [17] observe that in some settings transformation of non-events to events could be quite unlikely. In addition, they recommend the use of the analysis of fragility provided that three essential conditions are satisfied: (i) the clinical importance of a treatment; (ii) the statistical significance; (iii) the quantitative stability. By clinical importance the authors refer to the need of setting a minimum threshold δ (in terms of number of observed events) in order to be able to affirm that the treatment under scrutiny is effective. Statistical significance is the usual concept that derives from the use of a statistical test to verify the presence of an effect. With quantitative stability the authors refer to the need to measure the non-fragility of a result. This can be obtained, for example, by adopting the FI. In this case, the smallest number of events sufficient to change the significance to non-significance must be greater than the threshold δ which interprets the clinical importance.

Potter [6] suggests to dismantle the FI "as a complement to the *p*-value to summarize the statistical strength of evidence for a trial's result". This author generalizes to the case of comparison of two proportions the statement for which if two trials of different sizes have the same *p*-value, the smaller trial should provide more evidence for the experimental result [18, 19]. In particular, by fixing a *p*-value that leads to significance and using the Bayes factor, Potter verifies that more evidence to the alternative hypothesis should be given when the sample size is smaller. To justify this result, she states that "the smaller trial must have a larger treatment effect in order to match the larger trial's *p*-value, yet both trials control type I error at the same level". By contrast, calculating FI one would obtain that FI becomes larger with larger sample size. This conflict, however, can be solved using FQ, or relativizing the expression of FI taking $\sqrt{n}$ or $\sqrt[3]{n}$ [9]. In addition, Potter criticizes the use of FI in case of time-to-event outcomes, where a formal survival analysis should be performed. Also, in many cases the *p*-value is calculated through other tests than Fisher's exact test, while the latter is used to calculate the FI [20]. Thus, according to Potter, the FI should be dismantled since it has a weak mathematical basis, thus leading to fundamental inconsistences in its interpretation.

Baer et al. [7], critizing Potter's work on FI, put forward a reassembling of FI. These authors propose to unify FI and RFI in one index that takes a positive value if the experimental result is significant, and takes a negative value if the experimental result is not significant. To construct this index, both groups (i.e., experimental arm of the trial and control treatment) can be counterfactually modified adding or subtracting events, in agreement with Lin's definition [21]. Moreover, to respond to Walter et al.'s observation for which transformation of non-events to events could be quite unlikely, they propose the *incidence fragility index* which allows "only outcome modifications that are sufficiently likely", having fixed a probability measure to quantify the expression *sufficiently likely*. An exact algorithm to calculate the incidence fragility index is provided. These authors also study FI from a theoretical point of view [8], putting it into relation with the sample size and power of the test, responding to Potter's criticism [6, 22]. Baer et al. put forth a fragility index based test, where FI is interpreted as a random variable such that $P(FI > \varphi) \geq 1 - \tau$. The threshold φ is based on clinical appraisal, while τ is fixed as 0.5.

3 Measuring Strength of Clinical Trials

Let us consider a general setting of comparison of two proportions θ_1 and θ_2 of events in two groups G_1 and G_2, respectively:

$$H_0 : \theta_1 = \theta_2 \quad \text{vs.} \quad H_1 : \theta_1 \neq \theta_2. \tag{1}$$

Let us indicate with $\frac{x_1}{n_1}$ and $\frac{x_2}{n_2}$ the proportions of events verified in G_1 and G_2; without losing of generality we will assume: $\frac{x_1}{n_1} \leq \frac{x_2}{n_2}$. The value $\hat{\theta} = \frac{x_2}{n_2} - \frac{x_1}{n_1}$ is an estimate of $\theta = \theta_2 - \theta_1$ and

$$\hat{\sigma}^2 = \frac{x_1}{n_1^2}\left(1 - \frac{x_1}{n_1}\right) + \frac{x_2}{n_2^2}\left(1 - \frac{x_2}{n_2}\right) \tag{2}$$

is a suitable estimate of its variance. We assume that the difference of proportions can be approximated by a Normal random variable (r.v.).

3.1 *The Quantity Q*

We consider the quantity Q proposed by Quatto et al. [12] satisfying the Equation:

$$2\Phi\left(-\frac{|\frac{x_1}{n_1} - \frac{x_2}{n_2} + Q|}{\hat{\sigma}}\right) = \alpha \tag{3}$$

where Φ is the standard Normal cumulative distribution function. Expression (3) is equivalent to:

$$\left|\frac{x_1}{n_1} - \frac{x_2}{n_2} + Q\right| = \hat{\sigma} z_{1-\alpha/2}$$

where $z_{1-\alpha/2}$ is the percentile of order $1 - \alpha/2$ of the standardized Normal r.v. Z. In particular, for Q we propose the explicit formula:

$$Q = \frac{x_2}{n_2} - \frac{x_1}{n_1} - \hat{\sigma} z_{1-\alpha/2} = (z_{1-p/2} - z_{1-\alpha/2})\hat{\sigma} \tag{4}$$

where $z_{1-p/2} = \frac{\frac{x_2}{n_2} - \frac{x_1}{n_1}}{\hat{\sigma}} = \frac{\hat{\theta}}{\hat{\sigma}}$ and $p = 2\Phi\left(-\frac{\hat{\theta}}{\hat{\sigma}}\right)$ is the two-sided p-value for the null hypothesis (1). If the experiment leads to a significant result, expression (4) would take a positive value, while it would take a negative value in case of a non-significant result. More precisely, from (4) it follows $0 \leq Q \leq \frac{x_2}{n_2} - \frac{x_1}{n_1}$ in case of significance, and $-\hat{\sigma} z_{1-\alpha/2} \leq Q \leq 0$ in case of non-significance. In case of significance, from (4) we have:

$$\frac{x_1}{n_1} - \frac{x_2}{n_2} + Q = \frac{x_1 + n_1 Q}{n_1} - \frac{x_2}{n_2} = \frac{x_1}{n_1} - \frac{x_2 - n_2 Q}{n_2}.$$

Thus, Q represents the value that leads to subvert the state of significance, by adding $n_1 Q$ to the unit x_1 or subtracting $n_2 Q$ from x_2. Similar considerations can be put forth in case of non-significance.

3.2 Properties of Q

Through Q we can obtain the quantities $FI_1 = Qn_1$ and $FI_2 = Qn_2$. These represent the changes of the events that are necessary in G_1 or G_2 to subvert the experimental result. We consider $FI^* = Qmin(n_1, n_2)$, representing the minimal number of changes to subvert the experimental result. We remark the importance of the sign of Q; indeed, $FI^* > 0$ implies a significant result, while $FI^* < 0$ implies a non-significant result, in agreement with Baer et al. [8] who distinguish FI and RFI from the sign of the respective expressions. However, FI and RFI follow from p-values calculated through Fisher's exat test. FI^* follows from Normal approximations and can coincide or not with FI. Finally, as can be seen in formula (4), Q depends of the standard error, thus keeping into account the sample size when comparing different situations.

3.3 The Strength Index

In light of the previous results, we recall the normalized index SI introduced by Quatto et al. [12] given by:

$$SI = \begin{cases} \dfrac{Q}{\frac{x_2}{n_2} - \frac{x_1}{n_1}} = 1 - \dfrac{\hat{\sigma} z_{1-\alpha/2}}{\frac{x_2}{n_2} - \frac{x_1}{n_1}} = 1 - \dfrac{z_{1-\alpha/2}}{z_{1-p/2}} & : p \leq \alpha \\ \dfrac{Q}{-\hat{\sigma} z_{1-\alpha/2}} = \dfrac{\frac{x_2}{n_2} - \frac{x_1}{n_1}}{-\hat{\sigma} z_{1-\alpha/2}} + 1 = 1 - \dfrac{z_{1-p/2}}{z_{1-\alpha/2}} & : p > \alpha \end{cases} \tag{5}$$

To make SI operative, it is necessary to compare it with a suitable threshold. An intuitive value in the interval (0, 1) is 1/2. Let us set $\alpha = 0.05$ and let suppose to have obtained a significant experimental result. It would follow that this result could be classified as strong only if $p \leq 2\Phi(-2z_{1-\alpha/2}) \cong 0.0001$. Such value seems too conservative. We then recommend setting an alternative suitable threshold, following what has been proposed in the Bayesian literature concerning the credibility analysis. In that context, in case of a significant result, in order to attain the credibility threshold it is required that $p \leq 2\Phi\left(-\sqrt{\frac{1+\sqrt{5}}{2}}\right) \cong 0.0127$ as proposed by Matthews [23]. Alternatively, Held [24] has recommended setting the threshold $p \leq 2\Phi\left(-\sqrt{2}z_{1-\alpha/2}\right) \cong 0.0056$, in agreement with Benjamin et al's proposal lowering the p-value threshold from 0.05 to 0.005 [26]. The problem of a non-significant result is not presented. Herein, we suggest to follow Held's proposal, so that, from expression (5) it would follow:

$$\begin{cases} SI \geq 1 - \frac{1}{\sqrt{2}} \leftrightarrow \frac{\frac{x_2}{n_2} - \frac{x_1}{n_1}}{\hat{\sigma}} \geq \sqrt{2}z_{1-\alpha/2} \leftrightarrow p \leq 2\Phi(-\sqrt{2}z_{1-\alpha/2}) : p \leq \alpha \\ SI \geq 1 - \frac{1}{\sqrt{2}} \leftrightarrow \frac{\frac{x_2}{n_2} - \frac{x_1}{n_1}}{\hat{\sigma}} \leq \frac{z_{1-\alpha/2}}{\sqrt{2}} \leftrightarrow p \geq 2\Phi\left(-\frac{z_{1-\alpha/2}}{\sqrt{2}}\right) : p > \alpha \end{cases} \tag{6}$$

The index (5) can be extended to the case of unidirectional alternative hypothesis $H_1 : \theta_1 < \theta_2$. The threshold can be set as $1/2$, since in this case it is not too conservative. The expression of SI will be the same, but in (5) it will be necessary to substitute $z_{1-\alpha/2}$ with $z_{1-\alpha}$ and $z_{1-p/2}$ with z_{1-p}. Thus, (6) can be rewritten as:

$$\begin{cases} SI \geq \frac{1}{2} \leftrightarrow \frac{\frac{x_2}{n_2} - \frac{x_1}{n_1}}{\hat{\sigma}} \geq 2z_{1-\alpha} \leftrightarrow p \leq \Phi(2z_{1-\alpha}) : p \leq \alpha \\ SI \geq \frac{1}{2} \leftrightarrow \frac{\frac{x_2}{n_2} - \frac{x_1}{n_1}}{\hat{\sigma}} \leq \frac{z_{1-\alpha}}{2} \leftrightarrow p \geq \Phi\left(\frac{z_{1-\alpha}}{2}\right) : \alpha < p < 0.5 \end{cases}$$

putting $SI = 1$ if $p \geq 0.5$.

4 Illustrative Examples

Example 1 Vitamin D deficiency seems to be connected with various pathologies. While it is known that normal values of vitamin D may protect from bone-related diseases, it seems also protective, indirectly, against cardiovascular diseases. Since vitamin D could influence all-cause mortality, as well as mortality from specific causes, Neale et al. [27] studied the effect of the administration of vitamin D on all-cause mortality, in subjects in the age range 60–84. During a follow-up of 5 years, 10,661 subjects (G_1) have been treated with vitamin D, and 10,649 subjects (G_2) have been treated with placebo. 562 subjects died in G_1 and 538 subjects died in G_2. Using the Normal approximation, we calculate $\hat{\theta} = 0.0022$ and $\hat{\sigma} = 0.0503$ with $z_{1-p/2} = 0.711$ and $p = 0.478$. From (3) we get $Q = -0.0037$ and through (5), it follows $SI = 0.637$ which, being greater than the threshold, provides strength to the experimental non-significant result. We have $FI^* = -0.00375 \times 10649 = -40$

Table 1 An Example from Potter: n indicates the sample size, E_1 the number of events in G_1, E_2 the number of events in G_2, p the p-value, BF the Bayes factor

n	E_1	E_2	p	BF	Q	FI
100	26	54	0.0001	631.5	0.1498	14
200	113	151	0.0001	379	0.0990	20
500	398	44	0.0001	173.1	0.0471	24
100	82	93	0.03	1.8	0.0196	2
200	137	157	0.03	1.4	0.0014	2
500	252	287	0.03	0.9	0.0084	4

that represents the number of deaths that would have to be subtracted from the 538 deaths to achieve the change from non-significance to significance. Adopting the Chi-square test as an approximation of Fisher's exact test, we would obtain the same number of changes. This illustrative example allows to verify Lin's proposal [21] according to which the number of changes can be verified simultaneously in the treatment groups. Indeed, the change of state from non-significant to significant with $FI^* \cong -40$ can be obtained by subtracting 38 events from the treatment group and adding 1 event to the control group, or, equivalently, by subtracting 4 events from the treatment group and adding 37 events to the control group.

Example 2 We analyze some data already presented in Potter [6] where it is shown, according to the author, a fundamental shortcoming of the FI due to incoherent behavior with respect to sample size. Data are shown in Table 1. It can be noted that, with the same p-value, FI increases as the sample size increases, making less fragile the situation with larger sample size. The Bayes Factors, like Q, have an opposite trend, namely a trend that reflects the logic mentioned above.

Example 3 Valgimigli et al. [28] assessed the effect of Bivalirudin administered as part of percutaneous coronary intervention in 3,603 patients with an acute coronary syndrome. This was compared to unfractionated Heparin, administered to 3,603 patients. Considering mortality at 30 days as an outcome, in the former group 401 patients died, and in the second group 444 patients died. It follows $\hat{\theta} = 0.012$, and we obtain $\hat{\sigma} = 0.0076$. The z value is 1.578, thus $p = 0.1143$ (non significant). The experimental result is not supported by the analysis of strength. Indeed, using (5) we obtain SI = 0.183, which is below the threshold. It follows FI^*= -10, which represents the number of deaths that have to be subtracted from the 401 deaths to subvert the result from non-significant to significant.

5 Conclusions

The main goal of this article is to reinforce the traditional statistical analysis of RCTs with equal or unequal allocation obtained with the use of the p-value. The starting point is the introduction of a quantity, namely Q, which, according to its positive or negative sign, denotes a significant or non-significant result. Q allows to quantify the number FI^* of events to be added or subtracted to subvert the experimental result and leads to the normalized index SI, which, if it exceeds a predetermined threshold, makes the sample result strong or frail. One of the fundamental criticism of FI, namely that it is not associated with a threshold, is solved by invoking FI^*. In addition, also the criticism due to the fact that FI would grow as the sample size grows, can be solved by calculating the quantity Q. This is also valid using the Bayes factor, as suggested by Potter. In this way, both Q and the Bayes factor would contradict the so-called α postulate [29] which has been criticized by Royall [18]. Regarding another criticism of the FI, namely that it is negatively correlated with the p-value, this is also true for SI in the significance case, while there is a direct relationship in the non-significance case. This criticism does not hold since what matters is the comparison of SI with the threshold. Finally, an important advantage of Q is that it is not computationally demanding, since it can be obtained with a simple spreadsheet.

We should briefly comment on another putative shortcoming of FI, namely that it can be only calculated in case of dichotomous data. This does not appear as a critical problem in biomedical sciences, since, as underlined by Grimes [10] "dichotomous outcome designs are vital in randomized clinical trials, cohort studies, and observational data for ascertaining differences between experimental and control arms".

In conclusion, the analysis of fragility, herein revisited in terms of a strength analysis as proposed by Quatto et al. [12], seems to be a valuable tool for evaluating the strength of the experimental results both in RCTs and in other more general contexts where statistical inference is drawn referring to a p-value. This is in line and provides added value to the observation of Benjamini et al.: "in summary, p-values and significance tests, when properly applied and interpreted, increase the rigor of the conclusions drawn from data" [30]. In particular, we suggest to rely on the triplet Q, SI, and FI^*, which are founded upon solid mathematical and statistical bases, thus addressing one of the fundamental criticism raised against FI.

References

1. Feinstein, A.R.: The unit fragility index: an additional appraisal of statistical significance for a contrast of two proportions. J. Clin. Epidemiol. **43**(2), 201–209 (1990)
2. Walsh, M., Srinathan, S.K., McAuley, D.F., et al.: The statistical significance of randomized controlled trial results is frequently fragile: a case for a Fragility Index. J. Clin. Epidemiol. **67**(6), 622–628 (2014)
3. Ahmed, W., Fowler, R.A., McCredie, V.A.: Does sample size matter when interpreting the fragility index? Crit. Care Med. **44**(11), e1142–e1143 (2016)

4. Khan, M.S., Fonarow, G.C., Friede, T., et al.: Application of the reverse fragility index to statistically nonsignificant randomized clinical trial results. JAMA Netw. Open **3**(8), e2012469–e2012469 (2020)
5. Tignanelli, C.J., Napolitano, L.M.: The fragility index in randomized clinical trials as a means of optimizing patient care. JAMA Surg. **154**(1), 74–79 (2019)
6. Potter, G.E.: Dismantling the Fragility Index: A demonstration of statistical reasoning. Stat. Med. **39**(26), 3720–3731 (2020)
7. Baer, B.R., Gaudino, M., Charlson, M., Fremes, S.E., Wells, M.: Fragility indices for only sufficiently likely modifications. PNAS **118**(49), e2105254118 (2021)
8. Baer, B.R., Gaudino, M., Charlson, M., Fremes, S.E., Wells, M.: The fragility index can be used for sample size calculations in clinical trials. J. Clin. Epidemiol. **139**, 199–209 (2021)
9. Baer, B.R., Gaudino, M., Fremes, S.E., Charlson, M.: Letter to Editor. J Clin. Epidemiol. **142**, 317–318 (2022)
10. Grimes, D.R.: The Ellipse of Insignificance: A Refined Fragility Index for Ascertaining Robustness of Results in Dichotomous Outcome Trials. OSF Preprints, 28 Mar. 2022. Web. (2022)
11. Wasserstein, R.L., Lazar, N.A.: The ASA statement on p-values: context, process, and purpose. Am. Stat. **70**(2), 129–133 (2016)
12. Quatto, P., Ripamonti, E., Marasini, D.: Beyond the fragility index. Mimeo (2023)
13. Dettori, J.R., Norvell, D.C.: How fragile are the results of a trial? The fragility index. Glob. Spine J. **10**(7), 940–942 (2020)
14. Schröder, A., Muensterer, O.J., von Sochaczewski, C.O.: The fragility index may not be ideal for paediatric surgical conditions: the example of foetal endoscopic tracheal occlusion. Pediatr. Surg. Int. **37**, 967–969 (2021)
15. Condon, T.M., Sexton, R.W., Wells, A.J., To, M.-S.: The weakness of fragility index exposed in an analysis of the traumatic brain injury management guidelines: A meta-epidemiological and simulation study. PLoS ONE **15**(8), e0237879 (2020)
16. Niforatos, J.D., Zheutlin, A.R., Chaitoff, A., Pescatore, R.M.: The fragility index of practice changing clinical trials is low and highly correlated with P-values. J. Clin. Epidemiol. **119**, 140–142 (2020)
17. Walter, S.D., Thabane, L., Briel, M.: The fragility of trial results involves more than statistical significance alone. J. Clin. Epidemiol. **124**, 34–41 (2020)
18. Royall, R.M.: The effect of sample size on the meaning of significance tests. Am. Stat. **40**(4), 313–315 (1986)
19. Berger, J.O., Sellke, T.: Testing a Poin Null Hupothesis: the Irreconciliability of P values and Evidence. J Am. Stat. Ass. **82**(397), 112–122 (1987)
20. Del Paggio, J.C., Tannock, I.F.: The fragility of phase 3 trials supporting FDA-approved anticancer medicines: a retrospective analysis. Lancet Onc. **20**, 1065–69 (2019)
21. Lin, L.: Factors that impact fragility index and their visualizations. J. Eval. Clin. Pract. **27**, 356–364 (2021)
22. Potter, G.E.: Letter to the Editor. J. Clin. Epidemiol. **142**, 315–316 (2022)
23. Matthews, R.A.: Moving towards the post $p<0.05$ era via the analysis of credibility. Am. Stat. **73**(sup1), 202–212 (2019)
24. Held, L.: The assessment of intrinsic credibility and a new argument for $p<0.005$. R. Soc. Open Sci.**6**(3), 181534 (2019)
25. Recovery Collaborative Group.: Dexamethasone in hospitalized patients with Covid-19. New Eng. J. Med., **384**(8), 693-704 (2021)
26. Benjamin, D.J., Berger, J.O., Johannesson, M., Nosek, B.A., Wagenmakers, E.J., Berk, R., Johnson, V.E.: Redefine statistical significance. Nat. Hum. Behav. **2**(1), 6–10 (2018)
27. Neale, R.E., Baxter, C., Romero, B.D., McLeod, D.S., English, D.R., Armstrong, B.K., Waterhouse, M.: The D-health trial: a randomised controlled trial of the effect of vitamin D on mortality. Lancet Diabet. Endocrinol. **10**(2), 120–128 (2022)
28. Valgimigli, M., Frigoli, E., Leonardi, S., et al.: Bivalirudin or unfractionated heparin in acute coronary syndromes. N. Engl. J. Med. **373**, 997–1009 (2015)

29. Cornfield, J.: Sequential trials, sequential analysis and likelihood principle. Am. Stat. **20**(2), 18–23 (1966)
30. Benjamini, Y., et al.: ASA President's task force statement on statistical significance and replicability. Chance **34**(4), 10–11 (2021)

Robust Designs Against Data Loss: A General Approach

Roberto Fontana and Fabio Rapallo

Abstract We describe an algorithm that, given an initial design $\mathcal{F}_n$ of size n and a linear model with p parameters, provides a sequence $\mathcal{F}_n \supset \cdots \supset \mathcal{F}_{n-k} \supset \cdots \supset \mathcal{F}_p$ of nested robust designs. The sequence is obtained by removing each individual run of $\mathcal{F}_n$ until a p-run saturated design $\mathcal{F}_p$ is obtained. The potential impact of the algorithm on real applications is high, because it can be used in a wide spectrum of designs. The initial fraction $\mathcal{F}_n$ can be of any type and the output sequence can be used to organize the experimental activity. The experiments can start with the runs corresponding to $\mathcal{F}_p$ and then continue by adding one run after the other (from $\mathcal{F}_{n-k}$ to $\mathcal{F}_{n-k+1}$) until the initial design $\mathcal{F}_n$ is obtained. In this way, if for some unexpected reasons the experimental activity has to be interrupted before the end when only $n-k$ runs have been completed, the corresponding $\mathcal{F}_{n-k}$ will have a high value of robustness for $k \in \{1, \ldots, n-p\}$. The algorithm uses the circuit basis, a special representation of the kernel of a matrix with integer entries.

Keywords Fractional factorial designs · D-optimality · Incomplete designs

1 Introduction

Optimal designs and orthogonal fractional factorial designs are frequently used in many fields of applications, including medicine, engineering and agriculture. They offer a valuable tool for dealing with problems where there are many factors involved and each run is expensive. The literature on the subject is extremely rich. A non-

R. Fontana (✉)
Department of Mathematical Sciences, Politecnico di Torino, Corso Duca degli Abruzzi 24, 10129 Torino, Italy
e-mail: roberto.fontana@polito.it

F. Rapallo
Department of Economics, University of Genova, Via Vivaldi 5, 16126 Genova, Italy
e-mail: fabio.rapallo@unige.it

M. Bini et al. (eds.), *Advanced Methods in Statistics, Data Science and Related Applications*, Springer Proceedings in Mathematics & Statistics 467,
https://doi.org/10.1007/978-3-031-65699-6_10

exhaustive list of references includes [3] for design of experiments in general, [2, 7, 18] for optimal designs and [6, 14, 15] for orthogonal fractional factorial designs.

When searching for an optimal experimental design, the aim is to select a design which produces the best estimates of the relevant parameters for a given sample size. There are many criteria for choosing an optimal design for the problem under study. They include alphabetical design criteria, and among these D-optimality is one of the most commonly used in applications.

In this work we focus on the notion of robustness of a design [11]. Let us suppose that a given design has n runs and that the model to be estimated has p parameters, with $n > p$. The model-design pair determines the design matrix X. The notion of robustness is important mainly for two reasons. First, the robustness of the design can be interpreted as the probability that a randomly selected subset of p runs is a *saturated* design (i.e., the p parameters of the model can be estimated). A high value of robustness has practical importance. If during the experimental activity $n - p$ runs are lost (i.e., the corresponding response values are not available) the probability that the p-run final design is *saturated* is high. Second, a close connection between robustness and D-optimality for a large class of model matrices is proved in [11].

The aim of this contribution is to provide a general order-of-the-runs criterion based on the notion of robustness. It can be applied to a wide range of possible designs. All designs for qualitative factors can be considered, from factorial designs with two-level factors, to multi-level and mixed-level factors. Even in some cases of continuous factors the proposed criterion can be applied. Since the robustness is based on the combinatorial properties of the design matrix, the unique assumption is to have a design matrix with integer entries. In practice combinatorial algorithms are limited by the dimension of the designs under analysis. We will discuss this issue in the next sections.

The main result of this work is an algorithm that starts from an initial design $\mathcal{F}_n$ (of size $n > p$) and removes each individual run of $\mathcal{F}_n$ until a p-run *saturated* design $\mathcal{F}_p$ is obtained. The choice of which point is removed at each step k ($1 \leq k \leq n - p$) aims to find a $(n - k)$-run sub-fraction $\mathcal{F}_{n-k}$ of the initial design with the highest value of robustness. The output of the algorithm is a sequence of *robust* fractions $\mathcal{F}_n \supset \cdots \supset \mathcal{F}_{n-k} \supset \cdots \supset \mathcal{F}_p$. In practice, the experimental activity can start with the runs corresponding to $\mathcal{F}_p$ and then continues by adding one run after the other (from $\mathcal{F}_{n-k}$ to $\mathcal{F}_{n-k+1}$) until the initial design $\mathcal{F}_n$ is obtained. In this way, if for some unexpected reasons the experimental activity has to be interrupted before the end when only $n - k$ runs have been completed, the corresponding $\mathcal{F}_{n-k}$ will have a high value of robustness, $1 \leq k \leq n - p$. It is worth noting that the value of robustness of $\mathcal{F}_{n-k}$ is high *for each* k, $1 \leq k \leq n - p$ as shown in the simulation study. In the simulation study, some examples are illustrated to prove the effectiveness of the algorithm. The problem of partial availability of data and techniques to prevent possible loss of information are addressed in, e.g., [4, 5, 19], both in model-based and model-free frameworks. In the paper [8] a combinatorial approach is introduced for the analysis of orthogonal arrays with removed runs using aberrations and the Generalized Word-Length Pattern criterion.

The algorithm introduced here uses the circuit basis, i.e., a special representation of the kernel of a matrix with integer entries. By exploiting some combinatorial properties of such a basis, we obtain a sequence of nested designs with high performance in terms of robustness. This result is obtained with a unique computation of the circuit basis in the first step. The theory of robustness based on circuits is fully described in [11], while the estimability of saturated designs using circuits is studied in [10].

It is worth noting that there are no restrictions on the way in which the initial design $\mathcal{F}_n$ is determined. It can be an orthogonal fractional factorial design, a D-optimal design, or any other design defined according to the user's preferences. The only restriction applies to the design matrix that must have integer entries. It follows that ANOVA-type models for qualitative variables can be considered. Polynomial models for continuous variables can also be considered with the restriction that the entries of the design are rational numbers. The examples below show that, from the combinatorial point of view, in some cases quantitative factors are easily rewritten in the qualitative framework. The general case of quantitative factors where an approximation of the design matrix is needed falls outside the scope of the present work, and we will provide some reflections for this in the concluding remarks.

The material is organized as follows. In Sect. 2 the definition of circuit basis is introduced together with its main algebraic and combinatorial properties, and the connections with robustness of a design are reviewed. Section 3 is devoted to the description of the proposed algorithm and some computational remarks. A first example on a small design is illustrated with full details. In Sect. 4 some examples are presented and discussed. Finally, Sect. 5 contains some final comments and pointers for future work.

2 Circuits and Robustness

We consider a design $\mathcal{F}$ with n runs, chosen from a set $\mathcal{D}$ with N runs, $N > n$. For experiments with d discrete factors $X_1, \ldots, X_d$, the set $\mathcal{D}$ is usually represented as a Cartesian product such as

$$\{0, \ldots, s_1 - 1\} \times \cdots \times \{0, \ldots, s_d - 1\},$$

where $s_1, \ldots, s_d$ are the number of levels of the factors $X_1, \ldots, X_d$, respectively. However, for our theory the special coding of the factor levels and even the Cartesian product structure of the full-factorial design are irrelevant. We may simply assume that a subset of n runs has been selected from a large set labeled $\{1, \ldots, N\}$. In the language of fractional factorial designs, the design $\mathcal{F}$ is a fraction, while the large set $\mathcal{D}$ is a full-factorial design.

Given a full-factorial design $\mathcal{D}$ we consider a linear model on $\mathcal{D}$:

$$\mathbf{y} = X_{\mathcal{D}} \boldsymbol{\beta} + \boldsymbol{\varepsilon}, \tag{1}$$

where $\mathbf{y}$ is the vector containing the response variable, $X_{\mathcal{D}}$ is the model matrix, $\boldsymbol{\beta}$ is the vector of parameters, and $\boldsymbol{\varepsilon}$ is the error term. Without loss of generality, to simplify some algebraic issues of our theory, we assume that the matrix $X_{\mathcal{D}}$ is full-rank with dimension $N \times p$, where p is the number of estimable parameters.

When a fraction $\mathcal{F}$ is selected, the expression of the model in Eq. (1) becomes

$$\mathbf{y} = X_{\mathcal{F}}\boldsymbol{\beta} + \boldsymbol{\varepsilon}, \tag{2}$$

where the model matrix $X_{\mathcal{F}}$ has dimension $n \times p$ and is obtained from $X_{\mathcal{D}}$ by selecting only the rows pertaining the chosen runs.

As pointed out in the Introduction, we aim at defining an algorithm which gives "optimal" subsets of a given fraction. Here, we use the criterion of robustness. Following [12, 13], the robustness is defined in terms of saturated fractions.

Definition 1 Let $\mathcal{F}$ be a fraction with model matrix $X_{\mathcal{F}}$. The robustness of the fraction $\mathcal{F}$ under the model $X_{\mathcal{F}}$ is the proportion of saturated minimal fractions over the number of minimal fractions contained in $\mathcal{F}$:

$$r(X_{\mathcal{F}}) = \frac{\#\{\text{saturated } \mathcal{F}_p\}}{\binom{n}{p}}. \tag{3}$$

We observe that the number of runs of a minimal fraction is p and for a minimal fraction $\mathcal{F}$ the robustness can be either 0 or 1.

We now show how to use the circuits of the model matrix $X_{\mathcal{F}}$ to study the robustness of the fraction. Then, in the next section we will provide an algorithm to sequentially remove runs from $\mathcal{F}$ to maintain the robustness as high as possible.

To match the language of Combinatorics with the design theory, we work with the transposed of the model matrix, i.e., we consider the matrix $A_{\mathcal{F}} = X_{\mathcal{F}}^t$ and with a slight abuse of notation we still call it the model matrix. Note that working with $A_{\mathcal{F}}$ implies that the runs identify columns, while parameters identify rows.

In words, a circuit of $A_{\mathcal{F}}$ is an element $\mathbf{u}$ of $\ker(A_{\mathcal{F}})$ with integer entries and minimal support, where the support of a vector $\mathbf{u}$ is the set of indices i with $u_i \neq 0$. We denote by $\operatorname{supp}(\mathbf{u})$ the support of the vector $\mathbf{u}$.

Definition 2 Let $\mathbf{u} \in \mathbb{Z}^n$ be an n-dimensional integer vector. $\mathbf{u} \in \ker(A_{\mathcal{F}})$ is a circuit if the nonzero entries of $\mathbf{u}$ are relatively prime and there is no other vector $\mathbf{v} \in \ker(A_{\mathcal{F}})$ with $\operatorname{supp}(\mathbf{v}) \subset \operatorname{supp}(\mathbf{u})$.

The set of all the circuits of $A_{\mathcal{F}}$ is called the circuit basis of $A_{\mathcal{F}}$ and is denoted with $C(A_{\mathcal{F}})$. The circuit basis is always finite.

For a comprehensive introduction to circuits and its properties the reader can refer to [16, 20]. Here, the key issue for using the circuit basis as a special basis of $\ker(A_{\mathcal{F}})$ is given by the following property.

Proposition 1 *Let $\mathcal{F}'$ be a sub-fraction of $\mathcal{F}$, and decompose each circuit of $A_{\mathcal{F}}$ into $\mathbf{u} = (\mathbf{u}_{\mathcal{F}'}, \mathbf{u}_{\mathcal{F}-\mathcal{F}'})$. The circuits of $A_{\mathcal{F}'}$ are*

$$\{\mathbf{u}_{\mathcal{F}'} \ : \ \mathbf{u} \in C(A_{\mathcal{F}}), \mathrm{supp}(\mathbf{u}) \subseteq \mathcal{F}'\} .$$

Proposition 1 says that the circuit basis is the natural representation of the kernel $\ker(A_{\mathcal{F}})$ when we need to remove runs, because we don't need to recompute the basis of the kernel at each step. The first computation contains all the information needed to compute the robustness also for all possible sub-fractions.

Moreover, the minimality property established in Definition 2 is used in the following result, from [10], which characterize the saturated minimal fractions using the circuit basis.

Proposition 2 *Let $A_{\mathcal{F}}$ be a model matrix with circuit basis $C(A_{\mathcal{F}})$. A minimal sub-fraction $\mathcal{F}_p$, i.e., a fraction with p runs from $\mathcal{F}$, is saturated if and only if it does not contain any of the supports* $\mathrm{supp}(\mathbf{u})$ *for all* $\mathbf{u} \in C(A_{\mathcal{F}})$.

Exploiting Proposition 2, in [11] an algorithm for finding robust fractions using the design points of a candidate set $\mathcal{D}$, based on an exchange-type strategy is described. Without introducing all the technical details, the algorithm in [11] is based on two general principles: (i) remove the point contained in the largest number of circuits; (ii) remove the point contained in the smallest circuits. Such two rules are summarized in the definition of a loss function $L(P)$ defined as

$$L(P) = \sum_{\mathrm{supp}(\mathbf{u}) \supseteq \mathcal{F}, \mathrm{supp}(\mathbf{u}) \ni P} \binom{n - \#\mathrm{supp}(\mathbf{u})}{p - \#\mathrm{supp}(\mathbf{u})} . \tag{4}$$

Note that the sum is taken over all the circuits $\mathbf{u}$ in the fraction containing the point P.

The algorithm is formed by the following main steps: (a) Start from a fraction $\mathcal{F}$; (b) Remove from $\mathcal{F}$ the run with the highest loss function; (c) Add a new run from $\mathcal{D}$ not in $\mathcal{F}$; (d) Repeat steps (b)-(c) until no reduction in the number of circuits is possible.

To actually compute the circuit basis $C(A_{\mathcal{F}})$ for a given model matrix $A_{\mathcal{F}}$ there are several available packages and free software. For our computations we have used `4ti2`, see [1], a program for computing combinatorial objects like Markov bases, Graver bases, circuits, and more.

A common drawback of Algebraic Statistics tools is the limitation to small problems. The computation of the circuits does not make exception. The number of circuits for a model matrix on the full-factorial design increases fast with the number of runs and the computations are actually feasible only for small-sized problems. To give a rough idea, the circuits for a full-factorial 2^d design with main effects and first-order interactions is feasible only for $d \leq 6$. For the 2^d with only main effects, the circuit basis has 20 elements for $d = 3$; 1,348 elements for $d = 4$; 353,616 elements for $d = 5$.

As a running example, let us consider a $3 \cdot 2^2$ design with main effects and the interaction between the two binary factors. The full-factorial design has $N = 12$ points and there are $p = 6$ free parameters. To run the exchange-type algorithm described above, we compute the circuits of the model matrix on the full-factorial design and we obtain a circuit basis with 42 circuits: 18 circuits with support on 4 points and 24 circuits with support on 6 points. Now, to find a robust fraction with a fixed size, say $n = 8$, it is enough to run the algorithm above with an arbitrary starting fraction with 8 runs. One can start with a randomly selected fraction, or can select a fraction satisfying some given criteria, such as D-optimality. However, the algorithm introduced in this work does not need the computation of the circuit basis for the full-factorial design, but only the circuit basis for the starting fraction, and therefore it can be applied also in relatively large examples.

We make explicit here some computational remarks, and we will come back on these issues later after the introduction of the new algorithm in Sect. 3. First, the procedure above is based on the loss function defined in Eq. (4). Such formula does not count the exact number of non-estimable minimal fractions and thus it is not assured that for each sample size it returns a fraction with the highest possible robustness. In the simulations described in [11], for sample sizes near to minimal the performance is quite good, but for large fractions, where the number of circuits contained in more than one minimal fraction is not negligible, the algorithm may yield a fraction with low robustness. However, in the next section we will illustrate why the chosen loss function is a good choice for the selection of the runs to be removed. Second, the algorithm assumes that the circuit basis for the candidate set (in general the full-factorial design) is available. The computation of the circuit basis for large designs can be actually unfeasible and thus the practical applicability is usually limited to small cases. Third, the algorithm above works for a fixed sample size, and it yields non-nested fractions when applied with different sample sizes. This is due to the random addition of a new run, and to the presence of ties, i.e., runs with the same loss function to be removed.

3 Removing Runs from a Fraction

In this section we consider another version of the circuit-based algorithm for removing runs from a given design. While the procedure described in the previous section was essentially based on the first property of the circuits, summarized in Proposition 2, the algorithm below fully exploits the second property of the circuits, described in Proposition 1.

Given a fraction $\mathcal{F}$ and a model with (transposed) design matrix $A_{\mathcal{F}}$, we use the circuit basis $C(A_{\mathcal{F}})$ as a "geometric tool" to choose the order of the points to be removed from $\mathcal{F}$ with the goal of obtaining the best possible robustness of the sub-fractions. Note that in the loss function in Eq. (4) only the circuits with support on p points or less are involved. Thus, we systematically remove from the circuit basis the circuits with support on more than p points. Moreover, to avoid computational problems we assume that the design matrix $A_{\mathcal{F}}$ is full-rank. The algorithm works as follows:

1. Start from a fraction $\mathcal{F}$ with n runs, and compute the circuit basis $C(A_{\mathcal{F}})$;
2. Compute the loss function for the runs in $\mathcal{F}$ based on $C(A_{\mathcal{F}})$;
3. Remove from $\mathcal{F}$ the run with the highest loss function $L(P)$. In case of ties, randomize among the runs with the highest loss function, and define a sub-fraction $\mathcal{F}'$ with $n-1$ runs;
4. Iterate items 2 and 3 until the desired number of runs has been removed.

The validity of the algorithm rests on two facts. First, from Proposition 1, the circuit basis $C(A_{\mathcal{F}})$ computed for the fraction $\mathcal{F}$ is valid also for all the sub-fractions $\mathcal{F}' \subset \mathcal{F}$, and thus the computation of the circuits is limited to the initial step and only for the given starting fraction. Second, the chosen expression of the loss function helps in reaching robust fractions at each step. Indeed, if we consider a circuit $\mathbf{u}$ with support contained in $\mathcal{F}$, the number of non-estimable fractions containing the circuit $\mathbf{u}$ and the design point P is

$$\widetilde{L}(P,u) = \binom{n - \#\mathrm{supp}(\mathbf{u})}{p - \#\mathrm{supp}(\mathbf{u})}$$

if $P \in \mathbf{u}$ and

$$\widetilde{L}(P,u) = \binom{n - \#\mathrm{supp}(\mathbf{u}) - 1}{p - \#\mathrm{supp}(\mathbf{u}) - 1}$$

if $P \notin \mathbf{u}$. The sum over all the circuits would yield a loss function of the form

$$\widetilde{L}(P) = \sum_{\mathrm{supp}(\mathbf{u}) \supseteq \mathcal{F}, \mathrm{supp}(\mathbf{u}) \ni P} \binom{n - \#\mathrm{supp}(\mathbf{u})}{p - \#\mathrm{supp}(\mathbf{u})} + \sum_{\mathrm{supp}(\mathbf{u}) \supseteq \mathcal{F}, \mathrm{supp}(\mathbf{u}) \not\ni P} \binom{n - \#\mathrm{supp}(\mathbf{u}) - 1}{p - \#\mathrm{supp}(\mathbf{u}) - 1}. \quad (5)$$

However, in view of our greedy strategy, we aim primarily at reducing the number of circuits because the surviving circuits entail new non-estimable minimal fractions in the subsequent steps. The idea of looking at the circuits instead of the non-estimable fractions rely on the idea that the circuits are the causes of the non-estimability and thus we concentrate in the elimination of such causes. As a consequence, the two addends in Eq. (5) have a totally different meaning because the second addend adds a positive contribution to the loss function for circuits not containing the data point P. Thus, we only keep the first sum and we use the loss function defined in Eq. (4).

Ideally the algorithm can be iterated until a saturated fraction is reached. Remember that in this case the robustness of the fraction can be either 0 or 1. Although robustness is useful for small designs with run size near to the minimum p, the algorithm works for all run sizes from n to p. Even in the intermediate cases, where the complete enumeration of all the sub-fractions may be computationally difficult, the proposed algorithm is able to easily find robust sub-fractions because it works on the runs and not on the sub-fractions.

Let us illustrate now a very small example in order to show the applicability of the algorithm above. We consider again the $3 \cdot 2^2$ case with main effects and the interaction between the two binary factors. The problem is to find robust sub-fractions of a D-optimal design. We use here a lexicographic order of the factor levels. In this example, we start with the fraction $\mathcal{F}_9$ which has $n = 9$ runs:

$$\mathcal{F}_9 = \{(-1,-1,+1),(-1,+1,-1),(-1,+1,+1),(0,-1,-1),\\ (0,+1,-1),(0,+1,+1),(+1,-1,-1),(+1,-1,+1),(+1,+1,-1)\} \quad (6)$$

with robustness $r(X_{\mathcal{F}_9}) = 0.5952$. The model has $p = 6$ parameters, so we seek for sub-fractions with 8, 7 and 6 runs.

There are 7 circuits with support contained in $\mathcal{F}_9$: 3 circuits with support on 4 points, 4 circuits with support on 6 points. We point out that the relevant circuit basis now has only 7 circuits, while the circuits basis for the full-factorial design has 42 elements, as described in the previous section. The 7 circuits are listed below (where the columns are ordered according to the list in Eq. (6)):

0	0	0	1	-1	0	-1	0	1
0	1	-1	-1	0	1	1	0	-1
0	1	-1	0	-1	1	0	0	0
1	-1	0	-1	1	0	1	-1	0
1	-1	0	0	0	0	0	-1	1
1	0	-1	-1	0	1	1	-1	0
1	0	-1	0	-1	1	0	-1	1

In $\mathcal{F}_9$ the highest loss function $L(R)$ is reached by the 3 runs

$$(-1,+1,-1),(0,+1,-1),(+1,+1,-1).$$

We randomly choose the first run from the above list, thus defining the robust sub-fraction

$$\mathcal{F}_8 = \{(-1,-1,+1),(-1,+1,+1),(0,-1,-1),(0,+1,-1),\\ (0,+1,+1),(+1,-1,-1),(+1,-1,+1),(+1,+1,-1)\} \quad (7)$$

with robustness $r(X_{\mathcal{F}_8}) = 0.7143$.

Among the 7 circuits above, only 3 of them still survive (the columns are ordered according to the list in Eq. (7)):

0	0	1	-1	0	-1	0	1
1	-1	-1	0	1	1	-1	0
1	-1	0	-1	1	0	-1	1

To further reduce the sample size of the fraction we compute again the loss function of the 8 remaining points. The highest value is reached by the 4 runs:

$$(0,-1,-1),(0,+1,-1),(+1,-1,-1),(+1,+1,-1)\,.$$

We randomly choose the third run in this list, and we get

$$\begin{aligned}\mathcal{F}_7 = \{&(-1,-1,+1),(-1,+1,+1),(0,-1,-1),(0,+1,-1),\\ &(0,+1,+1),(+1,-1,+1),(+1,+1,-1)\}\end{aligned} \tag{8}$$

with robustness $r(X_{\mathcal{F}_7}) = 0.8571$.

Only 1 circuit has support contained in $\mathcal{F}_7$ (the columns are ordered according the the list in Eq. (8)):

```
1   -1   0   -1   1   -1   1
```

In the last step to reach the minimal fraction, we remove one of the 6 runs in the support of the last circuit. We randomly choose the last run and we obtain

$$\begin{aligned}\mathcal{F}_6 = \{&(-1,-1,+1),(-1,+1,+1),(0,-1,-1),\\ &(0,+1,-1),(0,+1,+1),(+1,-1,+1)\}\,.\end{aligned}$$

Of course this last fraction has robustness 1, because it is a minimal fraction and does not contain any support of the circuits.

In this example we can compare the robustness of the sub-fractions $\mathcal{F}_8, \mathcal{F}_7, \mathcal{F}_6$ with the robustness of all the possible sub-fractions with 8, 7, 6 runs respectively. There are 9 sub-fractions of $\mathcal{F}_9$ with 8 runs: 6 of them have robustness 0.5357; 3 of them have robustness 0.7143. There are 36 sub-fractions of $\mathcal{F}_9$ with 7 runs: 3 of them have robustness 0 and they are actually non-estimable; 24 of them have robustness 0.5714; 9 of them have robustness 0.8571. Finally, there are 84 sub-fractions of $\mathcal{F}_9$ with 6 runs: 34 of them have robustness 0; 50 of them have robustness 1. We observe that our procedure has identified the highest robustness in all the steps.

Before the illustration of the examples in the next section, some computational remarks are needed to clarify the special features of the circuits for finding robust designs using our algorithm. First, the sub-fraction property of the circuits in Proposition 1 operates here in two ways. On one hand, we don't need the computation of the circuits for the full-factorial design, but only for the starting design. This makes feasible the computations also in cases where the circuits for the full factorial design cannot be computed. Moreover, the circuit basis in the first step is still valid throughout the whole algorithm, and no further computations are needed. These features allow us to use the algorithm also in intermediate-sized examples: using a standard PC, the time needed for the computation of the circuits for all the examples discussed in this paper ranges from 0.01 s for the running example above to 0.55 s for the example in Sect. 4.1.2.

4 Examples and Applications

In Sect. 4.1 we evaluate the performance of the algorithm, that is the ability to produce a sequence of *robust* nested sub-fractions. We run a simulation study which considers different designs with both qualitative and quantitative factors. In Sect. 4.2, starting from the data available in a real application, we show the importance of choosing a robust fraction in terms of the ability to obtain a reliable estimate of β, the unknown vector of parameters, in the case some runs are lost.

4.1 *Performance of the Algorithm*

We consider two examples. The examples have been chosen to show the possibility to use the proposed approach in different contexts. In each example we consider a starting design with $n > p$ runs. Then we analyze the robustness of its sub-fractions which are obtained removing $1, \ldots, n-p$ points by the starting design. It follows that by removing k points ($1 \le k \le n-p$), we must consider $\binom{n}{k}$ sub-fractions of the starting design and, for computing the robustness of each sub-fraction, we must consider $\binom{k}{p}$ size-p fractions which are contained in it. In this way we obtain the exact distribution of the robustness of all the sub-fractions of size $n-k$ of the starting design. This distribution allows us to evaluate the goodness of the solutions proposed by the algorithm. In the second example the number $\binom{n}{k}\binom{k}{p}$ becomes too large, and thus we build an approximation of the distribution of the robustness by sampling. More specifically we consider $\min(\binom{n}{k}, 1000)$ sub-fractions and we evaluate the robustness of each sub-fraction by classifying $\min(\binom{k}{p}, 1000)$ fractions of size p which are contained in it as saturated or not.

4.1.1 Example 1: A Plackett–Burman Design

The first example considers five 2-level factors and a model with a constant term plus the 5 main effects. The number of degrees of freedom of the model is $p = 1 + 5 = 6$. The robustness of a Plackett–Burman design with $n = 12$ runs is analyzed. For this problem the circuit basis has 91 circuits, while the corresponding circuit basis for the full-factorial problem would consist of 44,560 circuits. The values of the robustness of the fractions which are obtained removing $k = 1, \ldots, n-p = 6$ points are computed and compared with the values of the robustness corresponding to the fractions found by the algorithm. The distributions of the robustness for each $k = 1, \ldots, 6$ are exact. The case $k = 0$ (i.e., no points removed) provides the robustness of the initial design. The results are summarized in Fig. 1.

Table 1 compares the values of the robustness of the fractions found by the algorithm (r_*) with the 75th, 90th and 95th percentile of the distributions of the robustness (p_{75}, p_{90}, p_{95} respectively). It is worth noting that for each number k of points

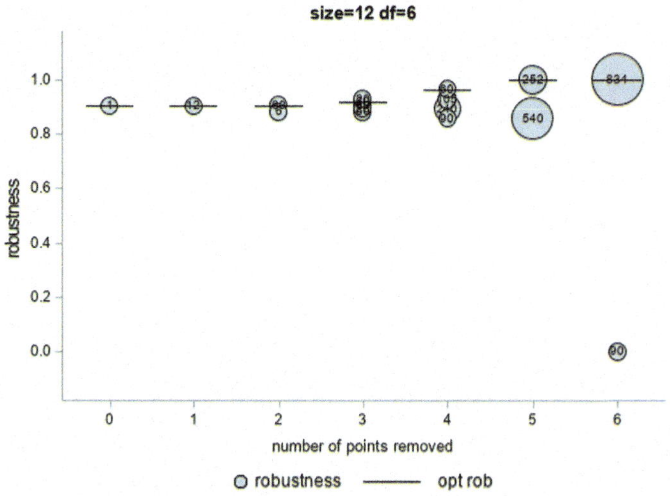

Fig. 1 Distribution of robustness *vs* number of points removed for the Plackett–Burman design in Sect. 4.1.1. For each sample size the robustness of the design selected by the proposed algorithm is represented with an horizontal line

Table 1 Example 1: Comparison of the output of the algorithm (r_*) with the 75th, 90th, and 95th percentile of the distribution of the robustness (p_{75}, p_{90}, p_{95} respectively). The value k is the number of points removed by the initial design. The value corresponding to the robustness of the initial design (r_0) is given at $k = 0$

k	p_{75}	p_{90}	p_{95}	r_*
0	$r_0 = 0.903$			
1	0.903	0.903	0.903	0.903
2	0.905	0.905	0.905	0.905
3	0.917	0.917	0.929	0.917
4	0.929	0.964	0.964	0.964
5	1	1	1	1
6	1	1	1	1

removed the algorithm provides values of robustness greater than the 75th percentile and apart from $k = 3$ equal to the 95th percentile.

4.1.2 Example 2: A Design with Quantitative Continuous Factors

The second example considers five continuous variables $x_i, i = 1, \ldots, 5$ which take values in the interval $[-1, +1]$. The model contains the intercept, five linear terms $x_i, i = 1, \ldots, 5$, five quadratic terms $x_i^2, i = 1, \ldots, 5$ and the four interaction terms $x_1x_2, x_1x_3, x_1x_4, x_1x_5$. The number of degrees of freedom of the model

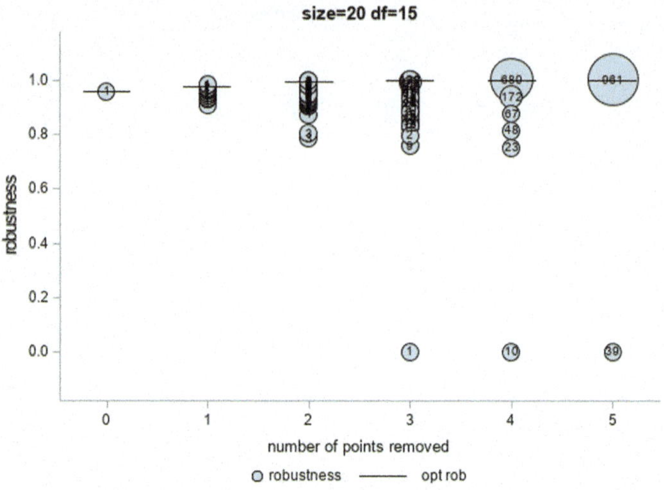

Fig. 2 Sampling distribution of robustness *vs* number of points removed for the 5-factor example in Sect. 4.1.2. For each sample size the robustness of the design selected by the proposed algorithm is represented with an horizontal line

is $p = 1 + 5 + 5 + 4 = 15$. For this problem the circuit basis has 276 circuits. The robustness of a D-optimal design with with $n = 20$ runs is analyzed. The 20-run D-optimal design has been obtained using as candidate set the full factorial design $\{-1, +1\}^5$. The values of the robustness of the fractions which are obtained removing $k = 1, \ldots, n - p = 5$ points are computed and compared with the values of the robustness corresponding to the fractions found by the algorithm. The distributions of the robustness for each $k = 1, \ldots, 5$ are obtained by sampling. The results are summarized in Fig. 2.

Table 2 compares the values of the robustness of the fractions found by the algorithm (r_*) with the 75th, 90th and 95th percentile of the distributions of the robustness (p_{75}, p_{90}, p_{95} respectively). It is worth noting that for each number k of points removed by the initial design the algorithm provides values of robustness greater than the 95th percentile.

4.2 *The Impact of Data Loss in a Real-Data Application*

We consider the application described in [17]. Briefly, a new technique for approximating the stress in pad-type nozzles attached to a spherical shell is presented. The stress values corresponding to a single replicated full factorial design with three 3-level factors (A, B, and C) are used for studying, using standard ANOVA, the effects of the factors on the membrane stress ratio. For showing the impact that our algorithm could have in practical applications we use the discontinous membrane stress

Table 2 Example 2: Comparison of the output of the algorithm (r_*) with the 75th, 90th, and 95th percentile of the distribution of the robustness (p_{75}, p_{90}, p_{95} respectively). The value k is the number of points removed by the initial design. The value corresponding to the robustness of the initial design (r_0) is given at $k = 0$

k	p_{75}	p_{90}	p_{95}	r_*
0	$r_0 = 0.954$			
1	0.971	0.975	0.977	0.978
2	0.979	0.989	0.991	0.994
3	0.993	0.996	1	1
4	1	1	1	1
5	1	1	1	1

Table 3 Values of the discontinous membrane stress y observed for each run of the full factorial design $\mathcal{D}$

Case no.	A	B	C	**y**	Case no.	A	B	C	**y**
1	−1	−1	−1	191.8	15	1	0	0	278.3
2	0	−1	−1	230.5	16	−1	1	0	153.7
3	1	−1	−1	269.6	17	0	1	0	213.3
4	−1	0	−1	159.3	18	1	1	0	236
5	0	0	−1	175.5	19	−1	−1	1	208.6
6	1	0	−1	186.3	20	0	−1	1	293.6
7	−1	1	−1	154.8	21	1	−1	1	369.8
8	0	1	−1	167.6	22	−1	0	1	170.5
9	1	1	−1	154.8	23	0	0	1	234
10	−1	−1	0	227.5	24	1	0	1	286
11	0	−1	0	295.1	25	−1	1	1	152.4
12	1	−1	0	369	26	0	1	1	206.1
13	−1	0	0	174.6	27	1	1	1	251.9
14	0	0	0	231.5					

(approximate analysis) as response. The full factorial design $\mathcal{D} = \{-1, 0, 1\}^3$ with the corresponding values of the response (**y**) are reported in Table 3.

As in [17], we consider a model with main effects and two-order interactions. The model can be written in matrix form as in Eq. (1), i.e.:

$$\mathbf{y} = X_{\mathcal{D}}\boldsymbol{\beta} + \boldsymbol{\varepsilon}.$$

The row $x_{(a,b,c)}$ of $X_{\mathcal{D}}$ corresponding to the design point $(a, b, c) \in \mathcal{D}$ is defined as

$$(1, x_a, x_b, x_c, x_{ab}, x_{ac}, x_{bc})$$

Table 4 For each number of available runs, $x = 23, 24, 25$, the mean error e_x, the number of estimable fractions E_x among all the sub-fractions of size x of $\mathcal{D}$, the relative frequency of the event $e_x^k > e_x$, and the relative frequency of non-estimable fractions $\overline{E}_x$ are reported

x	e_x	E_x	$\#(e_x^k > e_x)/E_x$	$\overline{E}_x/\binom{27}{x}$
23	0.299	16,821	0.838	0.0415
24	0.198	2,898	0.876	0.0092
25	0.269	351	0.567	0

where $x_y = (1_{y=-1} - 1_{y=1}, 1_{y=0} - 1_{y=1}), x_{yz} = x_y \otimes x_z, y, z \in \{a, b, c\}, \otimes$ denotes the Kronecker product, and 1_c is the indicator function that gives 1 if the condition c is true and 0 otherwise. The number of columns of $X_\mathcal{D}$ i.e., the number of degrees of freedom of the model is $p = 1 + 3 \cdot 2 + 3 \cdot 4 = 19$. The least-square estimate $\hat{\beta}$ of β is obtained as

$$\hat{\beta} = (X_\mathcal{D}^t X_\mathcal{D})^{-1} X_\mathcal{D}^t \mathbf{y}. \tag{9}$$

Using our algorithm we obtain a sequence of robust fractions $\mathcal{F}_{27} \equiv \mathcal{D} \supset \cdots \supset \mathcal{F}_{19}$. We focus on $\mathcal{F}_{25}, \mathcal{F}_{24}$, and $\mathcal{F}_{23}$ which are the fractions of $\mathcal{D}$ which correspond to the loss of 2, 3, and 4 runs, respectively. More specifically $\mathcal{F}_{25} = \overline{\mathcal{D}}_{\{18,1\}}$, $\mathcal{F}_{24} = \overline{\mathcal{D}}_{\{18,1,23\}}$, and $\mathcal{F}_{23} = \overline{\mathcal{D}}_{\{18,1,23,25\}}$, where $\overline{\mathcal{D}}_I$ means the full factorial design without the *caseno.* in the set I, where the *caseno.* of each run of the full factorial $\mathcal{D}$ is defined in Table 3.

Using $\mathcal{F}_x$ we compute the corresponding estimate $\hat{\beta}_x$ of β, $x = 23, 24, 25$. We measure how far $\hat{\beta}_x$ is from $\hat{\beta}$ (the estimate of β obtained using the full factorial design $\mathcal{D}$) using the mean error e_x defined as

$$e_x = \frac{1}{p} \sum_{i=1}^{p} \left| \frac{(\hat{\beta}_x)_i - (\hat{\beta})_i}{(\hat{\beta})_i} \right| = \frac{1}{19} \sum_{i=1}^{19} \left| \frac{(\hat{\beta}_x)_i - (\hat{\beta})_i}{(\hat{\beta})_i} \right|$$

where $(v)_i$ denotes the i-th component of the vector v.

For $x = 23, 24, 25$ we build all the fractions $\mathcal{A}_x^k$ of size x of $\mathcal{D}$, $k = 1, \ldots, \binom{27}{x}$. For each fraction $\mathcal{A}_x^k$ we compute $\hat{\beta}_x^k$, the least-square estimates of β using $\mathcal{A}_x^k$, and the corresponding errors e_x^k. It is worth noting (see Table 4) that

- the errors e_x are less than the median of e_x^k for all the x and for $x = 23, 24$ e_x are close to the 15-th percentiles of e_x^k, meaning that approximately 85% of fractions $\mathcal{A}_x^k$ provide mean errors e_x^k greater than e_x;
- for $x = 23$, and $x = 24$ there are $\overline{E}_{23} = 729$, and $\overline{E}_{24} = 27$ non-estimable fractions, and $E_{23} = 16{,}821$, and $E_{24} = 2{,}898$ estimable fractions, respectively (a fraction $\mathcal{F}$ is non-estimable when $\det(X_\mathcal{F}^t X_\mathcal{F}) = 0$). It follows that the probability that a randomly chosen fraction is non-estimable is $729/17{,}550 \approx 4.15\%$ for $x = 23$ and $27/2925 \approx 0.92\%$ for $x = 24$.

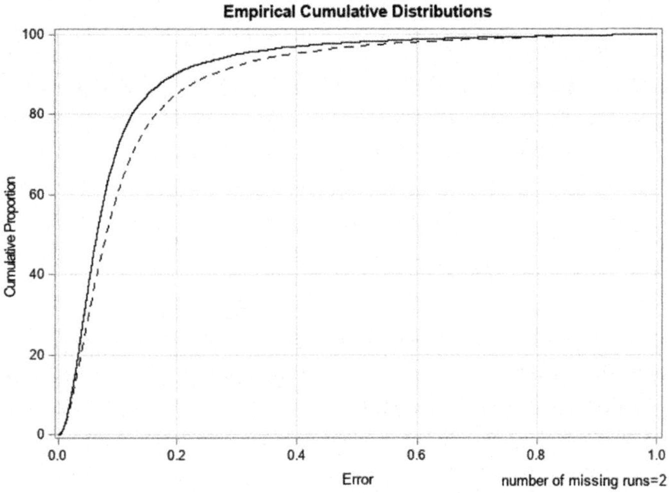

Fig. 3 Empirical distribution functions of $\{e_{i,25} : i = 1, \ldots, 10{,}000\}$ (solid line) and $\{\tilde{e}_{i,25} : i = 1, \ldots, 10{,}000\}$ (dashed line)

We run a simulation study for assessing the stability of the results. We considered $N = 10{,}000$ vectors $\mathbf{y_i}$ generated as

$$\mathbf{y_i} = X_{\mathcal{D}}\boldsymbol{\beta} + \boldsymbol{\varepsilon}_i,\ i = 1, \ldots, N$$

where $\boldsymbol{\beta}$ is taken equal to $\hat{\beta}$ as defined in Eq. (9), $\boldsymbol{\varepsilon}_i$ is a vector of independent standard normally distributed random variables, and $\boldsymbol{\varepsilon}_i$ and $\boldsymbol{\varepsilon}_j$ are independent when $i \neq j$. For each simulation i we compute $e_{i,x}$, the error obtained using the robust fraction $\mathcal{F}_x$ and $\tilde{e}_{i,x}$ the error obtained using a randomly chosen fraction of $\mathcal{D}$ of size $x, x = 23, 24, 25$. The empirical distribution functions of $\{e_{i,25} : i = 1, \ldots, 10{,}000\}$ and $\{\tilde{e}_{i,25} : i = 1, \ldots, 10{,}000\}$ computed using $x = 25$ runs (that is two runs have been lost) are reported in Fig. 3. It is evident that the errors obtained using the robust fraction $\mathcal{F}_x$ are smaller than those obtained using a random fraction of size x of $\mathcal{D}$ (e.g. the median of $\{e_{i,25} : i = 1, \ldots, 10{,}000\}$ is 0.0647 and the median of $\{\tilde{e}_{i,25} : i = 1, \ldots, 10{,}000\}$ is 0.0854). The goodness of the results is also confirmed for $x = 23$ and 24.

5 Final Remarks

The main result of this contribution is an algorithm for organizing the n runs of a given fraction $\mathcal{F}_n$ in such a way that, if for some reasons k of the n runs are lost, the remaining $n - k$ runs constitute a *robust* design, $1 \leq k \leq n - p$. As shown in the simulation study, the algorithm can provide very good designs in terms of robustness for all $k \in \{1, \ldots, n - p\}$. This means that k does not need to be defined at the design stage (it would have been extremely difficult to make a hypothesis on the number k of runs that could be lost before starting the execution of the experiments).

The algorithm can be used with any type of initial design. The starting design $\mathcal{F}_n$ can be an orthogonal fractional factorial design or a D-optimal design or any user-defined design. It is worth noting that the algorithm can work in many practical siyuations. The reason is that the circuits needed are those of the matrix $A_{\mathcal{F}_n}$ which has dimension $p \times n$ and, usually in the applications, both the number of parameters p and the size of the starting fraction $\mathcal{F}_n$ are not large.

One of the requirements is that the matrix $A_{\mathcal{F}_n}$ must have integer values. This is always the case for models with qualitative factors. Some preliminary work is needed for models with quantitative variables. For a model matrix with real numbers it is possible to build an approximate version of it with rational entries. The approximation can be built as accurately as required since the set of rational numbers $\mathbb{Q}$ is dense in the set of real number $\mathbb{R}$. Finally the model matrix can be transformed into a matrix with integer values simply by multiplying the rational matrix by a suitable integer constant. In some cases, as shown in the second example of Sect. 4, a D-optimal design used as a starting design for the algorithm contains only points with integer entries, and from the combinatorial point of view the problem reduces immediately to the qualitative case. Although the approximation of a real design matrix with a rational one is outside the scope of the present work, some first experiments in this direction are promising. For instance, we have considered the design matrix with non-rational entries presented in [9], p. 1674. Using 1, 2, or 3 decimal places to approximate the real entries, we obtain in all cases the same structure of the circuit basis.

Acknowledgements The authors are members of INdAM-GNAMPA.

References

1. 4ti2 team: 4ti2—a software package for algebraic, geometric and combinatorial problems on linear spaces (2018). https://4ti2.github.io
2. Atkinson, A., Donev, A., Tobias, R.: Optimum experimental designs, with SAS, vol. 34. Oxford University Press, Oxford, UK (2007)
3. Bailey, R.A.: Design of Comparative Experiments, vol. 25. Cambridge University Press, Cambridge, UK (2008)
4. Butler, N.A., Ramos, V.M.: Optimal additions to and deletions from two-level orthogonal arrays. J. R. Stat. Soc. Ser. B. Stat. Methodol. **69**(1), 51–61 (2007). https://doi.org/10.1111/j.1467-9868.2007.00576.x
5. Dey, A.: Robustness of block designs against missing data. Stat. Sin. **3**(1), 219–231 (1993)
6. Dey, A., Mukerjee, R.: Fractional Factorial Plans. Wiley, New York (2009)
7. Fedorov, V.V.: Theory of Optimal Experiments. Elsevier, United States (2013)
8. Fontana, R., Rapallo, F.: On the aberrations of mixed level orthogonal arrays with removed runs. Stat. Pap. **60**(2), 479–493 (2019). https://doi.org/10.1007/s00362-018-01069-5

9. Fontana, R., Rapallo, F.: Combinatorial analysis of factorial designs with ordered factors. In: Balzanella, A., Bini, M., Cavicchia, C., Verde, R. (eds.) SIS 2022 Book of the Short Papers, pp. 1670–1675. Pearson (2022)
10. Fontana, R., Rapallo, F., Rogantin, M.P.: A characterization of saturated designs for factorial experiments. J. Stat. Plan. Inference **147**, 205–211 (2014). https://doi.org/10.1016/j.jspi.2013.10.011
11. Fontana, R., Rapallo, F., Wynn, H.P.: Circuits for robust designs. Stat. Pap. **63**(5), 1537–1560 (2022). https://doi.org/10.1007/s00362-021-01285-6
12. Ghosh, S.: On robustness of designs against incomplete data. Sankhya: Indian J. Stat. Ser. B (1960–2002) **40**(3/4), 204–208 (1979)
13. Ghosh, S.: Robustness of bibd against the unavailability of data. J. Stat. Plan. Inference **6**(1), 29–32 (1982). https://doi.org/10.1016/0378-3758(82)90053-2
14. Hedayat, A.S., Sloane, N.J.A., Stufken, J.: Orthogonal Arrays: Theory and Applications. Springer, New York (2012)
15. Mukerjee, R., Wu, C.F.J.: A Modern Theory of Factorial Design. Springer, New York (2007)
16. Ohsugi, H., Hibi, T.: Toric ideals and their circuits. J. Commut. Algebra **5**(2), 309–322 (2013). https://doi.org/10.1216/JCA-2013-5-2-309
17. Oikawa, T., Oka, T.: A new technique for approximating the stress in pad-type nozzles attached to a spherical shell. J. Press. Vessel Technol. **109**(2), 188–192 (1987). https://doi.org/10.1115/1.3264894
18. Pukelsheim, F.: Optimal Design of Experiments. SIAM, New York (2006)
19. Street, D.J., Bird, E.M.: D-optimal orthogonal array minus t run designs. J. Stat. Theory Pract. **12**(3), 575–594 (2018). https://doi.org/10.1080/15598608.2018.1441081
20. Sturmfels, B.: Gröbner bases and convex polytopes. University Lecture Series, vol. 8. American Mathematical Society, Providence, RI (1996)

Bayesian Optimization for Cloud Resource Management Through Machine Learning

Bruno Guindani, Danilo Ardagna, and Alessandra Guglielmi

Abstract Optimal cloud configuration of recurring big data analytic jobs is a relevant and challenging task in the industry. To this end, Bayesian Optimization is a promising method for efficiently finding optimal or near-optimal configurations for such applications, which are often executed in the cloud. On the other hand, Machine Learning methods can provide useful knowledge about the application at hand thanks to the quality of their estimations. In this paper, we propose a hybrid algorithm that is based on Bayesian Optimization and integrates elements from Machine Learning techniques to tackle time-constrained optimization problems in a cloud computing setting. We consider a recurring job scenario, where unfeasible points are to be avoided by all means, as they are a waste of resources. In such a context, Machine Learning helps to convey valuable information about the violation of constraints. Experiments on big data applications have shown that our algorithm significantly reduces the amount of unfeasible executions with respect to a pure constrained BO approach.

Keywords Acquisition function · Cloud computing · Gaussian process

1 Introduction

Big data analytics are employed in several industrial fields to allow organizations and companies to make better decisions. The most suitable execution environment of big data analytic applications is a cluster of virtual machines (VMs) which allows the adjustment of the allocated resources (CPU, memory, disk, network) to match

B. Guindani (✉) · D. Ardagna · A. Guglielmi
Politecnico di Milano, Piazza Leonardo da Vinci 32, 20133 Milano, Italy
e-mail: bruno.guindani@polimi.it

D. Ardagna
e-mail: danilo.ardagna@polimi.it

A. Guglielmi
e-mail: alessandra.guglielmi@polimi.it

M. Bini et al. (eds.), *Advanced Methods in Statistics, Data Science and Related Applications*, Springer Proceedings in Mathematics & Statistics 467,
https://doi.org/10.1007/978-3-031-65699-6_24

the application current needs. Choosing the right cloud configuration to minimize execution times and reduce costs is essential to service quality and business competitiveness. This is especially crucial for recurring cloud jobs, i.e., applications that need to be executed multiple times, which constitute a significant part of the total amount of analytic jobs running in the cloud [2, 20]. However, variability in the progress and resource requirements of analytic jobs imply that choosing the best configuration for a broad spectrum of applications is a challenging process [2].

Bayesian Optimization (BO) has recently gained notoriety as a powerful tool to solve global optimization problems in which expensive black-box functions are involved; see the recent paper [15] or the popular tutorial paper [5]. BO is a sequential design strategy that requires few steps to get sufficiently close to the true optimum, while requiring no derivative information on the optimized function. Most commonly, it is initialized by choosing and evaluating a small handful of starting points, then fitting a Gaussian process (GP) on these points. This approach can be interpreted as assuming a prior distribution for the unknown infinite-dimensional parameter f, i.e., the function to be optimized. The posterior distribution of the fitted GP provides an estimate of both the function value at each point and the uncertainty around the estimate. BO then iteratively chooses new points at which to evaluate the function in a such a way to balance exploration (high uncertainty) and exploitation (best estimated function value); see, for instance [11]. The topic of constrained BO has also received attention in the literature [8, 11].

The goal of this work is to integrate Bayesian Optimization algorithms with Machine Learning (ML) techniques in the context of cloud computing optimization for recurring jobs. The former techniques have proven to be successful [2, 16] in exploring and finding optimal or near-optimal cloud configurations after a small amount of exploratory runs. On the other hand, the latter can provide useful information to be incorporated into the BO mechanism in several ways to improve its performance, for instance in the form of cheap estimates of target quantities to guide the exploration process.

This work builds on previous results found in [2], in which the *CherryPick* system has been successfully applied to benchmark applications on cloud computing frameworks such as Apache Spark. This system exploits pure constrained BO to find optimal cloud configurations. Our work is motivated by the belief that ML models can lend their estimation capabilities to BO to further improve its effectiveness. This topic has been explored in [13], which examines the performance of several ML models in carrying out prediction of execution times of Spark cloud jobs with different types of workloads. The hybrid BO algorithm we propose here is promising since it shows the usefulness of ML in the context of cloud computing configuration.

The setup of this paper is as follows. Section 2 describes the mathematical formulation of the problem, Sect. 3 presents our proposal of a BO algorithm, while Sect. 4 collects a few experimental results. We conclude the paper with a short discussion in Sect. 5.

2 Background and Mathematical Formulation

The goal of Bayesian Optimization is to estimate, using an iterative method, the minimum of a given function f, called objective function, by using as few iterations as possible. We focus here on the minimum of the function f, though we could consider the maximum with similar arguments. Specifically, we want to find $\widehat{x}$ such that

$$\widehat{x} = \arg\min_{x \in \mathcal{A}} f(x).$$

Strong assumptions on f or on the minimization domain $\mathcal{A}$ are not required, and BO algorithms are derivative-free, i.e., they do not require any knowledge about the derivatives of f. For these reasons, BO is often used to optimize expensive black-box objective functions (see [3]), that is, functions for which little to no information is available and whose evaluation has significant time, resource, and/or monetary costs.

We consider the mathematical formulation for our *constrained global optimization* problem similarly to [2]. Let $x \in \mathcal{A}$ denote the d-dimensional vector representing a configuration for the cloud job, including information such as the number of cores used for the job, with $\mathcal{A} \subset \mathbb{R}^d$ being the domain of all feasible configurations. The *objective function* to be minimized is the total cost $f(x) = P(x)T(x)$, where $T(x)$ is the unknown execution time and $P(x)$ is the price per unit (it is a known, deterministic function). We also assume the constraint that $T(x) \leq T_{max}$, where T_{max} is a given threshold. By making this model explicit in f, we obtain:

$$\begin{aligned} \min_{x \in \mathcal{A}} \; & f(x) = P(x)T(x) \\ \text{s.t.} \; & f(x) \leq P(x)\, T_{max}. \end{aligned} \tag{1}$$

In this paper, we assume the deterministic price function $P(x)$ as being proportional to the number of virtual machines or cores used by the application job, which is always included in the cloud configuration vector x. Other choices of the price function are possible.

The key idea of BO comes from the Bayesian approach to statistics, in which values taken by f are treated as random variables, and a *prior distribution* represents the a-priori information on the modeled phenomenon—in the case of BO, information on the location of the minimum. The prior distribution is then iteratively updated with information coming from the observed data, obtaining the *posterior distribution*. For the rest of the paper, we assume that observed data, i.e., the evaluations of f, are noise-free. This is justified by the analysis in [13] on the data considered for validation. In a more general scenario, data can be assumed to have independent, normally distributed additive noise with variance η^2. In this context, the parameter η^2 is usually estimated in a preliminary step.

In the BO framework, the Gaussian process (GP) is the preferred choice for the prior for f. This implies that for any $x \in \mathcal{A}$,

$$f(x) \sim \pi_x(\cdot) = \mathcal{N}(\mu_0(x), \sigma_0^2(x, x)).$$

The functions $\mu_0(\cdot)$ and $\sigma_0^2(\cdot, \cdot)$ are called mean and kernel functions, respectively, and are the GP model hyperparameters. These functions serve as the "initial guesses" on values of $f(\cdot)$ and on their uncertainty, and they will be iteratively updated with observed values. A constant mean function $\mu_0(\cdot) \equiv \mu_0$ is often adopted, whereas the choice of the kernel is more delicate, since it influences the smoothness of the process. Commonly used kernels include the squared exponential or Radial Basis Function and the Matérn kernel [19]. The former gives the GP an excessively large degree of smoothness, which is unrealistic in many practical scenarios. Therefore, in this work, we assume $\mu_0(\cdot) \equiv \mu_0$ and we use the Matérn kernel with smoothness parameter $\nu = 5/2$ (see [5]):

$$\sigma_0^2(x, x') := \frac{1}{2^{3/2}\Gamma(5/2)} \left(\sqrt{5}\|x - x'\|\right)^{5/2} K_{5/2}\left(\sqrt{5}\|x - x'\|\right). \tag{2}$$

In Eq. (2), $\|\cdot\|$ denotes the Euclidean norm, while K is the modified Bessel function of the second type [1]. As usual, $\Gamma(\cdot)$ is the gamma function.

Having observed values $H_n = \{(x_1, f(x_1)), \ldots, (x_n, f(x_n))\}$ of the objective function, one computes the posterior distribution of $f(x)$, for any x. This distribution is Gaussian as well, with posterior mean $\mu_n(\cdot)$ and variance $\sigma_n^2(\cdot)$, i.e.:

$$f(x)|H_n \sim \pi_x(\cdot|H_n) = \mathcal{N}(\mu_n(x), \sigma_n^2(x)),$$

and is computed by well-known properties of GPs (see [5]) as

$$\mu_n(x) = \mu_0(x) + \sigma_0^2(x, x_{1:n})^T \sigma_0^2(x_{1:n}, x_{1:n})^{-1}\big(f(x_{1:n}) - \mu_0(x_{1:n})\big), \tag{3}$$

$$\sigma_n^2(x) = \sigma_0^2(x, x) - \sigma_0^2(x, x_{1:n})^T \sigma_0^2(x_{1:n}, x_{1:n})^{-1}\sigma_0^2(x, x_{1:n}). \tag{4}$$

In Eqs. (3) and (4), $\sigma_0^2(x, x_{1:n})$ indicates the column vector of values of the $\sigma_0^2(\cdot, \cdot)$ function applied to pairs $(x, x_1), \ldots, (x, x_n)$, and similarly for $f(x_{1:n})$ and $\mu_0(x_{1:n})$. Analogously, $\sigma_0^2(x_{1:n}, x_{1:n})$ is the matrix of values of $\sigma_0^2(x_i, x_j)$ with $i, j = 1, \ldots, n$.

Bayesian Optimization is an iterative algorithm that obtains a new observation at each iteration by solving a proxy problem—the maximization of the *acquisition function* $g(x)$, which depends on the fitted GP model and measures the utility of evaluating the objective function at a given configuration x. This function is optimized at each round of the iterative algorithm, instead of directly optimizing the objective function itself, since it is available in closed form and inexpensive to evaluate. The acquisition function must strike a delicate balance—the exploration-exploitation trade-off. On the one hand, there are points to which large uncertainty is attached, for instance because they lie in a region of the domain which has not been explored yet. Choosing such points to evaluate the objective $f(\cdot)$ is appealing, especially early on in the optimization procedure, since this would allow a large decrease of the uncertainty on the position of the optimum. On the other hand, the algorithm does seek to find the optimum of the objective function, therefore it should also choose to evaluate points which most likely (according to the GP model) give small values of $f(\cdot)$. This is done by exploiting the information already available on the location of the optimum, especially in the late iterations of the algorithm. Convergence of the BO algorithm is guaranteed under mild conditions [3, 14].

Algorithm 1 Generic Bayesian Optimization algorithm

1: choose n_0 initial points
2: evaluate $f(\cdot)$ in the initial points, add evaluations to history H
3: **for** iterations $n = 1 : N$ **do**
4: update the current posterior distribution of the GP model with data in H
5: find point x_{n+1} which maximizes the acquisition function $g(\cdot)$ under the current model
6: evaluate $f(x_{n+1})$, add performed evaluation to H
7: **end for**
8: **return** estimated optimum $\widehat{x}$

The BO procedure is summarized in Algorithm 1. In step 1, a small number n_0 of initial points (i.e., cloud configurations in our specific application) are selected, usually 3–10, in order to initialize the algorithm. These points should be chosen so as to cover the maximum domain area possible, for instance using a Latin hypercube design [10]. We then evaluate these points (in practical terms, this means executing the application using these configurations), and we record the points and their evaluations (step 2). After the initialization phase, we enter the algorithm loop, where we update the posterior distribution (step 4), choose the next point x_{n+1} by maximizing the acquisition function (step 5), and evaluate it (step 6). The algorithm stops when the iteration budget N runs out.

Figure 1 shows how BO works. Specifically, in the top panel, the objective function to be minimized (the solid line) is estimated by the posterior mean function (the dashed line) and 95% credible interval (the highlighted area in the background) of the associated Gaussian process, after evaluating 3 points (the large dots). Both estimates are updated whenever a new observation is received, therefore getting more and more accurate over time. The vertical bell curve represents the Gaussian distribution which is assumed for each point x of the optimization domain; we recall that both the mean and variance of this distribution depend on x. Instead, the bottom panel shows the acquisition function given the current posterior distribution. Its values are lower in points which have already been sampled, because the utility of repeating an observation is generally smaller than evaluating a brand new point. The crossed point indicates the maximum of the acquisition function, i.e., the next point which will be evaluated. Note that such point is at the center of a region which has both large expected value and large variance, making it appealing in the exploration-exploitation trade-off.

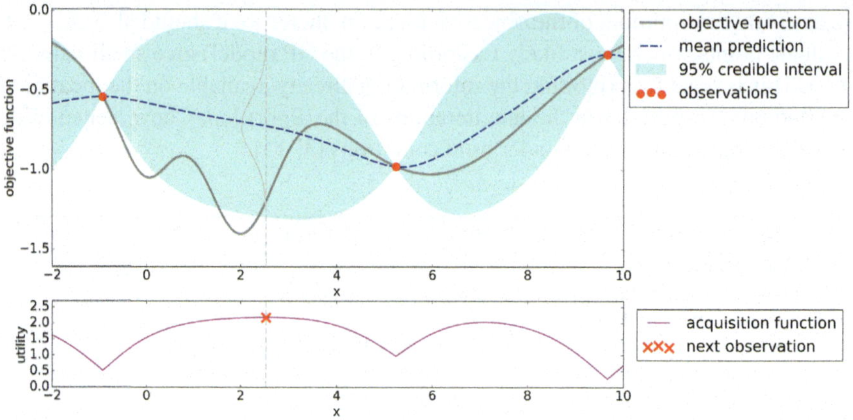

Fig. 1 Bayesian Optimization after 3 iterations. Top panel: the objective function and its Bayesian estimate. Bottom panel: acquisition function

In this paper, we compare different acquisition functions. All of them are based on two popular acquisition functions, which have been shown to be effective in the literature [6, 7, 18]. The *Expected Improvement* (EI) over the best value f_n^* found by the optimization process so far is defined as

$$EI_n(x) := \mathbb{E}_{\pi_x(\cdot|H_n)}[\max(f_n^* - f(x), 0)] \quad \text{with} \quad f_n^* = \min_{i \leq n} f(x_i).$$

The expectation is taken under the current posterior distribution $\pi(\cdot|H_n)$ of $f(x)$, given history H_n. We consider a generalization of EI to the constrained optimization setting—the *Expected Improvement with Constraints* (EIC) acquisition function [10, 17], which accounts for the probability of a point of satisfying the constraints (see Eq. (1)):

$$EIC_n(x) := EI_n(x) \cdot \mathbb{P}_{\pi_x(\cdot|H_n)}\big(f(x) \leq P(x)\, T_{max}\big). \tag{5}$$

Note that the second factor in Eq. (5), which is equal to the posterior probability of respecting the constraint, is used as a correction factor for the basic Expected Improvement acquisition function.

3 Our Hybrid Algorithm

As mentioned in the Introduction, ML estimation techniques can provide useful information to be incorporated into the Bayesian Optimization mechanism to improve its performance, for instance in the form of cheap estimates of target quantities to guide the exploration process.

Algorithm 2 Proposed algorithm

1: choose n_0 initial points
2: evaluate $f(\cdot)$ in the initial points, add evaluations to history H
3: **for** iterations $n = 1 : N$ **do**
4:     update the current posterior distribution of the GP model with data in H
5:     train model $\widehat{T}(\cdot)$ with data in H, to be used in $g(\cdot)$
6:     find point x_{n+1} which maximizes the acquisition function $g(\cdot)$ under the current model
7:     evaluate $f(x_{n+1})$, add performed evaluation to H
8:     update memory queue with x_{n+1}
9:     **if** stopping criteria are met **then**
10:         terminate the algorithm
11:     **end if**
12: **end for**
13: **return** estimated optimum $\widehat{x}$

We first summarize the complete procedure in Algorithm 2. Our algorithm is based on pure BO, but it integrates elements coming from ML techniques. We use a first-in-first-out memory queue for discrete features to prevent exploration of already visited values, similarly to the taboo search meta-heuristic methods [4]. In this memory queue, we save the last q points visited by the algorithm. Configurations currently in the queue are excluded from being selected again until they have shifted out of the queue, i.e., after q iterations. At each round, we train an ML model $\widehat{T}(\cdot)$ (step 5) with all points evaluated so far by the algorithm. Then, similarly to regular BO, we maximize the acquisition function of choice, which now incorporates the trained ML model. Note that we have used ML language (i.e., *we train an ML model* $\widehat{T}(\cdot)$) here, but these *ML models* consist of statistical estimation methods which have become very popular in the ML community. After that, we evaluate the newly chosen configuration (step 7) as usual, and we update the aforementioned memory queue (step 8). The algorithm continues until the evaluated execution time at the current iteration is sufficiently close to the time threshold: $T(x_n) \in [\alpha\, T_{max},\ T_{max}]$, with $\alpha \in (0, 1)$ (step 10). This is because our goal is to obtain a configuration that is compliant with the time threshold, but also uses as few resources as possible. Generally speaking, using more resources results in a lower execution time—meaning that a time which is just under the threshold likely consumes the least amount of resources for that configuration to be feasible. After termination of this algorithm, it is likely that we have found the true optimal configuration. Afterwards, we perform subsequent executions using such optimal or near-optimal configuration (recall that we are dealing with recurring jobs, whose periodic execution does not stop with the termination of the optimization procedure).

In the application of interest, our goal is ultimately to find optimal (or near-optimal) configurations which are also feasible, i.e., points x s.t. $T(x) \leq T_{max}$. Unfeasible points are to be avoided by all means, since they represent a waste of resources in a recurring job setting, providing additional unnecessary costs. Integrating ML models into the acquisition function is crucial in assessing the feasibility of the points under consideration. Hence, ML methods can prove to be useful additions to the acquisition function in a setting where evaluating the objective function $f(\cdot)$ is expensive. Indeed, a cheap estimation of values of $f(\cdot)$ can compensate for the scarcity of direct information on them. In our case, ML models can be used to convey valuable information about the violation of constraints. Specifically, we use ML models to compute $\widehat{T}(\cdot)$ and use this estimate to correct the acquisition function $EIC(\cdot)$.

We therefore propose the following acquisition functions:

- $g_A(x) = EIC(x)$: the original Expected Improvement with Constraints acquisition function [10, 17], used by *CherryPick* [2], which we use as baseline;
- $g_B(x) = g_A(x) \cdot \exp(-k\,\widehat{T}(x))$, the latter term being a [0, 1]-valued weight for g_A. This correction factor is called nascent minima distribution function [12], and it serves the purpose of turning $\widehat{T}(x)$ into an acquisition-like function. More precisely, the correction factor takes on values close to 1 if the estimated execution time $\widehat{T}(x)$ is small, thus making x a desirable point, while it is closer to 0 if such estimate is large;
- $g_C(x) = g_A(x) \cdot I_{\{\widehat{T}(x) \leq T_{max}\}}$, with I being the indicator function: the search is prevented in areas where the current estimated execution time violates the threshold T_{max}. Here we are using the model $\widehat{T}(x)$ to estimate the current feasible domain;
- $g_D(x) = g_A(x) \cdot \exp(-k\,\widehat{T}(x)) \cdot I_{\{\widehat{T}(x) \leq T_{max}\}}$: the combination of cases B and C.

Note that the original definition of the nascent minima distribution functions (used in variants B and D) includes the normalization constant $1/C_k$ as a multiplicative factor, with $C_k = \int_{\mathcal{A}} \exp(-k\,\widehat{T}(w))\,dw > 0$, as described in [12]. However, this value is independent of x, therefore we can omit it when maximizing the acquisition function in x.

Note that at each step of the algorithm, we *train* the ML regression model $\widehat{T}(\cdot)$, e.g., we fit the model with the data we have collected so far, obtaining a model which can estimate $T(x)$ for any $x \in \mathcal{A}$.

4 Experiments

We present experimental results using the techniques we have discussed in Sect. 3. We test variants B, C, and D of the algorithm, as well as pure constrained BO (represented by variant A), on two different big data applications, run with the Apache Spark analytics engine. The first is Query26 application from the TPC-DS industry benchmark, which we execute with input data size equal to 250 GB and time threshold equal to 150 seconds (s). The second is an application performing K-means clustering algorithm on a dataset with 15 million rows, under a time threshold equal to 330 s. In both cases, we optimize the total cost in Eq. (1) on the number of cores x, i.e., the minimization domain is uni-dimensional. The same three fixed initial points were used for all variants ($n_0 = 3$).

Figure 2 shows the comparison of pure constrained BO (top row) with our three variants (other rows) at each algorithm iteration, when applied to Query26. In the left

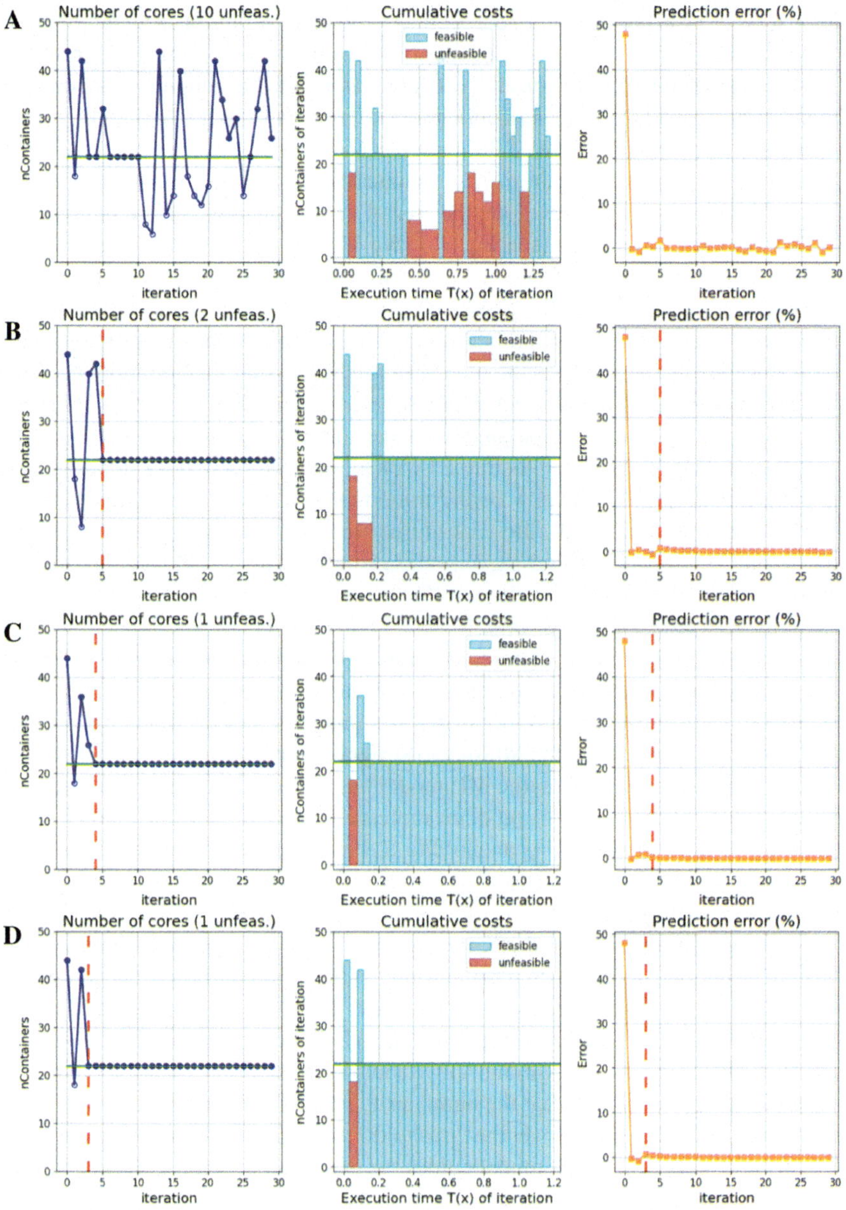

Fig. 2 Comparison of pure constrained BO (top row) with variants B, C, and D at each algorithm iteration for Query26. Left panel: number of cores x visited by the algorithm; center panel: cost of each configuration; right panel: percentage error of the ML model

panel, we represent the number of cores chosen by the algorithm at each iteration. The solid horizontal line is the true optimum of the constrained optimization problem. The vertical dashed line indicates the execution at which the stopping criterion kicks in, and after which the recurring application at hand sticks to the best configuration found by the algorithm. In the center panel, each bar represents a single execution. The width of a bar represents the execution time of the job, while its height represents the number of cores used. Therefore, the area of a bar is proportional to the cost $f(x)$ for that particular execution (see Eq. (1)). We highlight the bars corresponding to feasible and unfeasible configurations in different shades. The signed percentage errors of the ML model for the execution time are displayed in the right panel. In this case, we have used the Ridge regression method to estimate the execution time $T(x)$, since [13] show that it is the most accurate regression method in this context, with a mean absolute error smaller than 5%. In particular, $\widehat{T}(x) = x^\top \widehat{\beta}$, where $\widehat{\beta}$ is the estimate of the regression parameters under the Ridge regression method.

At each iteration, the model is trained with all data from previous iterations, and the error is evaluated on the new configuration chosen by the algorithm. We see in the right panel of Fig. 2 that after one initial iteration with a large error, our ML models quickly converge to errors very close to zero. This initial spike in the error can be explained by the large distance between the n_0 initialized points and the first point selected by the algorithm. Indeed, in this particular experiment, the initial points have a small number of cores. Therefore the algorithm, which is still in the exploration phase, selects a configuration with a large number of cores (as seen in the leftmost spike of the left panel), because it lies in a region of the domain that is still unexplored. For this reason, the ML model struggles to perform accurate estimations in the first iteration. After that, the ML model has accumulated enough information and consequently it is able to produce more accurate estimates, as shown in the following iterations in the right panel of Fig. 2. Finally, the termination criterion (see the vertical dotted line) correctly assesses the optimality of the configuration, and stops the exploration phase.

We show a similar plot for the K-means application in Fig. 3, for which we used the same ML model as Query26. Furthermore, Table 1 summarizes a few noteworthy average metrics from multiple runs of the algorithm, for both Query26 and K-means. It is clear that each of our algorithm variants (B–D) outperforms the constrained vanilla BO technique in [2] (variant A) with respect to multiple metrics. First of all, in Figs. 2 and 3, we can see that the number of "wasted" unfeasible executions, i.e., ones whose execution time is over the threshold T_{max}, decreases from 10 to 22 to just 1 or 2. In variants B–D, the ML model guides the exploration process towards feasible points, by excluding those that are likely to be unfeasible. These variants reach termination before the 6th iteration, whereas variant A does even struggle to reach convergence. On average (see Table 1), the total amount of unfeasible executions drops by two to three times. Furthermore, we measure the ratio of the total cost of unfeasible executions over the total cost of all executions (i.e., over the entire iteration budget), displayed as "ratio of unfeasible costs" in Table 1. With respect to variant A, this ratio also decreases by two to three times. Finally, the average cost of

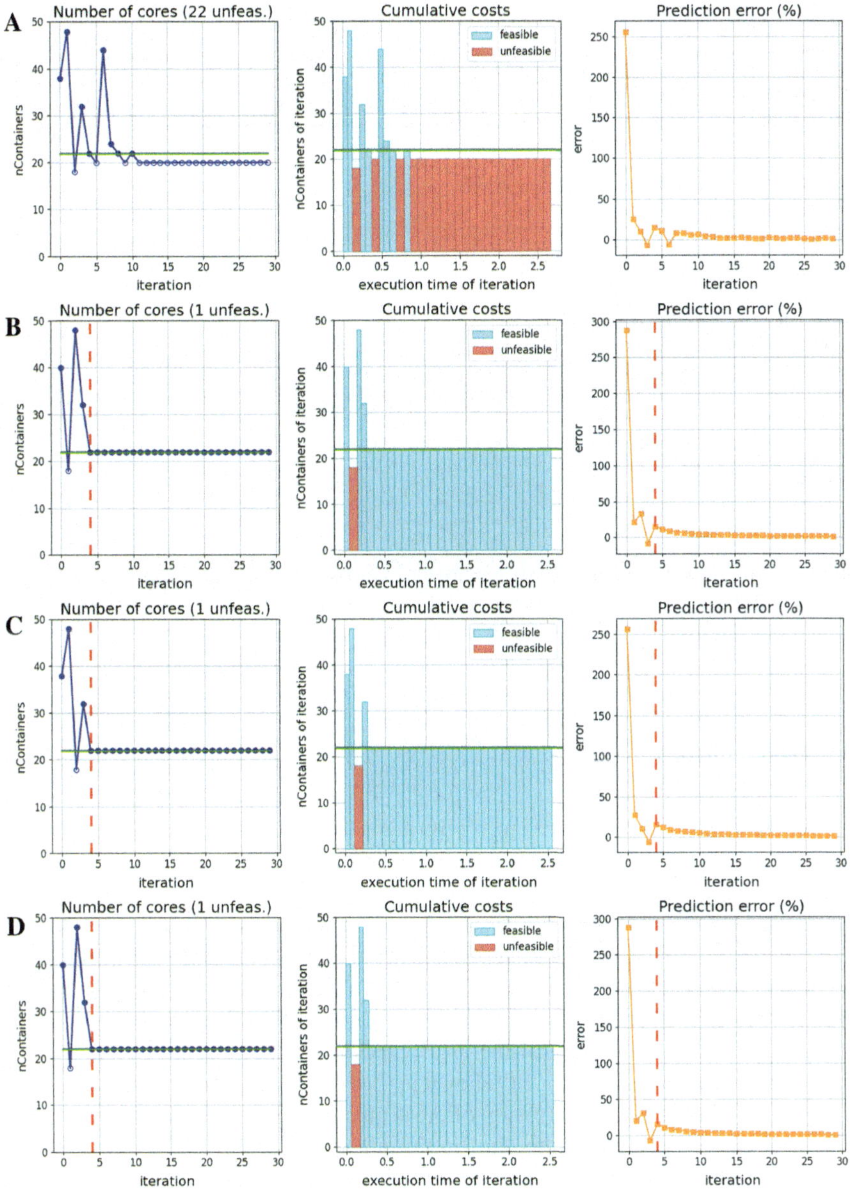

Fig. 3 Comparison of pure constrained BO (top row) with variants B, C, and D at each algorithm iteration for K-means. Left panel: number of cores x visited by the algorithm; center panel: cost of each configuration; right panel: percentage error of the ML model

Table 1 Measured metrics for variants A–D of the algorithm

Scenario	Var. A	**Var. B**	**Var. C**	**Var. D**
Query26				
Unfeasible executions	11.23	4.34	4.11	4.31
Ratio of unfeasible costs	0.36	0.14	0.13	0.14
Norm. mean feasible cost	1.00	0.87	0.87	0.88
K-means				
Unfeasible executions	10.52	5.74	5.93	5.96
Ratio of unfeasible costs	0.34	0.20	0.21	0.21
Norm. mean feasible cost	1.00	1.00	1.01	1.00

the feasible configuration (which we normalize over the cost of variant A in Table 1 for the sake of clarity) decreases by about 13% in Query26, while always remaining competitive with variant A in the K-means case.

5 Discussion

In this paper, we have presented a novel BO algorithm integrated with ML techniques, to find the optimal configuration for a recurring job running in the cloud, under a given time threshold. ML models help in the crucial task of recognizing configurations which lead to unfeasible executions, saving on unnecessary additional costs. Indeed, experiments on big data applications show that our algorithm significantly reduces the amount of unfeasible executions with respect to a pure constrained BO approach, as well as reducing the average cost of the configuration. Overall, each of our algorithm variants outperforms the state-of-the-art BO technique used as benchmark.

Convergence guarantees for BO shown in [3, 14] may no longer hold when introducing ML models in the expression of the acquisition function. However, we stress that in the context of cloud computing optimization, the reduction of exploration iterations is prioritized over the quality of the solution, which can also be near-optimal rather than optimal. Therefore, the issue of convergence is not relevant in this context. In such a setting, if the given iteration budget were limited enough, a traditional BO algorithm would not be able to reach convergence in the proper sense either. Besides this remark, note that our algorithm does have a sensible termination criterion, which assesses whether the current configuration is likely to be near-optimal, therefore taking the role of the former convergence guarantees.

A preliminary version of this manuscript has appeared as [9].

Acknowledgements The European Commission has partially funded this work under the Horizon 2020 Grant Agreement number 956137 LIGATE: LIgand Generator and portable drug discovery platform AT Exascale, as part of the EuroHPC Joint Undertaking.

References

1. Abramowitz, M., Stegun, I.A.: Handbook of Mathematical Functions with Formulas, Graphs, and Mathematical Tables, vol. 55. US Government Printing Office (1964)
2. Alipourfard, O., Liu, H., Chen, J., Venkataraman, S., Yu, M., Zhang, M.: Cherrypick: adaptively unearthing the best cloud configurations for big data analytics. In: USENIX Symposium on Networked Systems Design and Implementation (2017)
3. Brochu, E., Cora, V.M., De Freitas, N.: A tutorial on Bayesian optimization of expensive cost functions, with application to active user modeling and hierarchical reinforcement learning (2010). arXiv:1012.2599
4. Cvijović, D., Klinowski, J.: Taboo search: an approach to the multiple-minima problem for continuous functions. In: Handbook of Global Optimization, pp. 387–406. Springer (2002)
5. Frazier, P.I.: A tutorial on Bayesian optimization (2018). arXiv:1807.02811
6. Gardner, J.R., Kusner, M.J., Xu, Z.E., Weinberger, K.Q., Cunningham, J.P.: Bayesian optimization with inequality constraints. In: Proceedings of the 31st International Conference on International Conference on Machine Learning, vol. 32, pp. 937–945 (2014)
7. Gelbart, M.A., Snoek, J., Adams, R.P.: Bayesian optimization with unknown constraints. In: Proceedings of the Thirtieth Conference on Uncertainty in Artificial Intelligence, pp. 250–259 (2014)
8. Gramacy, R.B., Lee, H.K.H.: Optimization under unknown constraints. In: Bayesian Statistics 9. Oxford University Press (2011)
9. Guindani, B., Ardagna, D., Guglielmi, A.: Bayesian optimization with machine learning for big data applications in the cloud. In: SIS2022 - Book of Short Papers, pp. 1479–1484. Pearson (2022)
10. Jones, D.R., Schonlau, M., Welch, W.J.: Efficient global optimization of expensive black-box functions. J. Global optim. **13**(4), 455–492 (1998)
11. Letham, B., Karrer, B., Ottoni, G., Bakshy, E.: Constrained Bayesian optimization with noisy experiments. Bayesian Anal. **14**, 495–519 (2019)
12. Luo, X.: Minima distribution for global optimization (2018). arXiv:1812.03457
13. Maros, A., Murai, F., da Silva, A.P.C., Almeida, J.M., Lattuada, M., Gianniti, E., Hosseini, M., Ardagna, D.: Machine learning for performance prediction of spark cloud applications. In: IEEE International Conference on Cloud Computing, pp. 99–106 (2019)
14. Mockus, J.: Application of Bayesian approach to numerical methods of global and stochastic optimization. J. Global Optim. **4**(4), 347–365 (1994)
15. Pourmohamad, T., Lee, H.K.: Bayesian optimization via barrier functions. J. Comput. Graph. Stat. **31**(1), 74–83 (2022)
16. Reagen, B., Hernández-Lobato, J.M., Adolf, R., Gelbart, M., Whatmough, P., Wei, G.Y., Brooks, D.: A case for efficient accelerator design space exploration via Bayesian optimization. In: IEEE/ACM International Symposium on Low Power Electronics and Design, pp. 1–6. IEEE (2017)
17. Schonlau, M., Welch, W.J., Jones, D.R.: Global versus local search in constrained optimization of computer models. IMS Lecture Notes-Monograph Series, pp. 11–25 (1998)
18. Shahriari, B., Swersky, K., Wang, Z., Adams, R.P., De Freitas, N.: Taking the human out of the loop: a review of Bayesian optimization. Proc. IEEE **104**(1), 148–175 (2015)

19. Williams, C.K., Rasmussen, C.E.: Gaussian Processes for Machine Learning, vol. 2. MIT Press Cambridge, MA (2006)
20. Zhang, Y., Prekas, G., Fumarola, G.M., Fontoura, M., Goiri, I., Bianchini, R.: History-based harvesting of spare cycles and storage in large-scale datacenters. In: USENIX Symposium on Operating Systems Design and Implementation (2016)

The manufacturer's authorised representative in the EU is Springer
Nature Customer Service Centre GmbH, Europaplatz 3, 69115 Heidelberg,
Germany. If you have any concerns regarding our products, please
contact ProductSafety@springernature.com

Printed and bound by CPI Group (UK) Ltd, Croydon, CR0 4YY
01/10/2026
02225907-0010